Distillation Theory and Its Application to Optimal Design of Separation Units

Distillation Theory and Its Application to Optimal Design of Separation Units presents a clear, multidimensional, geometric representation of distillation theory that is valid for all types of distillation columns for all splits, column types, and mixtures. This representation answers such fundamental questions as:

- What are the feasible separation products for a given mixture?
- What minimum power is required to separate a given mixture?
- What minimum number of trays is necessary to separate a given mixture at a fixed-power input?

Methods of the general geometric theory of distillation, encoded in software, provide quick and reliable solutions to problems of flowsheet synthesis and to optimal design calculations. DistillDesigner software allows refinement and confirmation of the algorithms of optimal design. A sample of this software is available at www.petlyuk.com.

This book is intended for students and specialists in the design and operation of separation units in the chemical, pharmaceutical, food, wood, petrochemical, oil-refining, and natural gas industries, and for software designers.

Felix B. Petlyuk, Ph.D., D.Sc., has worked in the petrochemical engineering and oil-refining industries for more than 40 years. He currently works for the engineering firm ECT Service in Moscow.

Distillation Theory and Its Application to Optimal Design of Separation Units

F. B. Petlyuk

CAMBRIDGE
UNIVERSITY PRESS

PUBLISHED BY THE PRESS SYNDICATE OF THE UNIVERSITY OF CAMBRIDGE
The Pitt Building, Trumpington Street, Cambridge, United Kingdom

CAMBRIDGE UNIVERSITY PRESS
The Edinburgh Building, Cambridge CB2 2RU, UK
40 West 20th Street, New York, NY 10011-4211, USA
477 Williamstown Road, Port Melbourne, VIC 3207, Australia
Ruiz de Alarcón 13, 28014 Madrid, Spain
Dock House, The Waterfront, Cape Town 8001, South Africa

http://www.cambridge.org

© F. B. Petlyuk 2004

First published 2004

Printed in the United States of America

Typefaces Times Ten Roman 10/13 pt. and Gill Sans   *System* LaTeX 2_ε [TB]

A catalog record for this book is available from the British Library.

Library of Congress Cataloging in Publication Data
Petlyuk, F. B. (Felix B.), 1934–
    Distillation theory and its application to optimal design of separation units /
F. B. Petlyuk.
        p.   cm. – (Cambridge series in chemical engineering)
    Includes bibliographical references and index.
    ISBN 0-521-82092-8 (hardback)
    1. Distillation.   2. Distillation apparatus – Design and construction.
I. Title.   II. Series.
    TP156.D5P45   2004
    660′.28425 – dc22   2003058496

ISBN 0 521 82092 8 hardback

ac

Contents

Preface

This book is devoted to distillation theory and its application. Distillation is the most universal separation technique. Industrial distillation consumes a considerable part of the world power output. The distillation theory enables one to minimize power and capital costs and thus opens up new ways of designing economical separation units. The most important constituent of the distillation theory is the geometric approach, which reveals general rules governing the variation of component concentrations along the distillation column. In other words, it provides general rules for the arrangement of distillation trajectories in the so-called concentration space, in which every point represents some mixture composition. A considerable part of the book is concerned with these general rules, which are used as the basis in developing new methods and algorithms for the optimal design of separation units.

The geometric approach to distillation was put forward by the German scientists Ostwald and Schreinemakers in the early twentieth century. During the years that followed, it has been developed by scientists from various countries. However, until recently, the geometric approach found little use in the design of distillation units. The progress in this field was made by developing the pure computational approach, more specifically, ways of describing the liquid–vapor equilibrium and algorithms for solving sets of distillation equations. This approach has been fruitful: it has resulted in universal computer programs that enable one to design a distillation column (system) of any type for separation of any kind of mixture. However, the pure computational approach gives no answer to a number of fundamental questions that arise in the optimal design of distillation processes, particularly in the case of azeotropic distillation. These questions are the following: (1) What are the feasible separation products for a given mixture? In other words, what components can be present in or absent from the separation products? (2) What minimum power is required to separate a given mixture into the desired components? (3) What minimum number of trays is necessary to separate a given mixture into the desired components at a fixed-power input? Answers to these questions have been provided only by a general geometric theory of distillation.

Until recently, this theory had not advanced to a sufficient extent. Solutions were only obtained for particular cases. For many years, the author and his colleagues, relying on the results obtained by other researchers, have been putting a great deal of effort into elaborating general methods of the geometric theory to answer the fundamental questions listed above. An analysis of thermodynamically reversible distillation, the conception of "sharp" separation, the formulation of conditions under which distillation trajectories can tear-off from the boundaries of the concentration simplex, and the conditions of joining of column section trajectories have been particularly important steps in constructing the geometric theory of distillation. We have proposed a clear multidimensional geometric representation of distillation, which is valid for all types of distillation columns and complexes, for mixtures of any number of components and azeotropes, and for all splits. This representation provided answers to all the fundamental questions, which were previously enumerated. This success encouraged the author to write the present book.

The optimal design of a distillation plant includes the optimization of the sequence of the most economic columns and complexes for separation for a given mixture (flowsheet synthesis) and optimization of the operating and design parameters of these columns and complexes (optimal design calculations). Methods of the general geometric theory of distillation, encoded in software, provide quick and reliable solutions to both problems. The creation of this book necessitated the development of DistillDesigner software that allowed us to refine, check, and confirm the algorithms of optimal designing and also to provide for a significant portion of illustrations and exercises. The problems are solved neither by conventional "blind" methods nor by trial-and-error methods based on the designer's intuition. They are solved in a systematic way, and the solution has a geometric image so the designer can see that it is really optimal. The creation of the software product led, in its turn, to a revision of the general statements of the geometric distillation theory.

Furthermore, the book considers problems that are beyond the framework of the geometric theory of distillation but are still of importance from both the theoretical and practical standpoints.

Among these problems is the problem of maximizing energy savings by optimizing the type of separation unit and by maximizing heat recovery and the problem of the maximum yield of the most valuable products in the separation of thermolabile mixtures (e.g., the maximum yield of the light product in oil refining). Application of optimal design methods based on the general geometric theory of distillation and use of new, most economic distillation units and separation sequences bring the practice of separation to a much higher level.

This book is intended for a wide variety of specialists in the design and operation of separation units in the chemical, pharmaceutical, food, wood, petrochemical, oil-refining, and natural gas industries, and for those engaged in creating software for separation unit design. The circle of these specialists comprises software engineers, process designers, and industrial engineers. The software engineer will find new computational algorithms, the process designer will be provided with a useful guide in his or her search for economic engineering solutions, and the industrial engineer will find ways of reducing the process cost. This book can serve as a manual for students and postgraduates who want to refine their understanding of distillation.

The book has many illustrations, without which understanding of the geometric theory would be impossible. The visualization of trajectory location in the concentration space has great practical significance, as it allows the process designer to understand the main peculiarities of separation of each particular mixture. Developing the geometric theory of distillation necessitated the introduction of some new terms. Furthermore, for some concepts, there are no unique, commonly accepted terms. For these reasons, the book is supplemented with a short glossary, which is believed to be useful for the reader. For better understanding of the subject, each chapter has an introduction that presents the problems to be considered, their brief history, and a conclusion, which summarizes the basic results. Besides that, each chapter contains questions for review and exercises with DistillDesigner software. A sample of this software is available at www.petlyuk.com. The most important chapter for understanding the geometric theory of distillation is Chapter 5. The chapters preceding it are basically introductory, and those that follow speak mostly of the application of the theory.

Acknowledgments

The author is grateful to many people who have favored the creation of this book.

First, I express my gratitude to my closest assistant Roman Danilov whose participation was really indispensable. Together with him, I have developed the hitherto unrivaled software package that made it possible to check and put into practice the main ideas of this book. He also designed all the illustrations without which the book would not be comprehensible.

My debt of gratitude is to colleagues and research students who have taken part in numerous projects for decades: Victoria Avetyan, Vyacheslav Kiyevskiy, Maya Yampolskaya, Valentina Mashkova, Galina Inyayeva, Elizaveta Vinogradova, Zhanna Bril, Boris Isayev, Alexander Shafir, and Oleg Karpilovskiy.

My encounter with Professor Vladimir Platonov gave rise to my interest in distillation. Later acquaintance with Professor Leonid Serafimov led me to the investigation of the most complicated problems concerning azeotropic mixtures.

A number of scientists approved of my working on the book and favored it. I am grateful to Valeriy Kiva, Sigourd Skogestad, Arthur Westerberg, and Nikolay Kulov.

I am grateful to Andry Kalinenko and Vyacheslav Kiyevskiy, chiefs of the engineering firm ECT Service, where I have been working for a long time, for providing me with much support in developing new methods and writing this book.

I express my gratitude to Norsk University of Science and Technology for helping me when I was starting this book.

And I am thankful to my wife who made every effort so that my work would go on.

Nomenclature

A	separation work
A	stationary point of bond chain
A	vertex of product simplex
B	bottom stream (flow rate), kmol/sec
$C^{(k)}$	k-component boundary element of concentration simplex
C_n	concentration simplex for n-component mixture
d	dimension of trajectory bundle
D	overhead stream (flow rate), kmol/sec
E	entrainer stream (flow rate), kmol/sec
F	feed stream (flow rate), kmol/sec
h	enthalpy of liquid, kJ/kg or kcal/kg
H	enthalpy of vapor, kJ/kg or kcal/kg
h	heave key component
$i_D : i_B$	split in column (i_D and i_B – components of overhead and bottom products respectively)
$i : j$	split in section (i and j present and absent component of section product or pseudoproduct respectively)
K	equilibrium ratio
k	number of product components at sharp distillation
k	key component
k	key stationary point (pseudocomponent)
K_j^∞	equilibrium ratio of component j at infinite dilution
K^t	equilibrium ratio in tear-off point
l	light key component
L	liquid stream (flow rate), kmol/sec
m	number of product components at sharp distillation

m	number of stationary points of bond chain
n	number of components in a mixture
N	number of equilibrium stages
N^+ or N^-	stable or unstable node respectively
N_D^+ or N_D^-	stable or unstable node of overhead boundary element of concentration simplex or of distillation region respectively
N_B^+ or N_B^-	stable or unstable node of bottom boundary element of concentration simplex or of distillation region respectively
N_r^+ or N_r^-	stable or unstable node of rectifying trajectory bundle respectively
N_s^+ or N_s^-	stable or unstable node of stripping trajectory bundle respectively
N_e^+ or N_e^-	stable or unstable node of extractive trajectory bundle respectively
P	pressure, Pa
q	fraction of liquid in feed
Q	heat flow rate, kJ/sec or kcal/sec
qS	quasisaddle
R	reflux ratio
$R_{\min}$	minimum reflux ratio
$R_{\lim}^1$ or $R_{\lim}^2$	first or second boundary minimum reflux ratio respectively
$R_{\min}^t$ or $R_{\max}^t$	minimum or maximum reflux ratio for trajectory tear-off respectively
Reg_{ijk}	region
Reg_{ord}	component order region
$\text{Reg}_D^{(k)}$ or $\text{Reg}_B^{(k)}$ or $\text{Reg}_{D,E}^{(k)}$	k-component possible overhead or bottom or overhead-entrainer product region respectively
$\text{Reg}_D^{\,j}$ or $\text{Reg}_B^{\,j}$ or $\text{Reg}_{D,E}^{\,j}$	i-present components and j-absent components possible overhead or bottom or overhead-entrainer product region respectively
$\text{Reg}_{bound,D}^{\,j}$ or $\text{Reg}_{bound,B}^{\,j}$ or $\text{Reg}_{bound,D,E}^{\,j}$	boundary of possible overhead or bottom or overhead-entrainer product region respectively, i-present components, and j-absent components
$\text{Reg}_r^{t(k)}$ or $\text{Reg}_s^{t(k)}$ or $\text{Reg}_e^{t(k)}$	k-component tear-off region of rectifying or stripping or extractive section respectively
Reg^∞	distillation region at infinite reflux
$\text{Reg}_{bound,D}^\infty$, $\text{Reg}_{bound,B}^\infty$	top or bottom boundary element of distillation region at infinite reflux respectively
$\text{Reg}_{sep,r}^{\min,R}$, $\text{Reg}_{sep,s}^{\min,R}$	separatrix min-reflux region for rectifying or stripping section for given reflux R respectively

$\text{Reg}_{sep,r}^{sh,R}$, $\text{Reg}_{sep,s}^{sh,R}$	separatrix sharp split region for rectifying or stripping section for given reflux R respectively
$\text{Reg}_{w,r}^{R}$, $\text{Reg}_{w,s}^{R}$ $\text{Reg}_{w,e}^{R}$	rectifying or stripping or extractive section working region at given reflux R respectively
$\text{Reg}_{sh,r}^{i:j}$, $\text{Reg}_{sh,s}^{i:j}$, $\text{Reg}_{sh,e}^{i:j(E)}$	sharp split region for rectifying or stripping or extractive section for split $i:j$ respectively
$\text{Reg}_{rev,r}^{h}$, $\text{Reg}_{rev,s}^{l}$, $\text{Reg}_{rev,e}^{m}$	reversible distillation region for rectifying section with h heavy component or stripping section with l light component or extractive section with m middle component respectively
Reg_{att}	attraction region
Reg_{L-L}	two liquid phases region
Reg_{pitch}	region of pitchfork
Reg_{simp}	product simplex at infinite reflux
Reg_{sub}	subregion of distillation at infinite reflux
Reg_{tang}	tangential pinch region
S	reboil ratio
S	entropy
S	saddle
S^1	tear-off point of section trajectory at sharp split
S^2	tear-off point of section trajectory at minimum reflux
$S^1 - S^2 - N^+$	boundary element of trajectory bundle at sharp split
$S^2 - N^+$	boundary element of trajectory bundle at minimum reflux
SN	saddle-node
S_r or S_s or S_m	saddle point of rectifying or stripping or intermediate trajectory bundle respectively
T	temperature, K
V	vapor stream (flow rate), kmol/sec
x	mole fraction of liquid phase
x_{rev}^{t}	tear-off point of reversible distillation trajectory
x_D'	pseudoproduct point
x_f^{∞} or $x_f^{\min}$	composition on first plate under feed cross section at which number of stripping section plate is infinite or minimal respectively
x_{f-1}^{∞} or $x_{f-1}^{\min}$	composition on first plate above feed cross section at which number of rectifying section plate is infinite or minimal respectively
x_{rev}^{branch}	branch point of reversible distillation trajectory
(x_f^{sh}) or (x_{f-1}^{sh})	composition on first plate under or above feed cross section at sharp split respectively

$[x_f^{sh}]$ or $[x_{f-1}^{sh}]$	composition segment on first plate under or above feed cross section at sharp split respectively
y	mole fraction of vapor phase
z	mole fraction of liquid–vapor mixture
$1, 2, 3 \ldots$	components $1, 2, 3 \ldots$ respectively
$1, 2; 1, 3 \ldots$	mixtures of components 1 and 2; 1 and 3 ... respectively
1-2, 1-2-3 ...	boundary elements of concentration simplex
$12, 13 \ldots$	binary azeotropes of components 1 and 2; 1 and 3 ... respectively
$123, 124 \ldots$	ternary azeotropes of components 1, 2, and 3; 1, 2, and 4 ... respectively
$123, 132 \ldots$	regions of component order

Greek and Other Symbols

ε	component recovery
Δ	difference
λ	eigenvalue of distillation matrix
σ	excess reflux factor
∞	infinity
α	relative volatility
Σ	sum
θ	the root of an Underwood equation for both sections
φ or ψ	the root of an Underwood equation for rectifying or stripping section
η	product purity
η	thermodynamic efficiency
Δx_f^{sh} or Δx_{f-1}^{sh}	composition interval on plate under or above feed cross section
$\alpha_{12}, \alpha_{13} \ldots$	volatility of component 1 relative of component 2, of component 3 ...
$N^- \overset{S}{\Rightarrow} N^+$	distillation bundle included stationary points N^-, S, N^+
$x_{f-1} \Downarrow \Rightarrow x_f$	mixing in feed cross section
$\rightarrow$	bond, trajectory of distillation, one-dimensional trajectory bundle
$\Rightarrow$	set of all bonds (or of all distillation trajectories) of distillation bundle
$\Leftrightarrow$	flows between sections of distillation complex
$\Updownarrow$	decanter

Subscripts and Superscripts

az	azeotrop
ad	adiabatic

B	bottom product
con	condenser
D	overhead product
e	component of entrainer
E	entrainer
e	first plate under entrainer cross section
$e\text{-}1$	first plate above entrainer cross section
F	feed
f	first plate under feed cross section
$f\text{-}1$	first plate above feed cross section
h	heave key component h
Haz	heteroazeotrop
i	component of mixture
i, D	component i, which is present in product D
imp	impurity
int	intermediate condenser or reboiler
irr	irreversible
j	component j, which is absent on the boundary element of concentration simplex
j, DE	component j, which is absent in product D and entrainer E
j	plate of column
j	stationary point
k	component of mixture
k	plate of column
key	key component
l	light key component of mixture
$L1, L2$	first, second liquid phases
M	intermediate product
m	intermediate section
m	middle volatility component of mixture
new	new value at iterations
old	old value at iterations
$pinch$	pinch
pr	preferable
r	rectifying
reb	reboiler
rev	reversible
s	stripping
st	stationary point
t	tear-off point
$t1, t2$	first and second tear-off points of reversible distillation trajectories respectively
(k)	k-component boundary element of concentration simplex, k-component point, product point with k product components

w	working region, working trajectory
1,2,3 ...	component 1,2,3 ...; section 1,2,3 ...; feed 1,2; variant 1,2,3; column 1,2,3 ... respectively

Nomenclature to Figures

A	endpoint of tear-off segment of distillation trajectories
$A_1, A_2, A_3 ...$	vertexes of possible product composition regions
Az	azeotropes
boxed digits	component order regions
C-1, C-2 ...	columns
dash-dotted line	line of material balance
dashes	tray compositions on composition profiles
dotted line	trajectory of reversible distillation
dotty line	separatrix
double segment	possible composition of overhead product or trajectory tear-off segment of top section
thick black segment	possible composition of bottom product or trajectory tear-off segment of bottom section
gray segment	tear-off segment of extractive distillation trajectories
$F + E$	composition point of feed and entrainer mixture
F_0	composition point of initial feed
$F_1 + F_2$	composition point of mixture of feeds F_1 and F_2
H	height of column
HD	heave diesel oil
HN	heave naphta
LD	light diesel fuel
LN	light naphta
little black or white circle	stable or unstable node of concentration simplex respectively
little cross circle	saddle of concentration simplex
little cross square	bottom composition point
little square	overhead composition point
little triangle	feed composition point
short segment with arrow	tie-line liquid–vapor
st	steam
thick line	trajectory of distillation
thin line	equivolatility line
(1), (2) ...	column (1) or (2) respectively
(1), (2) ...	split (1) or (2) respectively
$\alpha_{12}, \alpha_{13} ...$	equivolatility line of components 1 and 2, 1 and 3 ... respectively

Phase Equilibrium and Its Geometric Presentation

1.1. Introduction

The process of distillation can be presented as consisting of numerous states of phase equilibrium between flows of liquid and vapor that have different compositions. Geometric analysis of the distillation process represented in the so-called *concentration space* (C) is the main instrument for understanding its regularities.

That is why, before we start the examination of the existing distillation process and its geometric interpretation, it is necessary to consider geometric interpretation, of the phase equilibrium. Numerous methods of calculating phase equilibrium are described in many monographs and manuals (see, e.g., Walas [1985]).

We will not repeat these descriptions but instead will examine only representation of equilibrium states and processes in concentration space.

1.2. Concentration Space

Molar composition of an n-component mixture is presented as an array that holds molar concentrations of all components:

$$x_i = \frac{m_i}{\sum m_i} \tag{1.1}$$

$$\sum x_i = 1 \tag{1.2}$$

where m_i is the amount of moles of the component i in the mixture.

Concentration space of an n-component mixture C_n is a space in which every point corresponds to a mixture of definite composition. *Dimensionality* of concentration space corresponds to the number of concentrations of components that can be fixed independently.

The $(n-1)$ concentration for an n-component mixture can be fixed independently because concentration of the nth component can be found from Eq. (1.2). That is why the dimensionality of the concentration space of binary mixture C_2 is one, of ternary mixture C_3 – two, of four-component mixture C_4 – tree, etc.

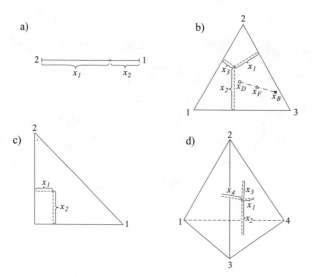

Figure 1.1. Concentration simplexes (a) for binary mixtures,
(b, c) for three-component mixtures. and (d) for four-component mixtures. x_1, x_2, x_3, x_4, concentrations of components.

Concentration space is the number of points representing all possible compositions of an n-component mixture. Concentration space of a binary mixture C_2 is
a segment of unit length; the ends correspond to pure components, and the inner
points correspond to mixtures of various compositions (Fig. 1.1a)

For a three-component mixture, it is convenient to present the composition
space C_3 as an equilateral triangle, the height of which equals one (Fig. 1.1b). The
triangle's vertexes represent pure components, the points within its sides, represent the binary constituents of the three-component mixture, and the inner points
of triangle represent the three-component mixture compositions. The lengths of
the perpendiculars to the triangle's sides correspond to the concentrations of the
components indicated by the opposite vertexes. The described system of coordinates, which bears the name of the system of uniform coordinates, was introduced
by Mobius and was further developed by Gibbs.

Another way to present a three-component mixture's composition space C_3
implies the use of an isosceles right-angle triangle (Fig. 1.1c), with a side equal
to one. In this method of representation the concentrations of components 1 and
2 are expressed by the length of perpendicular segments, as in the first case of
the composition's representation, and the concentration of the third component
is defined in accordance with the formula: $x_3 = 1 - (x_1 + x_2)$.

Four-component mixture composition can be represented by a point of an
equilateral tetrahedron C_4 (Fig. 1.1d). In this tetrahedron the vertexes represent
the pure components, the edges represent the binary constituents, and the faces
represent the three-component constituents.

In this book, we will often represent the mixture compositions corresponding
to the material balance (e.g., the compositions of feed flow and product flow of

the distillation column):

$$Fx_{iF} = Dx_{iD} + Bx_{iB} \tag{1.3}$$

$$(D + B)x_{iF} = Dx_{iD} + Bx_{iB} \tag{1.4}$$

$$D(x_{iF} - x_{iD}) = B(x_{iB} - x_{iF}) \tag{1.5}$$

Equation (1.5) represents the so-called *lever rule*: points x_{iF}, x_{iD}, and x_{iB} are located on one straight line, and the lengths of the segments $[x_{iF}, x_{iD}]$ and $[x_{iB}, x_{iF}]$ are inversely proportional to the flow rates D and B (Fig. 1.1b). Mixture with a component number $n \geq 5$ cannot be represented clearly. However, we will apply the terms simplex of dimensionality $(n - 1)$ for a concentration space of n-component mixture C_n, hyperfaces C_{n-1} of this simplex for $(n - 1)$-component constituents of this mixture, etc.

1.3. Phase Equilibrium of Binary Mixtures

An equilibrium between liquid and vapor is usually described as follows:

$$y_i = K_i x_i \tag{1.6}$$

where y_i and x_i are equilibrium compositions of vapor and liquid, respectively, and K_i is the liquid–vapor phase equilibrium coefficient.

To understand the mutual behavior of the components depending on the degree of the mixture's nonideality caused by the difference in the components' molecular properties, it is better to use graphs $y_1 - x_1$, $T - x_1$, $T - y_1$, $K_1 - x_1$, and $K_2 - x_1$ (Fig. 1.2). In Fig. 1.2, the degree of nonideality increases from a to h: a is an ideal mixture, b is a nonideal mixture with an *inflection on the curve* $y_1 - x_1$ (a and b are zeotropic mixtures), c is a mixture with a so-called *tangential azeotrope* (curve $y_1 - x_1$ touches the diagonal in the point $x_1 = 1$), d is an *azeotropic mixture* with minimum temperature, e is a mixture with a so-called *inner tangential azeotrope*, f is a *mixture with two azeotropes*, g is a *heteroazeotropic mixture*, and h is an *azeotropic mixture with two liquid phases*. *Azeotrope* is a binary or multicomponent mixture composition for which the values of phase equilibrium coefficients for all components are equal to one:

$$K_i^{Az} = 1 \qquad (i = 1, 2, \ldots n) \tag{1.7}$$

Heteroazeotrope is an overall composition of a mixture with two liquid phases for which the values of the overall coefficients of phase equilibrium for all components are equal to one:

$$K_{ov,i}^{Haz} = 1 \qquad (i = 1, 2, \ldots n) \tag{1.8}$$

where $K_{ov,i} = y_i/x_{ov,i}$, $x_{ov,i} = x_i^{(1)}a + x_i^{(2)}(1 - a)$, a is the portion of the first liquid phase in the whole liquid, and $x_i^{(1)}$ and $x_i^{(2)}$ are the concentrations of the ith component in first and second liquid phases correspondingly.

In this book, we will see that the previously discussed features are of great importance. Even b case results in serious abnormalities of the distillation process.

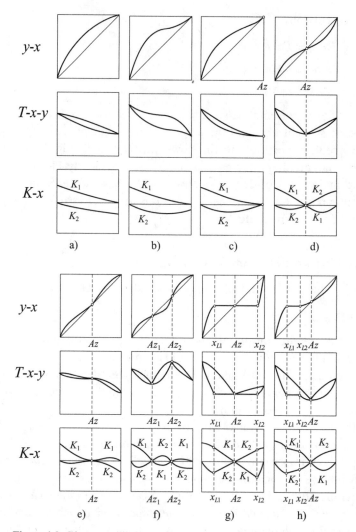

Figure 1.2. Phase equilibrium of binary mixtures: (a) ideal mixture; (b) nonideal mixture; (c) tangential azeotropic mixture ($x_{1,\,Az}=1$); (d) azeotropic mixture; (e) mixture with internal tangential azeotrope ($0 < x_{1,\,Az} < 1$); (f) mixture with two azeotropes Az_1 and Az_2; (g) heteroazeotropic mixture; and (h) azeotropic mixture with two liquid phases ($y-x$, $T-x-y$, and $K-x$ diagrams). Az, azeotropic or heteroazeotropic point; x_{L1} and x_{L2}, compositions of liquid phases.

The appearance of azeotropes makes the separation of the mixture into pure components impossible without special procedure application.

Further increase in nonideality and transition to heteroazeotropes makes it again possible to separate mixtures, not using just a distillation column, but a *column with decanter* complex. Cases *e* and *f* occur, but very seldom; therefore, we will not consider them further.

In the azeotrope point, $K_1 = K_2 = 1$. For a *tangential azeotrope*, $x_1^{Az} = 1$ or $x_1^{Az} = 0$. It might seem that a tangential azeotrope is no obstacle for separation.

However, later in this book, we will see that if $x_1^{Az} = 1$, it is impossible to get component 1 with a high degree of purity, and if $x_1^{Az} = 0$, it is impossible to get component 2 with a high degree of purity.

1.4.   Phase Diagrams of Three-Component Mixtures

Three-component mixtures represent the simplest type of multicomponent mixtures. The majority of multicomponent mixture peculiarities become apparent in three-component mixtures. This is why the three-component mixtures are best studied. Liquid–vapor equilibrium in the concentration triangle C_3 is represented by a vector connecting a point of liquid composition with a point of equilibrium vapor composition $x \rightarrow y$. This vector is called a *liquid–vapor tie-line*. The opposite vector $y \rightarrow x$ (vapor–liquid) is called a *vapor–liquid tie-line*. The tie-lines field in the concentration triangle characterizes phase equilibrium in each of its points.

However, tie-lines can cross each other. That is why, for phase equilibrium characteristics in the concentration space, it is convenient to use another kind of line, the so-called *residue curves*. Let's consider a process of *open evaporation* (*simple distillation*) illustrated in Fig. 1.3.

Let's assume that the initial amount of liquid in a flask makes L moles and the liquid has a composition x_i ($i = 1, 2, \ldots n$). After the evaporation of a small amount of liquid ΔL, vapor with a composition y_i ($i = 1, 2, \ldots n$), will be formed which represents an equilibrium of the remaining liquid, the amount of which is equal to $L - \Delta L$ moles and the composition is $x_i + \Delta x_i$.

The material balance for i component is:

$$Lx_i = (\Delta L)y_i + (L - \Delta L)(x_i + \Delta x_i) \tag{1.9}$$

In limit at $\Delta L \rightarrow 0$,

$$Ldx_i/dL = x_i - y_i \tag{1.10}$$

Figure 1.3. Open evaporation process (open distillation). x, y, composition of liquid and equilibrium vapor phases; L, amount of liquid; dL, infinitesimal amount of evaporated liquid.

Denoting $dt = dL/L$, we will get the equation of a residue curve:

$$dx_i/dt = x_i - y_i \qquad (i = 1, 2, \ldots n) \tag{1.11}$$

The residue curve represents the change in a mixture composition during the open evaporation process. Each point of this line corresponds to a certain moment of time and to a portion of evaporated liquid.

From Eq. (1.11), it results that in each point of a residue curve a liquid–vapor tie-line is tangent to this line. The residue curves are convenient for the description of phase equilibrium because as these lines are continuous and noncrossing.

These lines were used for the first time to describe phase behavior of three-component azeotropic mixtures at the beginning of the twentieth century (Ostwald, 1900; Schreinemakers, 1901). Later, the residue curves of three-component azeotropic mixtures were studied in the works of Reinders & De Minjer (1940a, 1940b) for the azeotropic mixture acetone–chloroform–benzene and more widely in the works by Bushmakin & Kish (1957a, 1957b). Gurikov (1958) developed the first classification of three-component mixtures residue curve diagrams. In the works of Zharov (1967, 1968a, 1968b) and Serafimov (1969) the residue curve diagrams analysis and classification were applied for four-component and multicomponent mixtures. Several years later, these works were summarized in a monograph by Zharov & Serafimov (1975). In recent years, other versions of residue curve diagram classifications were developed (Matsuyama & Nishimura, 1977; Doherty & Caldarola, 1985).

Points of pure components and azeotropes are *stationary or singular points of residue curve bundles*. At these points, the value dx_i/dt in Eq. (1.11) becomes equal to zero. A stationary point at which all residue curves come to an end is called a *stable node* (the temperature increases in the direction of this point). A specific point at which all residue curves start is called an *unstable node* (the temperature

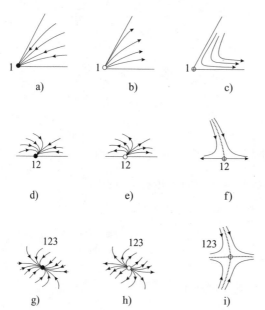

Figure 1.4. Types of stationary points of three-component mixtures: (a) one-component stable node, (b) one-component unstable node, (c) one-component saddle, (d) two-component stable node, (e) two-component unstable node, (f) two-component saddle, (g) three-component stable node, (h) three-component unstable node, and (i) three-component saddle. Arrows, direction of residium curves.

increases in the opposite direction of this point). The rest of stationary points are called *saddles* (Fig. 1.4).

A stationary point type is defined by the *proper values* of Yakobian from Eq. (1.11). For a stable node, both proper values are negative, $\lambda_1 < 0$ and $\lambda_2 < 0$; for an unstable node, both proper values are positive, $\lambda_1 > 0$ and $\lambda_2 > 0$; and for a saddle, one proper value is negative, $\lambda_1 < 0$, and the second is positive, $\lambda_2 > 0$.

For a distillation process not only the stationary point type, but also the behavior of the residue curve in the vicinity of the stationary point is of special importance. If the residue curves in the vicinity of the specific point are tangent to any straight line (singular line) (Fig. 1.4a, b, d, e, g, h), the location of this straight line is of great importance. A special point type and behavior of residue curves in its vicinity are called stationary point *local characteristics*.

The whole concentration space can be filled with one or more residue curve bundles. Each residue curve bundle has its own initial point (unstable node) and its own final point (stable node). Various bundles differ from each other by initial or final points.

The boundaries separating one bundle from another are specific residue curves that are called the *separatrixes of saddle stationary points*. In contrast to the other residue curves, the separatrixes begin or come to an end, not in the node points but in the saddle points. A characteristic feature of a separatrix is that in any vicinity of its every point, no matter how small it is, there are points belonging to two different bundles of residue curves. The concentration space for ideal mixtures is filled with one bundle of residue curves. Various types of azeotropic mixtures differ from each other by a set of stationary points of various types and by the various sequence of boiling temperatures in the stationary points.

The first topological equation that connects a possible number of stationary points of various types for three-component mixtures (*N*, node; *S*, saddle; upper index is the number of components in a stationary point) was deduced (Gurikov, 1958):

$$2(N^3 - S^3) + N^2 - S^2 + N^1 = 2 \tag{1.12}$$

Figure 1.5 shows mainly physically valuable *types of three-component azeotropic mixtures* deduced by Gurikov (1958) by means of systematic application of Eq. (1.12). In Fig. 1.5, one and the same structure cover a certain type of mixture and an antipodal type in which stable nodes are replaced by unstable ones and vice versa (i.e., the direction of residue curves is opposite). Besides that, the separatrixes are shown by the straight lines. Let's note that the later classifications of three-component mixture types (Matsuyama & Nishimura, 1977; Doherty & Caldarola, 1985) contain considerably greater number of types, but many of these types are not different in principle because these classifications assume light, medium, and heavy volatile components to be the fixed vertexes of the concentration triangle.

Types of azeotropic mixture and separatrixes arrangements are also called mixture *nonlocal characteristics*.

The part of the concentration space filled with one residue curve bundle is called a *distillation region* Reg^∞ (Schreinemakers, 1901). A distillation region $\mathrm{Reg}_{(3)}^\infty$ has

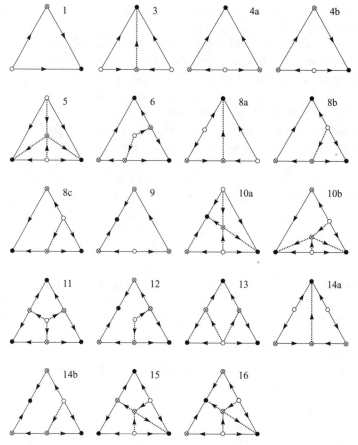

Figure 1.5. Types of three-component mixtures according to Gurikov (1958). Arrows, direction of residium curves (bonds); dotted lines, separatrixes.

boundary elements that include the separatrixes, segments of the concentration triangle sides $\text{Reg}_{(2)}^{\infty}$, and stationary points $\text{Reg}_{(1)}^{\infty}$ referring to this region. A distillation region of a three-component mixture $\text{Reg}_{(3)}^{\infty}$ is *two-dimensional*; separatrixes, and segments of the concentration triangle sides $\text{Reg}_{(2)}^{\infty}$ are *one-dimensional*; and stationary points $\text{Reg}_{(1)}^{\infty}$ have *zero dimensionality*. Distillation regions and their boundary elements are also called concentration space *structural elements*.

Besides these structural elements, concentration space has other structural elements that are of great importance for a distillation process under various modes.

1.5.   Residue Curve Bundles of Four-Component Mixtures

The structure of residue curve bundles of four-component mixtures is significantly more complex and diverse than that of three-component mixtures. This is due to the fact that each four-component mixture consists of four three-component constituents. Therefore, the number of types of four-component mixtures is enormous. In addition to that, four-component mixtures can have four-component node and saddle azeotropes. In contrast to three-component mixtures, the enormous

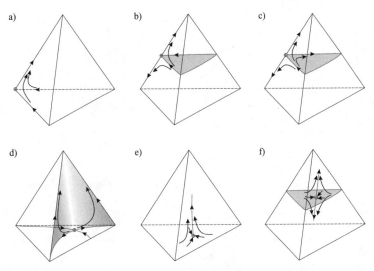

Figure 1.6. Types of saddle points of four-component mixtures: (a) one-component saddle, (b, c) two-component saddle, (d, e) three-component saddle, and (f) four-component saddle. Separatrix surfaces are shaded. Arrows, direction of residium curves; dotty lines, separatrixes.

number of four-component mixture structures makes their overall sorting out practically impossible. However, a topological equation for four-component mixtures similar to Eq. (1.12) was obtained (Zharov & Serafimov, 1975).

To understand the peculiarities of location of residue curve bundles of four-component mixtures, let's consider their behavior in the vicinity of saddle points (Fig. 1.6) and the nonlocal characteristics of the residue curve bundles using separate examples of the four-component mixture structures (Fig. 1.7). In Fig. 1.7, the *separating surfaces* of the residue curve bundles representing the *two-dimensional bundles* $Reg^{\infty}_{(3)}$ are shaded. Considering the nonlocal characteristics of the residue curve bundles, the simplest of such characteristics refers to each pair of stationary points. A pair of stationary points can be connected or not connected by the residue curve. To be brief, let's call the line of distillation that connects a pair of stationary points a *bond* (link) – it will be designated by the arrow ($\rightarrow$) that is directed toward the side of the temperature increase (Petlyuk, Kievskii, & Serafimov, 1975a, 1975b, 1977, 1979). For example, in Fig. 1.7a, $12 \rightarrow 23$. In the same figure, points 1 and 2 are not bonded.

The totality of all bonds characterizes the mixture's structure. The bond serves as the *elementary nonlocal characteristic* of the residue curve bundle structure. Bonds form bond chains. The *bond chains* of maximum length connect the unstable node N^- and the stable node N^+ of the distillation region Reg^{∞}. Let's call a polyhedron formed by all stationary points of one maximum-length bond chain and containing all components of the mixture a *distillation subregion* Reg_{sub}.

The distillation region Reg^{∞} is a polyhedron formed by all stationary points of the totality of all maximum-length bond chains connecting the same unstable node of the composition space with the same stable node (it will be designated $\Rightarrow$). The examples of distillation regions Reg^{∞} are $12 \Rightarrow 4$, $12 \Rightarrow 2$ (at Fig. 1.7a),

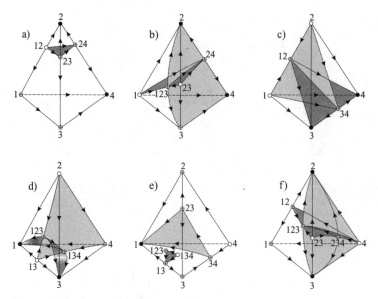

Figure 1.7. The examples of four-component structures (bonds and distillation regions Reg^{∞}). Separatrix surfaces are shaded. Arrows, direction of residium curves; dotty lines, separatrixes.

$1 \Rightarrow 4$, $1 \Rightarrow 2$, $23 \Rightarrow 4$, $23 \Rightarrow 2$ (at Fig. 1.7b), $1 \Rightarrow 3$, $1 \Rightarrow 4$, $2 \Rightarrow 3$, $2 \Rightarrow 4$ (at Fig. 1.7c), $13 \Rightarrow 1$, $13 \Rightarrow 3$, $2 \Rightarrow 1$, $2 \Rightarrow 3$ (at Fig. 1.7d), $134 \Rightarrow 1$, $4 \Rightarrow 1$ (at Fig. 1.7e), $23 \Rightarrow 2$, $23 \Rightarrow 3$, $4 \Rightarrow 2$, and $4 \Rightarrow 3$ (at Fig. 1.7f).

The examples of distillation subregions Reg_{sub} are $12 \to 23 \to 3 \to 4$, $12 \to 1 \to 3 \to 4$, and $12 \to 23 \to 24 \to 4$ (Fig. 1.7a). In this case, the distillation region Reg^{∞} is $12 \Rightarrow 4$ ($\text{Reg}_{sub} \in \text{Reg}^{\infty}$), or

$$
\begin{array}{ccccc}
\uparrow & \to & 1 & \to & \downarrow \\
12 & \to & 23 & \to & 3 \to 4 \\
\downarrow & \to & 24 & \to \uparrow &
\end{array}
$$

As we will see in Chapter 3, the distillation region and subregion characterize those possible product compositions that can be produced from the given feedstock composition by distillation under one of the most important modes, in particular, under the *infinite reflux mode*.

A *bond, bond chain, distillation subregion, and region are the nonlocal structural elements* of the azeotropic mixture concentration space.

1.6. Matrix Description of the Multicomponent Mixture Residue Curve Structure

The structure of the residue curve bundles can be obviously represented only for binary, three-, and four-component mixtures. For mixtures with more components, it is impossible. However, practice needs make necessary the analysis of the bundle structure with any number of components. This problem can be solved by means of a structure matrix description (Petlyuk et al., 1975a, 1975b).

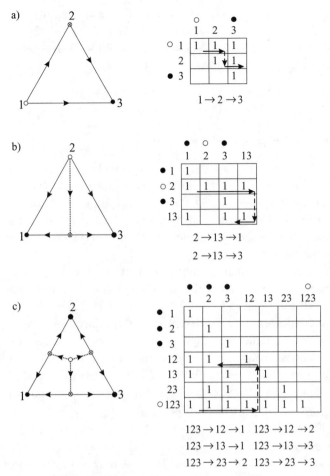

Figure 1.8. The examples of three-component structures and their structural matrices. Arrows, direction of residium curves; dotted lines, separatrixes; thick line with arrow, bond; dotty lines, transfer to next bond.

By the *structural matrix* of the azeotropic mixture concentration space, we will name a square matrix, the columns and lines of which correspond to the stationary points and the elements of which $a_{ij} = 1$, if there is a bond directed from stationary point i to stationary point j ($a_{ij} = 0$, if such a bond is missing). For the purpose of obviousness, some examples of three-component mixture structural matrices are shown in Fig. 1.8.

Each line of a structural matrix corresponds to the ith stationary point and each column to the jth one. Diagonal elements $a_{ij} = 1$ (it is accepted conditionally that each specific point is bonded to itself). The components are labeled 1, 2, 3; binary azeotropes are designated by two-digit numbers, 12, 13, 23; and the ternary azeotrope by a three-digit number, 123. Zero column corresponds to an unstable node N^- and zero line to the stable N^+ one (except for the diagonal elements). Structural matrices provide an opportunity to easily single out all maximum-length bond chains (i.e., all the distillation subregions). For example, in Fig. 1.8c, the distillation subregions Reg_{sub} are as follows: $123 \to 12 \to 1, 123 \to 13 \to 1, 123 \to 12 \to 2, 123 \to 23 \to 2, 123 \to 23 \to 3$, and $123 \to 13 \to 3$. Respectively, the distillation regions Reg_∞ are as follows: $123 \Rightarrow 1, 123 \Rightarrow 2, 123 \Rightarrow 3$, or

$$\begin{array}{ccc} \uparrow \quad \to \quad 13 \quad \to \quad \downarrow, & \uparrow \quad \to \quad 23 \quad \to \quad \downarrow, & \uparrow \quad \to \quad 13 \quad \to \quad \downarrow \\ 123 \to 12 \to 1 & 123 \to 12 \to 2 & 123 \to 23 \to 3 \end{array}$$

1.7.   Lines, Surfaces, and Hypersurfaces $K_i = K_j$

In Sections 1.3 to 1.5, the residue curve bundles, which characterize the direction of liquid–vapor tie-lines in each point of the concentration space (i.e., the phase equilibrium field), were considered. As stated previously, such characteristics of the phase equilibrium field and structural elements related to it (bonds, distillation regions, and subregions) are the most important for one of the distillation modes, in particular, for the infinite reflux mode.

However, the liquid–vapor phase equilibrium field has other important characteristics that become apparent under other distillation modes, in particular, under *reversible distillation* and usual (adiabatic) *distillation with finite reflux.*

To such characteristics are referred, first of all, lines, surfaces, and hypersurfaces of the phase equilibrium coefficients equality ($K_i = K_j$). For the purpose of brevity, we will name these lines, surfaces, and hypersurfaces as *α-lines, α-surfaces, and α-hypersurfaces* (or *univolatility lines, surfaces, and hypersurfaces:* $\alpha_{ij} = K_i/K_j = 1$).

Univolatility α-lines, α-surfaces, and α-hypersurfaces divide the concentration simplex into *regions of order of components* Reg_{ord}^{ijk} (in Reg_{ord}^{ijk} $K_i > K_j > K_k$) (Petlyuk & Serafimov, 1983).

The totality of several regions of components' order for which one and the same component appear to be the most light volatile ($K_l = \max_i K_i$) or the most heavy volatile ($K_h = \min_i K_i$) was named as a region of reversible distillation $\text{Reg}_{rev, s}^{l}$ or $\text{Reg}_{rev, r}^{h}$ (Petlyuk, 1978). Such a name can be explained by the crucial meaning of these regions for possibly realizing of reversible distillation (see Chapter 4).

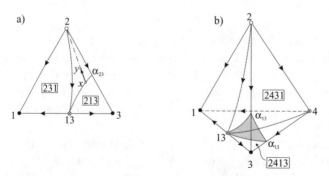

Figure 1.9. α-lines, α-surfaces (shaded), and regions of order of components Reg_{ord} for (a) three-component and (b) four-component mixtures. *231, 213, 2431*, and *2413*, regions of component order $\text{Reg}_{ord}^{2,3,1}$, $\text{Reg}_{ord}^{2,1,3}$, $\text{Reg}_{ord}^{2,4,3,1}$, and $\text{Reg}_{ord}^{2,4,1,3}$; $x \to y$, tie-line liquid–vapor for point x on α-line; arrows, direction of residium curves; dotty lines, separatrixes.

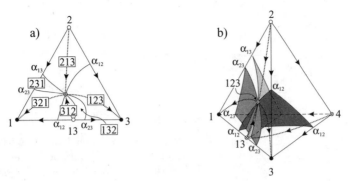

Figure 1.10. α-lines and α-surfaces (shaded) caused by ternary azeotropes for (a) three-component and (b) four-component mixtures. Ar-
rows, direction of residium curves; $\underset{2,1,3}{213}, \underset{1,2,3}{123}, \underset{1,3,2}{132}, \underset{3,1,2}{312}, \underset{3,2,1}{321}, \underset{2,3,1}{231}$, regions
of component order $\mathrm{Reg}_{ord}^{2,1,3}, \mathrm{Reg}_{ord}^{1,2,3}, \mathrm{Reg}_{ord}^{1,3,2}, \mathrm{Reg}_{ord}^{3,1,2}, \mathrm{Reg}_{ord}^{3,2,1}, \mathrm{Reg}_{ord}^{2,3,1}$;
dotty lines, separatrixes.

It is obvious that a binary azeotrope, in the point of which $K_i = K_j = 1$, generates
an α-line, surface, or hypersurface in the concentration space (Fig. 1.9).

In Fig. 1.9a, azeotrope 13 gives rise to an α_{13}-line (on α_{13}-line $K_1 = K_3$), which
crosses edge 2–3 in α_{13}-point and divides the concentration triangle into two re-
gions Reg_{ord}^{ijk}, where the order of components is $231(\mathrm{Reg}_{ord}^{2,3,1})$ and $213(\mathrm{Reg}_{ord}^{2,1,3})$.

In Fig. 1.9b, azeotrope 13 gives rise to α_{13}-surface, which crosses edges 1–3
and 3–4 in α_{13}-points and divides the concentration tetrahedron into two regions
Reg_{ord}^{ijk}, where the order of components is $2431(\mathrm{Reg}_{ord}^{2,4,3,1})$ and $2413(\mathrm{Reg}_{ord}^{2,4,1,3})$.

The ternary azeotrope, in the point of which $K_i = K_j = K_k$, gives rise to three
α-lines in the concentration triangle (Fig. 1.10a). In the concentration tetrahedron,
it gives rise to three α-surfaces in the points of which $\alpha_{ij} = 1$, $\alpha_{ik} = 1$, and $\alpha_{jk} = 1$
(Fig. 1.10b).

Let's note that the ternary azeotrope gives rise to six regions of order of compo-
nents Reg_{ord}^{ijk} and six α-points along the composition triangle contour, the indices
of which are repeated in every pair of indices while passing around the contour.
By means of phase equilibrium model, it is not difficult to define all the α-points
on the sides of the concentration triangle or on the edges of the concentration
tetrahedron or concentration simplex of a greater dimensionality, if the number
of components is greater than four ($n > 4$).

For example, for the diagram shown in Fig. 1.9a, the graphs of dependence of the
phase equilibrium coefficients of the components along the composition triangle
contour are shown in Fig. 1.11. Because one of the components is missing on
each side, its phase equilibrium coefficient is calculated under the infinite dilution
(K^∞). The graphs kindred to the one given in Fig. 1.11 allow all the α-points to be
defined. This allows the ternary azeotrope availability to be predicted. The ternary
azeotrope should exist only if there are six α-point indices, which are repeated with
every pair of indices along the concentration triangle contour.

It is characteristic of all points of α-lines that the liquid–vapor tie-lines in these
points are directed along the straight lines passing through that vertex of the
concentration triangle, the number of which is missing in the index of α-line.

a)
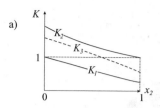

b)
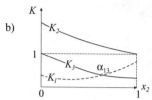

Figure 1.11. Dependences $K - x$ on the sides of the concentration triangle for mixture in Fig. 1.9a: (a) side 1–2, (b) side 2–3, (c) side 1–3. Thick lines, $K - x$ for present on side components; dotted lines, $K - x$ for absent on side components ($K^\infty - x$).

c)
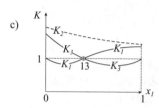

Indeed, if $\alpha_{ij} = K_i/K_j = 1$, then $y_i/y_j = x_i/x_j$ (i.e., points $[x_i, x_j]$ and $[y_i, y_j]$ lie on the straight line that passes through vertex k [$k \neq i$, $k \neq j$]). For example, in the points of α_{13}-line in Fig. 1.9a, the liquid–vapor tie-lines are directed to vertex 2.

In the concentration tetrahedron, all points of α-surfaces are characterized by the property that the liquid–vapor tie-lines in these points are directed along the straight lines passing through that edge of the concentration tetrahedron, which connects the vertexes whose numbers are missing in the index of α-surface. For example, in the points of α_{13}-surface in Fig. 1.9b, the liquid–vapor tie-lines are directed to edge 2–4.

In the concentration tetrahedron, the ternary azeotrope gives rise not only to three α-surfaces, but also to one specific α-line in the points of which not two but three components of the phase equilibrium coefficients are equal to each other. We will call the line a *three-index α-line*. For example, in Fig. 1.10b, the ternary azeotrope 123 gives rise to the α_{123}-line, which crosses the face 1–3–4 in the α_{123}-point (it isn't shown).

It is characteristic of all points of the three-index α-line that the liquid–vapor tie-lines in these points are directed along the straight lines passing through that vertex of the concentration tetrahedron, the number of which is missing in the index of α-line.

For example, in Fig. 1.10a in the points of the α_{123}-line, the liquid–vapor tie-lines are directed to vertex 4. Let's note that the α_{123}-line is a line of intersection of all three α-surfaces (α_{12}, α_{13}, and α_{23}).

The quaternary azeotrope gives rise to six α-surfaces in the concentration tetrahedron (the number of combinations is every two from four). Each α-surface gives rise to regions of order of components (i.e., the quaternary azeotrope gives rise to twelve regions of order of components $\underset{ijk}{\text{Reg}_{ord}}$).

Along with it, each α-surface crosses three edges of the concentration tetrahedron and forms three α-points on the edges (points of intersection of α-surface with the edges). Therefore, the quaternary azeotrope gives rise to eighteen α-points on the concentration tetrahedron edges. The availability of these α-points can be a sign important for practice, which allows the existence of a quaternary azeotrope to be predicted.

In this section, we have considered the characteristics of the phase equilibrium coefficients field related to the phase equilibrium coefficients ratio of various components: α-points, lines, surfaces and hypersurfaces, regions of identical order of components $\underset{ijk}{\text{Reg}_{ord}}$, and regions of reversible distillation $\text{Reg}^h_{rev,\,r}$ or $\text{Reg}^l_{rev,\,s}$.

The regions of reversible distillation and regions of the identical order of components are especially significant for the analysis of possible cases of separation by distillation.

As will be seen in the next chapters, the arrangement of the regions of the boundary elements of the concentration simplex where the missing components have the highest or lowest value of the phase equilibrium coefficients is the most significant. To perform this task, it is enough to determine the sequence of α-points on all edges of the concentration simplex, as well as the order of components within the segments between these points (Petlyuk et al., 1985), just the way it is done in Fig. 1.11.

1.8. Liquid–Liquid–Vapor Phase Diagrams

To separate mixtures in which components are characterized by a *limited intersolubility*, not only is liquid–vapor equilibrium of great importance (as it was considered throughout the previous sections), but *liquid–liquid equilibrium* is also important.

Figure 1.12 shows a *liquid–liquid–vapor* phase diagram of isopropyl alcohol (1)–benzene (2)–water (3) mixture. Figure 1.12 shows the critical point of liquid–liquid equilibrium (*cr*) in which the compositions of two equilibrium liquid phases are identical. The thin line shows the vapor line for the region of two liquid phases Reg_{L-L}. On this line, there are points of compositions of vapor that

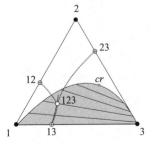

Figure 1.12. Liquid–liquid–vapor phase diagram for benzene (1)–isopropil alcohol(2)–water(3) mixture. Region of two liquid phases Reg_{L-L} is shaded. *cr*, critical point; dotty lines, separatrixes; thin lines, liquid–liquid tie-lines, vapor line.

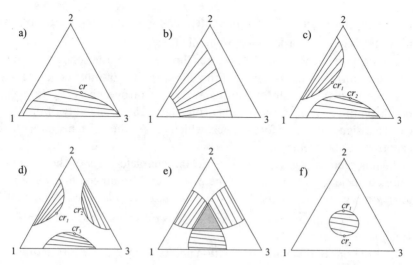

Figure 1.13. Some types of liquid–liquid phase diagrams for three-component mixtures. Region of three liquid phases Reg_{L-L-L} is shaded. Thin line, tie-line liquid–liquid; cr, critical points.

are in equilibrium with both liquid phases. In the segment between the binary and ternary heteroazeotrope, the vapor line almost coincides with the boundary between the distillation regions Reg^∞.

In this particular case, the vapor line is completely located in the region of two liquid phases Reg_{L-L}. However, in other cases this line can go out of two liquid phases' region boundary Reg_{bound}^{L-L} (binodal line).

A liquid–liquid–vapor phase diagram for a ternary mixture is a combination of liquid–vapor and liquid–liquid phase diagrams. Vast collections of liquid–liquid phase diagrams are available, which show the extent to which the behavior of mixtures with two or more liquid phases can differ. Figure 1.13 shows some types of liquid–liquid phase diagrams.

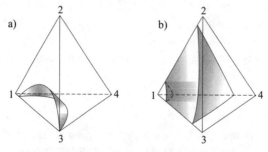

Figure 1.14. Examples of liquid–liquid phase diagrams for four-component mixtures: (a) with one binary two-phase liquid constituent (binodal surface Reg_{bound}^{L-L} is shaded), and (b) with two binary two-phase liquid constituents (region of two liquid phases Reg_{L-L} is shaded).

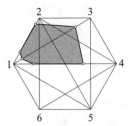

Figure 1.15. Liquid–liquid phase diagram for six-component mixture (points indicate the compositions of equilibrium liquid phases of binary constituents).

We can see that the phase diagrams differ by the number of one- and two-phase liquid regions Reg_{L-L}. In Fig. 1.13e, a system with a three-phase liquid region Reg_{L-L-L} is shown among others.

The most common types of phase diagrams are those represented in Figs. 1.13a and b with one two-phase region. In Figs. 1.14a and b, the examples of liquid–liquid phase diagrams for four-component mixtures with one *two-phase region* are shown.

For a clear representation of liquid–liquid equilibrium in multicomponent systems with one two-phase liquid region Reg_{L-L}, it is possible to use the graph in Fig. 1.15. From Fig. 1.15, it is clear that component 1 is a heteroforming agent (in practice, it is water that plays this role in most cases). Components 1–2, 1–3, and 1–4 form two liquid phases. The rest of the components do not form liquid phases between each other. In such a way, the description of liquid–liquid phase diagrams for multicomponent mixtures with one two-phase region Reg_{L-L} is rather simple.

1.9.   Conclusion

The variety of possible compositions of a multicomponent mixture can be represented as a multidimensional simplex. Liquid–vapor phase equilibrium is a factor determining the distillation process results. Depending on the nonideality degree, the following types of mixtures – differing in their behavior during the distillation process – can be distinguished: ideal, nonideal zeotropic, mixtures with a tangent azeotrope, azeotropic mixtures, and heteroazeotropic mixtures. Residue curves and their bundles characterize phase behavior of mixtures in each point of the concentration simplex. Residue curve bundles define the possible cases of mixture separation in one of the distillation-limiting modes, in particular, the infinite reflux mode. Residue curve bundles split the concentration simplex into distillation regions Reg^{∞}, separated by the lines, surfaces, or hypersurfaces.

Bonds between the stationary points (points of the components and azeotropes) and distillation subregions Reg_{sub} are the structural elements of the distillation regions. Residue curve bundle structure of multicomponent mixtures can be described with the help of a structural matrix that reflects the bonds available between the stationary points.

Lines, surfaces, and hypersurfaces of equal phase equilibrium coefficients of two components split the concentration simplex into regions of identical order of components Reg_{ord}^{ijk} ($K_i > K_j > K_k$) that define the possible causes of separation under the finite reflux mode.

1.10.   Questions

1.   What is the essence of the "lever rule"?

2.   What is an azeotrope?

3.   What is a heteroazeotrope?

4.   Name the stationary points of types of residue curve bundles?

5.   Define a bond, a distillation subregion Reg_{sub}, and a region Reg^∞.

6.   Draw a phase diagram for a three-component mixture with two binary azeotropes of minimum boiling temperatures and fill in the structural matrix for this phase diagram.

7.   Name the distillation regions Reg^∞ and subregions Reg_{sub} for a phase diagram mentioned in item 6.

8.   Indicate the arrangement of the straight lines coming through liquid–vapor tie-lines on the α-line points in the concentration triangle.

9.   Perform the same assignment as in item 8, but for the points of the α-surfaces in the composition tetrahedron.

10. How is it possible to define the availability of a ternary azeotrope, given only the phase equilibrium coefficients of all components from the contour of the concentration triangle?

1.11.   Exercises with Software

1.   For a mixture of acetone(1)–methanol(2)–chloroform(3)–ethanol(4), draw the bonds between components and azeotropes in the concentration tetrahedron, as well as the boundaries of each distillation region.

2.   For a mixture of acetone(1)–benzene(2)–chloroform(3)–toluene(4), draw the α-lines and α-surfaces in the concentration tetrahedron.

References

Bushmakin, I. N., & Kish, I. N. (1957a). Rectification Investigations of a Ternary System Having an Azeotrope of the Saddle-Point Type. *J. Appl. Chem.*, 30, 401–12 (Rus.).

Bushmakin, I. N., & Kish, I. N. (1957b). Separating Lines of Distillation and Rectification of Ternary Systems. *J. Appl. Chem.*, 30, 595–606 (Rus.).

Doherty, M. F., & Caldarola, G. A. (1985). Design and Synthesis of Homogeneous Aseotropic Distillations. 3. The Sequencing of Columns for Azeotropic and Extractive Distillations. *Ind. Eng. Chem. Fundam.*, 24, 474–85.

Gurikov, Yu. V. (1958). Some Questions Concerning the Structure of Two-Phase Liquid–Vapor Equilibrium Diagrams of Ternary Homogeneous Solutions. *J. Phys. Chem.*, 32, 1980–96 (Rus.).

Matsuyama, H., & Nishimura, H. (1977). Topological and Thermodynamic Classification of Ternary Vapor–Liquid Equilibria. *J. Chem. Eng. Japan.*, 10, 181–7.

Ostwald, W. (1900). Dampfdrucke ternarer Gemische, Abhandlungen der Mathematisch-Physischen Classe der Konige Sachsischen. *Gesellschaft der Wissenschaften*, 25, 413–53 (Germ.).

Petlyuk, F. B. (1978). Thermodynamically Reversible Fractionation Process for Multicomponent Azeotropic Mixtures. *Theor. Found. Chem. Eng.*, 12, 270–6.

Petlyuk, F. B., Kievskii, V. Ya., & Serafimov, L. A. (1975a). Thermodynamic and Topological Analysis of Phase Diagrams of Polyazeotropic Mixtures. 1. Determination of Distillation Regions Using a Computer. *J. Phys. Chem.*, 49, 1834–5 (Rus.).

Petlyuk, F. B., Kievskii, V. Ya., & Serafimov, L. A. (1975b). Thermodynamic and Topological Analysis of Phase Diagrams of Polyazeotropic Mixtures. 2. Algorithm for Construction of Structural Graphs for Azeotropic Ternary Mixtures. *J. Phys. Chem.*, 49, 1836–7 (Rus.).

Petlyuk, F. B., Kievskii, V. Ya., & Serafimov, L. A. (1977). Method for Isolation of Regions of Rectification Polyazeotropic Mixtures Using an Electronic Computer. *Theor. Found. Chem. Eng.*, 11, 1–7.

Petlyuk, F. B., Kievskii, V. Ya., & Serafimov, L. A. (1979). Determination of Product Compositions for Polyazeotropic Mixtures Distillation. *Theor. Found. Chem. Eng.*, 13, 643–9.

Petlyuk, F. B., Zaranova, D. A., Isaev, B. A., & Serafimov, L. A. (1985). The Presynthesis and Determination of Possible Separation Sequences of Azeotropic Mixtures. *Theor. Found. Chem. Eng.*, 19, 514–24.

Reinders, W., & De Minjer, C. H. (1940a). Vapour–Liquid Equilibria in Ternary Systems. 1. The System Acetone–Chloroform–Benzene. *Rec. Trav. Chim. Pays-Bas.*, 59, 392–400.

Reinders, W., & De Minjer, C. H. (1940b). Vapour–Liquid Equilibria in Ternary Systems. 2. The Course of the Distillation Lines in the System Acetone–Chloroform–Benzene. *Rec. Trav. Chim. Pays-Bas.*, 59, 401–406.

Schreinemakers, F. A. H. (1901). Dampfdrucke ternarer Gemische. *J. Phys. Chem.*, 36, 413–49 (Germ.).

Serafimov, L. A. (1969). The Azeotropic Rule and the Classification of Multicomponent Mixtures. 4. *N*-Component Mixtures. *J. Phys. Chem.*, 43, 981–3 (Rus.).

Walas, S. M. (1985). *Phase Equilibria in Chemical Engineering*. Boston: Butterworth.

Zharov, V. T. (1967). Free Evaporation of Homogeneous Multicomponent Solutions. *J. Phys. Chem.*, 41, 1539–55 (Rus.).

Zharov, V. T. (1968a). Free Evaporation of Homogeneous Multicomponent. Solutions. 2. Four-Component Systems. *J. Phys. Chem.*, 42, 58–70 (Rus.).

Zharov, V. T. (1968b). Free Evaporation of Homogeneous Multicomponent Solutions. 3. Behavior of Distillation Lines Near Singular Points. *J. Phys. Chem.*, 42, 195–211 (Rus.).

Zharov, V. T., & Serafimov, L. A. (1975). *Physico-Chemical Foundations of Bath Open Distillation and Distillation*. Leningrad: Khimiya (Rus.).

2

Basic Concepts of Distillation

2.1. Purpose and Process Essence of Distillation

Distillation is the oldest and the most universal process of chemical technology and other branches of industry incorporating separation of mixtures.

Practically, all natural substances and substances produced in the chemical reactors are mixtures that do not have the properties required for using them in techniques and for household needs. These mixtures should be separated into components or groups of components.

Distillation has substantial advantages over other processes applied in order to separate a mixture, such as extraction, crystallization, semipermeable membranes, etc. As a rule, it is the most cost-effective process, so it may be used for mixtures with very diverse properties. This process is based on the fact that the composition of the boiling liquid and that of the vapor over it differ. Thus, if the boiling temperature is low (e.g., air separation), it is necessary to use low-temperature refrigerants and conduct the process at a higher pressure. If it is high (e.g., in separation of heavy oil fractions or metals), high-temperature heat carriers or fire preheating have to be used and the process is run under vacuum.

If the composition of the boiling liquid and that of the vapor over it are quite close (e.g., xylene isomers or isotope separation), there is substantial energy consumption, which results in high capital costs.

It is impossible to conduct the distillation process in the case of azeotropic composition (i.e., if the composition of the boiling liquid and that of the vapor are identical).

If the detrimental chemical reactions take place at the boiling temperature of mixture (i.e., the mixture is thermolabile), it is also impossible to run the distillation process.

So if we use the basic theory of distillation and the methods that follow from it, the cost of separation of the mixtures of substances with close boiling temperatures can be decreased, and the problems of azeotropic and thermolabile mixtures separation may also be substantially overcome. On the whole, in this book, special attention will be paid to the problems mentioned above.

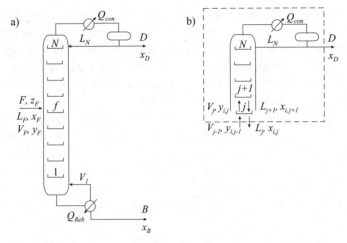

Figure 2.1. (a) A distillation column with condenser, reboiler, and reflux capacity; (b) control volume (dotted line) for obtaining material balance equations for the top section.

2.1.1. Description of Distillation Process

To begin with, let's consider a schematic diagram of a distillation column with a condenser and a reboiler (Fig. 2.1a).

The column feed, in the common case, is vapor–liquid mixture of flow rate F and with concentrations of components z_{iF}, where i is the component number, the vapor feed of flow rate V_F with concentrations of components y_{iF}, and the liquid feed of flow rate L_F with composition x_{iF}.

The reboiler serves to make a vapor flow, which goes upward along the column length; the condenser serves to make a liquid flow, which flows down from tray to tray from top to bottom. On tray j, the liquid flow from tray $j + 1$ meets the vapor flow from tray $j - 1$. These flows are not equilibrium and, therefore, a mass exchange takes place on the tray; a part of the lightest components converts from liquid into the vapor phase, and a part of the heaviest components (of higher boiling temperatures) converts from vapor into the liquid phase.

Although on real trays complete equilibrium between liquid and vapor is never reached, in the world practice a model of a *theoretical tray* (Sorel, 1893) for which this equilibrium is achieved ($1, 1 \div 2$ real trays correspond to one theoretical tray) is accepted. Due to the mass exchange between liquid and vapor, the composition on the trays varies along the column length – on the overhead trays, there is a high concentration of light components and, on the bottom trays, there is a high concentration of heavy components. Thus, a separation of the initial mixture occurs. The vapor rising from the column is condensed in the condenser. One part of the formed liquid is removed as an overhead product (distillate is the amount D of composition x_{iD}), and the other part comes back into the column (reflux in the amount L_N of the same composition x_{iD}). Such a condenser, which condenses all vapor from the column, is called a total condenser. Sometimes a partial condenser is applied, when the distillate is removed in form of vapor; and sometimes a mixed condenser is applied, when one part of the distillate is liquid and the other part is vapor.

Liquid from the bottom of the column goes to the reboiler where it is partially evaporated (reboilers of this kind are called partial). The vapor in amount V_1, of composition y_{i1} comes back to the column, and the remaining liquid in amount B of composition x_{iB} is removed as bottom product.

Ratio $R = L_N/D$ is called the reflux ratio and ratio $S = V_1/B$ is called the reboil ratio. In the reboiler, there is a input of heat in amount Q_R, and, in the condenser, there is a removal of heat in amount Q_{con}.

Thus, distillation is a two-phase (liquid–vapor) multistage counterflow potentially equilibrium process (in some cases – in cases of heteroazeotropic distillation – three phases may occur on the trays: two liquid phases and one vapor phase).

2.1.2. System of Algebraic Equations of Distillation

The distillation process is described by a system of algebraic equations, for the deduction of which let's consider a closed- loop covering, for example, the column overhead beginning from tray j (Fig. 2.1b).

The equation of component material balance:

$$V_{j-1}y_{i,j-1} = L_j x_{ij} + D x_{iD} \tag{2.1}$$

The equation of heat balance:

$$V_{j-1}H_{j-1} = L_j h_j + D h_D + Q_{con} \tag{2.2}$$

The equations of phase equilibrium (for "theoretical" tray):

$$y_{i,j} = K_{ij} x_{ij} \tag{2.3}$$

The summation equations:

$$\sum y_{ij} = 1, \quad \sum x_{ij} = 1 \tag{2.4}$$

Here, $K_{ij} = f(T, P, x_1 \ldots x_n, y_1 \ldots y_n)$ is a coefficient of phase equilibrium. $H_j = \varphi(T, P, y_1 \ldots y_n)$ and $h_j = \psi(T, P, x_1 \ldots x_n)$ are the enthalpies of vapor and liquid, respectively.

At first sight, the system of Eqs. (2.1) $\div$ (2.4) appears to be rather simple, but it is necessary to bear in mind that the equation of phase equilibrium [Eq. (2.3)] together with the equations of summation [Eq. (2.4)] are always nonlinear, even in the case of the so-called ideal mixtures, with $\alpha_{ih} = K_i/K_h = const$ (the component relative volatilities are not influenced by temperature and composition).

In real mixtures, functions $K_{i,j}$ have rather complicated form (especially for azeotropic and heteroazeotropic mixtures).

Sometimes the system [Eq. (2.1) $\div$ (2.4)] is simplified with the rejection of the heat balance equation [Eq. (2.2)] and with the adoption of the flows L_j and V_j constancy within each column section (the term *section* refers to the part of a column between the flow inlet and outlet points).

The system [Eq. (2.1) ÷ (2.4)] may have a large number of equations. First, the number of theoretical trays N may be enormous. Second, number of components n may also be very large. For example, petroleum contains thousands of components, which actually, for practical reasons, will be combined into tens of pseudocomponents (or fractions).

The system [Eq. (2.1) ÷ (2.4)] may be solved only by iteration, and the solution is not always immediately obtained, so it requires a high degree of initial approximation. As a result of the system [Eq. (2.1) ÷ (2.4)] solution at the preset number of theoretical trays in each section, we get not only the compositions of products x_{iD} and x_{iB}, but also the compositions on all trays x_{ij} and y_{ij} − profiles of concentrations along the column length, or distillation trajectories, that come to be the basic subject of this book.

2.2. Geometric Interpretation of Binary Distillation: Reflux and the Number of Trays

2.2.1. McCabe-Thiele Diagram

Geometric interpretation is extremely important for the understanding of distillation process. In this relation, binary distillation gives us particularly large possibilities. Only for binary distillation are we able to show in a flat diagram the composition of both liquid and vapor (curves $y_1 - x_1$).

This gives us an opportunity to understand easily some general regularities of the distillation: the dependence of the required number of trays upon the reflux ratio for a preset separation (preset purity of products), as well as the fact that under a preset separation the reflux ratio and the number of trays cannot be less than some minimum values ($R_{\min}$ and $n_{\min}$).

For this purpose, let's use *diagram* $y_1 - x_1$ (McCabe & Thiele, 1925) with the so-called operation lines applied (Fig. 2.2).

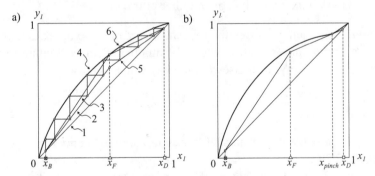

Figure 2.2. McCabe-Thiele diagram for (a) ideal and (b) nonideal mixtures. 1, operating line at infinite reflux. 2, operating line at finite more minimum reflux; 3, operating line at minimum reflux; 4, equilibrium line; 5, composition of liquid and vapor flow that meet on tray; 6, composition of liquid and vapor flow that leave from tray; x_{pinch}, point of tangential pinch.

Let's transform Eq. (2.1), dividing all its members by $V - D$ and taking into account that $V = L + D$ and $L/D = R$:

$$y_{j-1} = [R/(R+1)]x_j + [1/(R+1)]x_D \qquad (2.5)$$

While deriving the Eq. (2.5), we omitted index i, as for the binary mixture the concentration of one component defines the composition. Besides, we omitted index j for V and L flows, accepting their constancy along the section lengths.

Equation (2.5) represents the material balance equation and is called an *operating line equation*.

If we make $x_j = x_D$ substitution in Eq. (2.5), we get $y_D = x_D$. This means that at $x = x_D$, the operation line crosses the diagonal line. Operation line has a tangent of slope angle to the axis of abscissas $R/R + 1$. It allows us to draw the operating lines of sections (e.g., they are given for liquid feed in Fig. 2.2a).

The steps drawn between the section operating lines and the equilibrium curve illustrate the compositions on the trays: points on the operating lines correspond according to Eq. (2.5) to the composition of liquid from the jth tray, which meets the composition of vapor from the tray below and points on the equilibrium curve correspond to the compositions of liquid and vapor leaving the jth tray. Figure 2.2a allows a number of important conclusions:

1. The operating line of rectifying section has two ultimate positions: it may coincide with the diagonal line when $R = \infty$; it may be in a point $x_j = x_F$ reach the equilibrium curve as it is shown in Fig. 2.2a (line 3 corresponds to minimum reflux $R = R_{min}$).
2. When $R = \infty$ (infinite reflux mode), the number of theoretical trays is minimum, i.e., $n = n_{min}$ (the steps between the equilibrium curve and the diagonal line are the largest ones).
3. When $R = R_{min}$ (minimum reflux mode), the number of stages is infinite (in the feed point, the step between stages becomes equal to zero – this is an area of constant concentrations or pinch).
4. With the reflux increasing, the number of trays decreases.

For multicomponent mixtures, the regularities are more complex. But as a rule, there is some minimum R-value at which the number of stages is infinite, and the required number of trays decreases when the R-value increases.

2.2.2. Influences of Nonideality

Now let's see how the nonideality of binary mixtures influences the distillation process (Fig. 2.2b).

From Fig. 2.2b it is clear that, in this particular case, the infinite number of steps and, respectively, the area of constant concentrations appear in the point of tangency of the top section operation line to the equilibrium curve (this point is indicated as x_{pinch}), but not in the feed point. Such an area of constant concentrations is called a *tangential pinch*.

Now, if we consider the *tangential azeotrope mixture* at $x_1 = 1$ (Fig. 1.2c), then for $x_D = 1$ the operation line, tangent to a equilibrium curve, should coincide with the diagonal line (i.e., $R_{min} = \infty$). This makes the production of high-purity component 1 practically impossible.

If we take the azeotropic mixture (Fig. 1.2d), we see that for points x_D and x_B divided by an azeotrope, the distillation process become impossible.

2.3. Geometric Interpretation of Multicomponent Mixture Distillation: Splits

Geometric interpretation of distillation process of binary mixtures has been decisive in understanding the subject and the basic principles of the distillation units design development. Geometric interpretation of multicomponent mixtures distillation is also important for deep insight into the pattern of the multicomponent mixture distillation and better understanding of the methods of design of the units used for the separation of these mixtures.

As Chapter 1 introduced, the composition of a ternary (three-component) mixture is symbolized by a point in the concentration triangle, and the composition of a quaternary mixture is also symbolized by a point in the concentration tetrahedron. The curve of points illustrating the compositions on the distillation column trays is a trajectory of the distillation process within the composition space. The regularities of these trajectories arrangement are the essence of the multicomponent mixture distillation general geometric theory forming the foundation for an optimum design.

Unfortunately, it is impossible to visualize the trajectories of distillation of mixtures with five or more components. However, it will not prevent us from investigating the regularities of multicomponent mixture distillation because we have already observed all these regularities while analyzing the arrangement trajectories of the quaternary mixture distillation.

Let's go on now with the term *split*. Under the split for the preset feed composition x_{iF}, we understand the set of components of each product of separation.

Under the sharp split, we understand such a case when both product points belong to the boundary elements of the composition space (i.e., each product contains only a part of the feed components).

As far as the aim of distillation is, more often, the separation of the mixture into pure components, we are mostly interested in the *sharp splits*.

The rest of the splits we call the *nonsharp splits*.

For a three-component ideal mixture (here and further on component 1 is the lightest, component 2 is the intermediate, and component 3 is the heaviest), an example of sharp split is $1 : 2, 3$ (i.e., $x_{2D} = 0$, $x_{3D} = 0$, $x_{1B} = 0$ – point x_{iD} belongs to vertex 1, point x_{iB} belongs to side 2-3 of triangle). This split has got an additional name – *direct split* (the lightest component is separated from the remaining ones).

For a four-component ideal mixture at $K_1 > K_2 > K_3 > K_4$, the direct split is $1 : 2, 3, 4$. The indirect splits of ternary and quaternary mixtures are $1, 2 : 3$ or $1, 2, 3 : 4$, respectively.

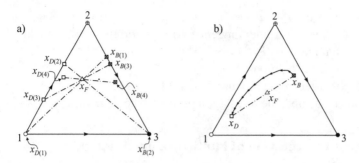

Figure 2.3. Possible splits ($x_{D(1)} : x_{B(1)}$, $x_{D(2)} : x_{B(2)}$, $x_{D(3)} : x_{B(3)}$, $x_{D(4)} :$ $x_{B(4)}$) for (a) three-component ideal mixture and (b) a concentration profile under infinite reflux. Segments with arrows represent liquid–vapor tie-lines.

For four-component mixture there also exists a *intermediate split*: 1, 2 : 3, 4.

Finally, the sharp splits include the *splits with the components to be distributed*: 1,2 : 2,3 (for three-component mixture), and 1,2,3 : 2,3,4; 1,2 : 2,3,4; 1,2,3 : 3,4 (for four-component mixture).

In Fig. 2.3a, split $x_{D(1)} : x_{B(1)}$ is a direct split, $x_{D(2)} : x_{B(2)}$ is a indirect split, $x_{D(3)} :$ $x_{B(3)}$ is a split with a component to be distributed, $x_{D(4)} : x_{B(4)}$ is a nonsharp split.

The sharp splits may be carried out only in infinite columns, but they are of prime importance for the geometric theory of distillation. In the real columns it is possible to obtain the products being as close as you like to the products of sharp splits.

2.4.   Trajectory Bundles Under Infinite Reflux: Distillation Diagrams

In the case of infinite reflux when $R = \infty$, Eq. (2.5) will be as follows (Thormann, 1928:77):

$$y_{j+1} = x_j \tag{2.6}$$

The distillation process under infinite reflux is described by Eqs. (2.3) and (2.6) and is illustrated in the composition space by the trajectory being the *interconnected tie-lines* (the end of one tie-line serves as a beginning of another tie-line when moving upward from the column bottom).

A *concentration profile* ($x_B \rightarrow x_D$) under infinite reflux for an ideal mixture is illustrated in Fig. 2.3b (point F of feed composition is shown as well). The broken line may be substituted with a continuous curve (*c-line*). For any point taken on this line as a tie-line beginning, the tie-line end point is located on the same line (Zharov & Serafimov, 1975). Such a substitution is especially convenient to extend a *c-line* beyond the product points up to the unstable and stable node of the concentration space and to get over to *c-line bundles* (1 ⇒ 3; Fig. 2.4a), (1 ⇒ 2, 3 ⇒ 2; Fig. 2.4b) from an individual *c-line*.

For an ideal mixture, the whole concentration triangle is filled with one *bundle of trajectories* (the trajectory direction toward the temperature increases as is shown

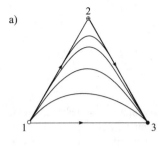

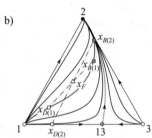

Figure 2.4. Trajectory bundles under infinite reflux for (a) three-component ideal and (b) azeotropic mixtures. $x_{D(1)}$: $x_{B(1)}, x_{D(2)} : x_{B(2)}$, possible splits; solid lines, trajectories; dotty line, separatrix under infinite reflux.

by the arrows). All the trajectories begin in vertex 1 and terminate in vertex 3 and round vertex 2. At the vertexes, the tie-line length becomes equal to zero. Thus, the vertexes are the stationary trajectory points.

If all the trajectories coming from the *stationary point*, in this case, such stationary point is called the *unstable node* N^- (vertex 1). The stationary point to which the trajectories get in is called the *stable node* N^+ (vertex 3). At last, the stationary point that all trajectories bend around is called a *saddle point S* (vertex 2).

In Fig. 2.4b, another example of the trajectory bundles is shown (let's call the picture of trajectory bundles a *distillation diagram*), but already for a three-component azeotropic mixture: acetone(1)-benzene(2)-chloroform(3).

In this case, we have two trajectory bundles, differing by their unstable nodes and separated from each other with a specific trajectory, which begins not at the unstable node, but in a saddle (azeotrope 13 of maximum temperature) and is called the *separatrix*.

The distillation diagram illustrates the arrangement of trajectories to be the profile of concrete column concentrations (Fig. 2.4b). It is enough to choose two points, for example, points $x_{D(1)}$ and $x_{B(1)}$ or $x_{D(2)}$ and $x_{B(2)}$, on the one trajectory and to meet the requirement of the material balance (all points x_D, x_F, and x_B should lay within the one straight line) and to state that a part of trajectories between points $x_{D(1)}$ and $x_{B(1)}$ or $x_{D(2)}$ and $x_{B(2)}$ serves to be a concentration profile of possible distillation column under the infinite reflux.

2.5. Trajectory Bundles Under Finite Reflux

To return to Eqs. (2.3) and (2.5) for the rectifying section and to fix x_{iD} and R parameters, we obtain a number of points x_{ij} by solving this system from the upper tray.

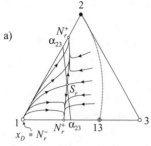

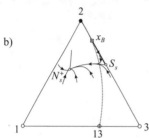

Figure 2.5. Trajectory bundles under finite reflux of acetone(1)-benzene(2)-chloroform(3) azeotropic mixture for (a) rectifying and (b) stripping section. Solid lines with arrows, trajectories; solid line, α-line; dotty line, separatrix under infinite reflux; big circles, stationary points under infinite reflux; little circles, stationary points under finite reflux.

The concentration profile of the rectifying section under reflux R and an overhead product composition x_{iD} will be represented by broken lines, the lengths of which come through points x_{ij}, which are found by means of solving Eqs. (2.3) and (2.5).

In a similar way, with the help of the Eq. (2.3) and the equation of the material balance:

$$x_{j+1} = [S/(S+1)]y_j + [1/(S+1)]x_B \qquad (2.7)$$

we create a trajectory for a stripping section.

Just as in the case of the infinite reflux, the broken lines can be replaced by the continuous curves. The distillation trajectories under the finite reflux, first, are different for two column sections and, second, have the composition points of the corresponding product (x_{iD} or x_{iB}) as parameters as well as reflux ratio or reboil ratio (R or S).

The trajectory bundles of the rectifying and stripping sections for the azeotrope mixture: acetone(1)-benzene(2)-chloroform(3) under $R = 2.5$, $S = 1.4$ are illustrated in Figs. 2.5a and 2.5b, respectively, while the product compositions are $x_{1D} = 1$, $x_{2D} = 0$, $x_{3D} = 0$, and $x_{1B} = 0$, $x_{2B} = 0.85$, $x_{3B} = 0.15$.

The trajectories of Figs. 2.5a and 2.5b were constructed in the following way: in the case of fixed R and x_D or S and x_B, an arbitrary point in the triangle x was chosen and the calculation was performed from this point to bottom in accordance with Eqs. (2.3) and (2.5) for the rectifying section and from this point to top in accordance with the Eqs. (2.3) and (2.7) for the stripping section.

The trajectory starting in product point x_D or x_B and ending in the point corresponding to the feed tray is the only one of the whole bundle. It is the profile of concentrations of the column section.

The given trajectory belongs to some trajectory bundle bounded by its fixed points (points N_r^-, S_r and N_r^+ of Fig. 2.5a and points S_s and N_s^+ of Fig. 2.5b), the separatrixes of the saddle points S and the sides of the concentration triangle.

Knowledge about the regularities of the trajectory bundles arrangement under the finite reflux provides an opportunity to develop the reliable and fast-acting algorithm to fulfill design calculations of distillation to determine the required number of trays for each section.

2.6. Minimum Reflux Mode: Fractionation Classes

Knowledge about the distillation process regularities under minimum reflux is the background of the distillation theory, and this mode analyzing is the most important stage of the distillation column design.

As it has been already mentioned above, at minimum reflux a column has infinite number of steps ($N = \infty$) (i.e., the trajectory passes through one or more stationary points of the bundle). In the column, these stationary points will correspond to the so-called zones of constant concentrations that are identical at adjacent trays.

2.6.1. Binary Distillation

Let's consider now the binary distillation (Fig. 2.6a,b). Having a set value of parameter D/F, we start to increase R from 0 up to ∞ in the infinite column.

In Fig. 2.6a, $R_1 = 0$, $R_3 > R_2 > R_1$. With the increase of R, while maintaining D/F ratio, points x_{iD} and x_{iB} become remote from point x_F, maintaining the constant concentration area of both sections in the feed cross-section. Such a mode is called the *first class of fractionation*. Its specific feature is that the feed composition and the compositions in the areas of constant concentrations of both sections, adjoining the feed tray, coincide.

In the case of $R = R_3$, point x_D coincides with the vertex 1 ($x_{1D} = 1$, $x_{2D} = 0$). Such a mode is a boundary one for the first class of fractionation. Under this

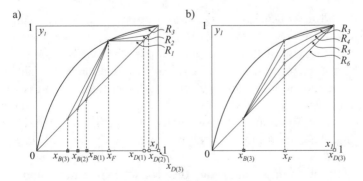

Figure 2.6. Operating lines for (a) first and (b) third class of fractionation for given feed x_F. $R_6 > R_5 > R_4 > R_3 > R_2 > R_1$, splits $x_{D(1)} : x_{B(1)}$ at R_1, $x_{D(2)} : x_{B(2)}$ at R_2, $x_{D(3)} : x_{B(3)}$ at R_3, at R_4, at R_5, and at R_6.

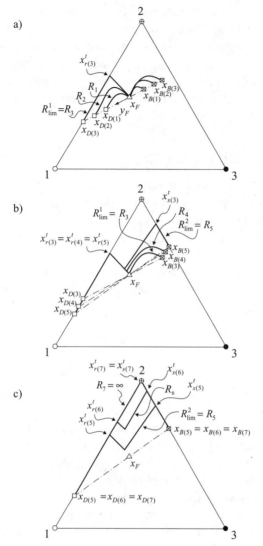

Figure 2.7. A location of product points and trajectories under minimum reflux for given three-component feed x_F: (a) first class of fractionation, (b) second class of fractionation, (c) third class of fractionation. $R_1 < R_2 < R_3 < R_4 < R_5 < R_6 < R_7 = \infty$; splits $x_{D(1)} : x_{B(1)}$ at R_1, $x_{D(2)} : x_{B(2)}$ at R_2, $x_{D(3)} : x_{B(3)}$ at $R_3 = R^1_{\lim}$, $x_{D(4)} : x_{B(4)}$ at R_4, $x_{D(5)} : x_{B(5)}$ at $R_5 = R^2_{\lim}$, at R_6 and $R_7 = \infty$, x^t_r and x^t_s — tear-off points of rectifying and stripping section trajectories.

mode, the second area of constant concentrations appears in the rectifying section (column overhead at $x_1 = 1$).

With further increase of R, we immediately pass to the *third class of fractionation*. For binary mixtures, the second class of fractionation is unavailable. The third class of fractionation is characterized by the fact that, in the case of R increase, the compositions of the separation products are not changed and the areas of constant concentrations in feed cross-section disappear (Fig. 2.6b). In the case of R changing, the compositions on the trays will change as well (in Fig. 2.6b, $R_6 = \infty$, $R_6 > R_5 > R_4 > R_3$).

We have come to an important result: the product compositions under infinite reflux and under a significantly large finite reflux (the third class of fractionation) are identical.

a)
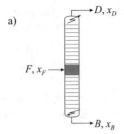

b)
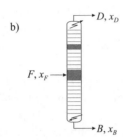

Figure 2.8. A location of zones of constant concentration (pinches) in columns for distillation of three-component mixtures under minimum reflux: (a) $R < R_{\lim}^1$ (first class of fractionation), (b) $R_{\lim}^1 < R < R_{\lim}^2$ (second class of fractionation), (c) $R > R_{\lim}^2$ (third class of fractionation); pinches are shaded.

c)
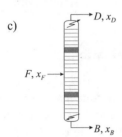

2.6.2. Distillation of Three-Component Mixtures

For mixtures with $n \geq 3$ side by side with the first and third classes of fractionation, an intermediate class – the *second class* – exists.

Let's consider the change of compositions of three-component ideal mixture products in the concentration triangle (Fig. 2.7) under the same conditions as before for the binary one. With the increase of R in the first fractionation class, points x_D and x_B are moving in opposite directions and transferred along the straight line passing through the "vapor–liquid" feed tie-line $x_F \to y_F$ (Fig. 2.7a). The zones of constant concentrations of the column are in the feed cross-section (Fig. 2.8a).

In the case of $R_{\lim}^1 = R_3$ (boundary mode of the first fractionation class), point x_D reaches side 1-2. At this time, the trajectory of distillation of the rectifying section (Fig. 2.7a) is situated along side 1-2 from point x_D up to the tear off point x_r^t, and later it comes inside the concentration triangle up to point x_F. Under these conditions, the trajectory of the stripping section is located completely inside the concentration triangle. The zones of constant concentrations of the column are given in Fig. 2.8b.

In the case of further R increase (the *second fractionation class*), point x_D is traveling along side 1-2 toward vertex 1 and reaches its limiting position $x_{D(5)}$ at fixed D/F parameter, when $R = R_{lim}^2 = R_5$ (boundary mode of the second fractionation class). The tear-off point x_r^t at the second fractionation class conserves its composition and the point of bottom product is traveling along the straight line being parallel to side 1-2, and when $R = R_{lim}^2$, it reaches side 2-3 (Fig. 2.7b).

In the case of further R increase (the third fractionation class), the compositions x_D and x_B do not change and the tear-off points x_r^t and x_s^t travel along sides 1-2 and 2-3 toward vertex 2 until they join in this vertex (Fig. 2.7c) at $R = R_7 = \infty$.

Under the conditions of the second fractionation class, the compositions of products change, but the composition on the feed cross-section differs from the composition of the feed.

In the majority of cases, the product compositions under the infinite reflux coincide with the compositions of the product under a mode on the verge of the second and the third classes of fractionation.

Thus, the analysis of possible compositions of the product under infinite reflux is of practical importance and appreciably easier than the analysis under finite reflux.

2.7.   Adiabatic, Nonadiabatic, and Reversible Distillation

Hitherto, we considered the columns characterized by the fact that heat was brought in only in the reboiler and removed only in the condenser. Therefore, we can call the method of heat feeding and removing the *adiabatic distillation*. The method is the most simple and, therefore, the most prevalent one, but thermodynamically (from the standpoint of the second law of thermodynamics) nonoptimum due to the high temperature of heat feeding and the low temperature of heat removing. In practice, it often requires the high-temperature (i.e., the more expensive) heat carriers and/or the high-priced low-temperature refrigerants. Reduction of the excessive costs is the problem of primary importance when separating the wide-boiling mixtures (e.g., the petroleum). In the case of crude separation, the portion of heat is removed in the middle area of the upper section by the so-called pumparound when the portion of heat is transferred to the petroleum to be separated (Fig. 2.9). Therefore, we can call this method *nonadiabatic distillation*.

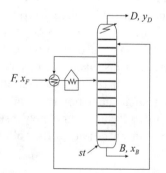

Figure 2.9. A column with pumparound (nonadiabatic distillation) and live steam.

a)

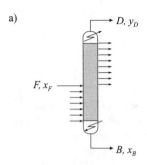

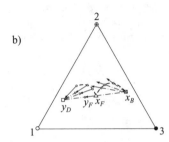

b)

Figure 2.10. A thermodynamically reversible distillation: (a) an infinite column with heat input and output (segments with arrows) at any cross-section of column, (b) a trajectory of reversible distillation. Segments with arrows, liquid–vapor tie-lines for certain cross-section of column (little circles).

If the column is infinite, and if the heat is fed in and removed on each tray by infinitely small portions in such a way that the internal flows of vapor and liquid are variable, and in Eq. (2.1) the concentrations of vapor and liquid flows meeting with each other, $y_{i,j-1}$ and $x_{i,j}$, are equilibrium for all i and j, then such process of distillation will be thermodynamically reversible or it will be equilibrium in each cross-section.

The thermodynamically *reversible distillation* is a hypothetical process in an infinite column in which heat is fed in or removed to each tray at zero temperature differences, there are no heat losses, there is no pressure drop along the column length, and there is no nonequilibrium in all points, including feed point and points of vapor supply from the reboiler and reflux from the condenser.

For the reversible distillation, the following condition is implemented (the second law of thermodynamics):

$$\sum Q_j / T_j = (S_F - S_D - S_B), \tag{2.8}$$

where Q_j is the heat that is fed or removed on tray j at temperature T_j; S_F, S_D, and S_B are the entropies of feed, overhead, and bottom products, respectively.

For the reversible distillation, the operation line should coincide with the equilibrium line for the binary mixture in the McCabe-Thiele diagram.

The scheme of the reversible process is shown in Fig. 2.10a. Figure 2.10b illustrates a trajectory of the reversible distillation for three-component ideal mixture.

In each trajectory point of the stripping section, the liquid–vapor tie-line continuation passes through point x_B and, in each trajectory point of the rectifying section, the liquid–vapor tie-line continuation passes through point y_D (this arises

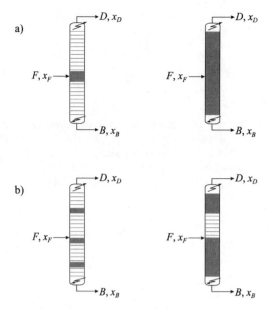

Figure 2.11. A location of pinches (shaded) in column for adiabatic distillation at minimum reflux and reversible distillation for equal product composition: (a) first class of fractionation ($R < R_{lim}^1$) and reversible distillation, (b) second class of fractionation ($R = R_{lim}^2$) and partially reversible distillation.

from the equilibrium conditions between y_{j-1} and x_j and conditions of the material balance for columns' ends).

Hence, it appears that the maximum possible locations of points y_D and x_B are the points of the straight-line intersection, which passes through the feed tie-line with sides 1-2 and 2-3.

For the adiabatic column in the first class of fractionation, the product compositions coincide with the product compositions for the reversible distillation (Fig. 2.11a).

Generally speaking, for the first and second fractionation classes under the minimum reflux mode, the points of compositions in the zones of constant concentrations (i.e., stationary points of the trajectory bundles) should be arranged at the trajectories of reversible distillation built for the product points. It follows from the conditions of the material balance and the phase equilibrium in the zones of constant concentrations. Figure 2.11b illustrates the *partially reversible process* (it is reversible in the column parts that are from the constant concentration zones for the minimum reflux mode up to the column ends).

In the case of the reflux ratio alteration and conservation of the product composition, the stationary points of trajectory bundle are traveling along the reversible distillation trajectories built for a given product, so the trajectories may be called *lines of stationarity*. Thus, the analysis of the reversible distillation trajectory arrangement in the concentration simplex is decisive in general geometric theory of distillation.

The analysis of temperature alteration, as well as the vapor and liquid flows along the trajectory of the reversible distillation, is the basis of the methods for nonadiabatic distillation unit design; the background for developing the new

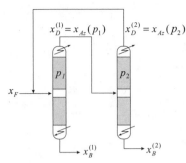

Figure 2.12. A sequence with recycle for separation of bynary azeotropic mixtures under two pressures. p_1 and p_2, pressures in columns.

cost-effective schemes of adiabatic distillation is the analysis of scheme of the vapor and liquid flows between sections of the reversible distillation column.

2.8. Separation of Azeotropic Mixtures by Distillation Under Two Pressures or Heteroazeotropic and Extractive Distillation

On many occasions, general geometric theory of distillation allows development of flowsheets of multicomponent azeotropic mixture separation without using such special methods as distillation under two pressures or heteroazeotropic and extractive distillation with entrainers (i.e., with additional components injected into the unit).

In some other cases, these methods have to be used. For example, if a binary mixture contains an azeotrope whose composition is highly dependent on pressure, it is possible to separate this mixture into two columns, *operating under different pressures*, according to the flowsheet with recycle (Fig. 2.12).

In separating the azeotropic mixtures, we are most frequently obliged to apply an *entrainer* (i.e., to introduce an additional component into the separable mixture

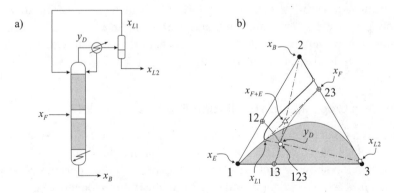

Figure 2.13. (a) A heteroazeotropic column with decanter, (b) the distillation trajectory of heteroazeotropic column for separation of benzene(1)-isopropil alcohol(2)-water(3) mixture (benzene-entrainer). x_{L1} and x_{L2}, two liquid phases; x_F, initial feed; x_{F+E}, total feed into column; region of two liquid phases is shaded.

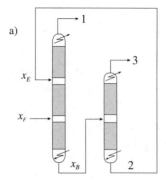
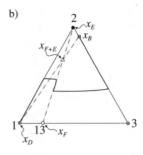

Figure 2.14. (a) A sequence with recycle for extractive distillation (first column, extractive column; second column, column of entrainer recovery); (b) the distillation trajectory of extractive column for separation of acetone(1)-water (2)-methanol(3) mixture (water-entrainer). x_F, initial feed; x_{F+E}, total feed into first column.

that converts, for example, a binary mixture into a three-component mixture with a quite definite structure type, a more easily separable one).

The *heteroazeotropic distillation* (Fig. 2.13a) is applicable only for the mixtures that have a heterogeneous region in the liquid phase Reg_{L-L}.

Figure 2.13b shows a structure of concentration space of mixture benzene (entrainer)(1)-isopropyl alcohol(2)-water(3). The mixture has a ternary azeotrope and three binary azeotropes.

Due to the entrainer and the *decanter*, it is possible to separate binary azeotropic mixture x_F into products x_B (almost pure isopropyl alcohol) and x_D (contaminated water), which may be purified easily in the second column. Point x_D lies not on the distillation trajectory but on the liquid–liquid tie line ($x_D \equiv x_{L2}$).

Figure 2.14a shows a flowsheet of the column of extractive distillation and, in Fig. 2.14b, an example of acetone(1)-water(entrainer)(2)-methanol(3) mixture with section trajectories is shown. This mixture, which is impossible to separate sharply into acetone (x_D) and methanol-water mixture (x_B) in the single-feed column, may be separated into these products in the column with an extractive section located between two feed inlets.

2.9.   Is Process Opposite to Distillation Process Possible?

If there is a reversible distillation process, then there should be also an *opposite process*, which may be called a process of the distilled flow mixing.

Figure 2.15b shows a McCabe-Thiele diagram with the operation lines of the *opposite process of distillation*, and Fig. 2.15a illustrates a scheme of this process.

In the opposite process, a light component is introduced into the column bottom, a heavy component is introduced into the column overhead, and a mixture is removed from the column middle part. In the reboiler, the heat is fed at low temperature T_D and, from the condenser, the heat is removed at high temperature T_B. Thus, the opposite process acts as the heat pump, which transfers heat from

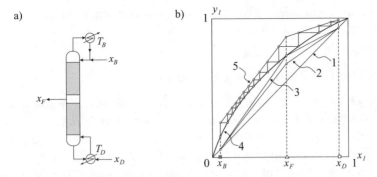

Figure 2.15. (a) A column for the process that is opposite to the distillation process, (b) McCabe-Thiele diagram (operation lines: at infinite reflux, 1; at finite reflux, 2; at minimum reflux, 3; at reversible distillation, 4; at opposite process, 5; little circles, tray composition).

the low temperature level T_D to the high temperature level T_B; however, in this case it does not require mechanical work. This heat transfer occurs due to the fact that mixture x_F has an entropy higher than the total of entropies of flows x_B and x_D (Petlyuk et al., 1984).

It is natural that the temperature in this process increases not from top to bottom as in distillation, but from bottom to top.

In the case of the extractive distillation (Fig. 2.14), when increasing the entrainer flow rate $E \to \infty$, in the extractive section the conditions will come close to those of the top section of the opposite process.

That is why in an extractive section the temperature from top to bottom should not only obligatorily rise, but it may also fall.

2.10. Mixtures with Limited and Unlimited Separability

Under unlimited separability, we understand possible separation from the mixture of any component or a component group (fractions) with any purity (i.e., with the preset negligibly small amounts of other impurity components).

There are two classes of mixtures that are not characterized by unlimited separability: the azeotropic mixtures and the thermolabile ones.

In Section 2.8, we discuss several methods of azeotropic mixtures separation. Thus, the azeotropic mixtures are characterized by a limited separability, while processed in an individual distillation column, and complete separability of these mixtures will be achievable in the case of selection of special schemes consisting of several columns with recycles.

During the separation of thermolabile mixtures, for example, the petroleum, that is, mixtures to be separated at a high temperature at which detrimental chemical reactions take place, the problem is the maximum possible extraction of light components from these mixtures without the excess of specific temperature level. Vacuum separation in the presence of steam is used to decrease the temperature.

Moreover, the application of specific separation schemes and stripping columns, in particular, is of primary importance.

2.11.   The Problem of Designing Distillation Units

The main object of the distillation theory is the development of new methods of designing the units for separation of mixtures into required products at minimum costs. Two basic interconnected stages of the conceptual design of distillation units are (1) synthesis of optimum flowsheet; and (2) determination of the best reflux number, quantity of trays in the column sections, and optimum recycling flows.

The synthesis of optimum sequences for the multicomponent azeotropic mixture is the issue of the distillation theory. Geometric theory of distillation overcomes the principal part of this problem – the determination of possible splits for each potential distillation column that may be included into the synthesized sequence. The best feasible sequences selection is carried out on the basis of the criteria of a minimum number of columns, as well as minimum liquid and vapor flows, under the minimum reflux mode.

The analysis of the minimum reflux mode is used at the stage of sequence selection, as well as at the stage of determination of optimum reflux ratios and the quantity of column trays. The geometric theory of distillation makes it possible to develop the general methods of calculation of minimum and more reflux mode.

For the mixtures characterized by infinite separability, the necessity for determination of possible splits no longer arises because they are known a priori.

Therefore, the problem of how to make the best sequence selection from the many probable ones still exists.

The maximum recovery of the most valuable components from the feed (e.g., the separation of light oil products from the crude oil) is considered to be the criterion of primary importance when choosing the best sequence used for thermolabile mixtures processing.

The second stage, including the selection of the best reflux numbers and the quantity of column section trays, will be the important one. The geometric distillation theory makes it possible to determine the feasible compositions that are to be in trays above and below the feed cross-section, then make the design calculations of the trajectory of sections and determine the best ratio of section tray numbers. The new algorithms allow for an increase in the design quality; and apart from that, they make it possible to lower the separation costs and to practically exclude the human participation in the process of calculation.

2.12.   Questions

1.   What is the "theoretical tray"?
2.   What types of equations are used to calculate the distillation?
3.   What are the distinctions in the representation of minimum, and more finite and infinite reflux modes in the diagram of McCabe and Thiele?

4. How many trays should be there in the column under minimum reflux mode?
5. Enumerate the types of sharp separation.
6. How many trays should be there in the column in the case of sharp distillation?
7. What is the trajectory of distillation?
8. What is the distillation trajectory bundle?
9. What is the stationary point of the distillation trajectory bundle?
10. What is the arrangement of the distillation trajectory bundles under infinite and finite reflux modes dependent on?
11. Where are the product points located in the first fractionation class?
12. Where are the stationary points of the distillation trajectory bundles located?
13. Where are the operating lines of the process opposite to the process of distillation in the diagram of McCabe and Thiele located?

References

McCabe, W. L., & Thiele, E. W. (1925). Graphical Design of Fractionating Columns. *Ind. Eng. Chem.*, 17, 606–11.

Petlyuk, F. B., Serafimov, L. A., Timofeev, V. S., & Maiskii, V. I. (1984). *Method of Heat and Mass Exchange Between Liquids with Different Boiling Temperatures.* Patent USSR No. 1,074, 555 (Rus.).

Sorel, E. (1893). *La Rectification de l'Alcohol.* Paris: Gauthier-Villars. (French).

Thormann, K. (1928). *Destillieren und Rektifizieren.* Leipzig: Verlag von Otto Spamer (Germ.).

Zharov, V. T., & Serafimov, L. A. (1975). *Physico-Chemical Foundations of Bath Open Distillation and Distillation.* Leningrad: Khimiya (Rus.).

3

Trajectories of Distillation in Infinite Columns Under Infinite Reflux

3.1. Introduction

Our main purpose is to understand which column sequences can be used in order to get the necessary products. This task is called the *task of sequencing* (synthesis). The sequencing task is being solved in consecutive order beginning with the first distillation column, where initial n-component mixture comes to. For each column, it is necessary to determine feasible splits.

As a rule, our task is to separate this mixture in a few distillation columns into pure components (perhaps, with the addition of subsidiary components – entrainers, or using, besides distillation, other methods of separation). That is why we are first interested in the sharp splits in each column, when each product of the column contains a number of components smaller than the feeding of the column. The finite number of sharp splits makes determining the sharp splits quite clear and definite.

The important advantage of mixture separability analysis for each sharp split consists of the fact that this analysis, as is shown in this and three later chapters, can be realized with the help of simple formalistic rules without calculation of distillation. A split is feasible if in the concentration space there is trajectory of distillation satisfying the distillation equations for each stage and if this trajectory connects product points. That is why to deduct conditions (rules) of separability it is necessary to study regularities of distillation trajectories location in concentration space.

Because sharp separation is not always feasible for azeotropic mixtures, we also consider the best semisharp splits, when one of the products contains a smaller number of components than the feeding and when the possible product point of the second product is the farthest from the product point of the first product in the concentration space.

The investigation of regularities of distillation trajectories' location in infinite columns under infinite reflux is directed to the solution of the task of determination of possible splits.

The mode of infinite reflux is interesting for us not only as one of limit distillation conditions, but also mainly as a mode to which splits achievable in real columns at finite but quite big reflux correspond. These splits are ones of distillation for border mode between the second and third classes of fractioning.

The question of the reflux at which these splits are achievable in real columns and of how, along with that, the distillation trajectory is located in the concentration space is discussed in Chapter 5. Here, we investigate only the splits themselves.

Often the splits for zeotropic mixtures are ones of sharp separation without distributed components. At practice, these splits are the most widespread because they are the sequences with the smallest number of columns ($n - 1$ column for n-component mixture, if each component is a purpose product) that correspond to them.

For azeotropic mixtures, not all the practically interesting splits are feasible at the infinite reflux. However, the sequencing should have the infinite reflux mode as its starting point because these splits are the easiest to realize at finite reflux. That is why we start systematic examination of distillation trajectories with the infinite reflux rate. It is also proved to be correct because the regularities of trajectories' locations for this mode are the simplest.

The analogy with the process of open evaporation favored the fact that this mode was investigated earlier than the others. Systematic examination of distillation at the infinite reflux was initially carried out in works (Zharov & Serafimov, 1975; Balashov & Serafimov, 1984). The analysis of infinite reflux mode in the infinite columns was made (Petlyuk, 1979; Petlyuk, Kievskii, & Serafimov, 1977; Petlyuk & Serafimov, 1983) that allowed general regularities of separation to be defined for the mixtures with any number of components and azeotropes. A number of important investigations was realized (Doherty, 1985; Doherty & Caldarola, 1985; Laroche et al., 1992; Bekiaris et al., 1993; Safrit & Westerberg, 1997; Rooks et al., 1998) and others.

3.2. Analogy Between Residue Curves and Distillation Trajectories Under Infinite Reflux

Investigations of *residue curves* have been conducted for over 100 years, beginning Ostwald (1900) and Schreinemakers (1901). Later, close correspondence between residue curves (i.e., curves of mixture composition change in time at the open evaporation) and *distillation trajectories at infinite reflux* (i.e., lines of mixture composition change at the plates of the column from top to bottom) was ascertained.

The similarity and the difference of these lines are defined by their equations:

$$dx_i/dt = y_i - x_i = x_i(K_i - 1) \tag{3.1}$$

(For residue curves, see Chapter 1.)

$$x_i^{(k+1)} = y_i^{(k)} = K_i x_i^{(k)} \tag{3.2}$$

[For distillation trajectories at the infinite reflux, see Chapter 2 Thormann (1928).]

a)

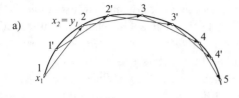

b)

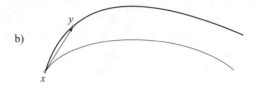

Figure 3.1. Conjugated tie-line liquid–vapor, c-line, and residue curve: (a) conjugated tie-line liquid–vapor and c-line, (b) c-line and residue curve, and (c) intersection c-line and residue curves. $1 \rightarrow 2 \rightarrow 3 \rightarrow 4 \rightarrow 5$ and $1' \rightarrow 2' \rightarrow 3' \rightarrow 4'$, two conjugated tie-lines liquid–vapor on one c-line; thick lines, c-lines; thin lines, residue curves.

c)

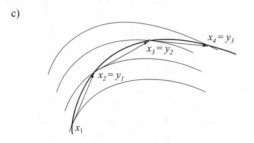

The distillation trajectory under infinite reflux is a *line of conjugated liquid–vapor tie-lines*, each of which corresponds to one of the column plates, in accordance with Eq. (3.2). In the works (Zharov, 1968; Zharov & Serafimov, 1975), the broken line of conjugate liquid–vapor tie-lines is replaced with a *continuous c-line*, for which the liquid–vapor tie-lines are chords (Fig. 3.1a). At the same time it follows from Eq. (3.1) that the liquid–vapor tie-line is a tangent to the residue curve. Therefore, liquid–vapor tie-line, on the one hand, is a tangent to residue curve and, on the other hand, is a chord of the *c*-line. This fact determines the similarity and the difference between the residue curves and the *c*-lines (see Fig. 3.1b). In Fig. 3.1c, it is shown that the distillation trajectory under the infinite reflux (*c*-line) crosses the set of residue curves.

At the same time in the vicinity of stationary points of concentration space (pure components and azeotropes), where the length of the liquid–vapor tie-lines becomes close to zero, the *c*-lines and the residue curves coincide (i.e., the local characteristics of stationary points of *c*-lines bundles and residue curves bundles are the same). That means that topologically the *c*-lines bundles and the residue curves bundles are identical; the structure of the concentration space is also the same: bonds between stationary points, regions Reg^{∞} and subregions Reg_{sub} of distillation and boundaries between them. In other words, distillation regions Reg^{∞} and subregions Reg_{sub} under infinite reflux and boundaries between them, on the one hand, and open evaporation regions and subregions and boundaries between them, on the other hand, contain the same sets of stationary points. However, the exact location of boundaries between the *c*-lines bundles and the residue curves bundles inside the concentration space is different, as can be seen in Figs. 3.1b and 3.1c.

Taking into consideration the aforesaid, sections of Chapter 1 referring to residue curves bundles, to the structural elements of these bundles, and to the matrix description of the concentration space structure are completely valid regarding distillation trajectories under the infinite reflux.

In literature, several different terms for distillation regions at the infinite reflux are used: simple distillation regions, basic regions of distillation, and regions of closed distillation. We use a longer but more exact term – *distillation region at the infinite reflux* (for the sake of briefness, we sometimes use just *distillation region* – Reg^∞).

3.3. Distillation Trajectories of Finite and Infinite Columns at Set Feed Composition

3.3.1. Dimensionality of Product Composition Regions for Finite and Infinite Columns

From Eq. (3.2) and the equation of material balance of the column, a simple mathematic model follows:

$$x_D = x_B K_i^{(1)} K_i^{(2)} \dots K_i^{(N)} \qquad (i = 1, 2 \dots n) \tag{3.3}$$

$$x_D(D/F) + x_B(1 - D/F) = z_F \qquad (i = 1, 2 \dots n) \tag{3.4}$$

From the system of Eqs. (3.3) ÷ (3.4), it follows that *at a given feed composition* z_F and at a fixed field of phase equilibrium coefficients, $K_i = f_i(T, P, x_1, \dots x_n)$ separation products compositions x_D and x_B depend on only two parameters – relative withdrawal of one of the products D/F and amount of theoretical plates N. At infinite reflux, the location of feeding plate does not influence the compositions of distillation products nor profile of concentrations. This is quite understandable – the external flow coming to the feeding plate is infinitely small in comparison with internal flows in the column.

Let's assume at the beginning that for a set composition of feeding z_F unique distillation products compositions x_D and x_B (uniqueness of stationary state) correspond to one set of parameters D/F and N. This assumption is not always carried into effect (see Section 3.7), but in the majority of cases it is. If it is fulfilled, at all feasible values of parameters D/F and N, all the points x_D (and also x_B) form in the concentration simplex of any dimensionality a two dimensional region (*possible product composition regions at fixed feeding composition are two dimensional* because the coordinates of points of these regions depend on two parameters). In particular, for three-component mixtures this region is part of the concentration triangle.

Obviously, at a finite number of stages, the distillation trajectory under the infinite reflux should lie in one of the c-lines and cannot pass through a stationary point of the concentration simplex, start or end in it. At the infinite number of stages, on the contrary, trajectory should go through a stationary point or at least start (end) in it.

As far as c-lines cannot cross each other and boundary elements of concentration simplex are filled with their c-lines bundles, c-lines cannot pass from the internal space of the simplex to its boundary element. Therefore, the distillation trajectories at the infinite reflux can lie completely inside the concentration simplex or inside its boundary elements.

As it follows from the aforesaid, *at a finite number of separation stages, both product points should lie on one c-line inside the concentration simplex. If the number of separation stages is infinite, the following variants of the product point location are feasible: (1) one product point lies inside the concentration simplex and the second one coincides with one of the nodal stationary points; and (2) both product points lie on two different boundary elements of the concentration simplex, and the distillation trajectory goes through their common point, which is a saddle stationary point of the concentration simplex, or goes through two saddle stationary points belonging to these boundary elements.*

3.3.2. Product Composition Regions for Ideal Three-Component Mixtures

Let us examine how the location of the product points should change for three-component mixtures under the infinite reflux at a set value of parameter D/F with the increase of stages number (Figs. 3.2a,b,c). With the increase of N, one of the product points moves toward the node (Fig. 3.2a), or both product points move toward sides of the concentration triangle (Fig. 3.2c), or one of the product points moves toward the node and another one moves toward the side (Fig. 3.2b).

The split (Fig. 3.2a) corresponds to the condition $D/F < x_{F1}$, the split (Fig. 3.2b) corresponds to the condition $D/F = x_{F1}$, and that in Fig. 3.2c corresponds to the condition $x_{F1} < D/F < (x_{F1} + x_{F2})$.

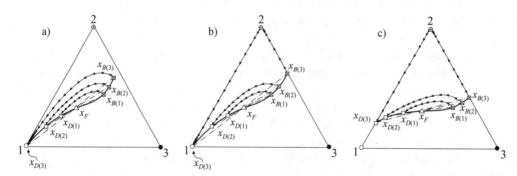

Figure 3.2. Product points and distillation trajectories under infinite reflux for different number of trays: (a) semisharp split, (b) sharp direct split, and (c) split with distributed component. Ideal mixture $(K_1 > K_2 > K_3)$, $x_{D(1)}, x_{D(2)}, x_{D(3)}, x_{B(1)}, x_{B(2)}, x_{B(3)}$, product points for different number of trays, $x_F = \text{const}$, $D/F = \text{const}$; short segments with arrows, conjugated tie-lines liquid–vapor (distillation trajectories under infinite reflux); thick solid lines, lines product composition for different number of trays.

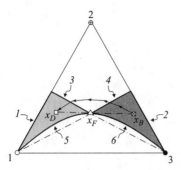

Figure 3.3. Product regions (shaded, bottom region darker shaded) under infinite reflux for given x_F and different number of trays and D/F. Ideal ternary mixture ($K_1 > K_2 > K_3$), line $1 - x_D$ at $N = \infty$ and $x_{F1} < D/F < (x_{F1} + x_{F2})$, line $2 - x_B$ at $N = \infty$ and $x_{F1} < D/F < (x_{F1} + x_{F2})$, line $3 - x_D$ at $N = \infty$ and $(x_{F1} + x_{F2}) < D/F < 1$, line $4 - x_B$ at $N = \infty$ and $0 < D/F < x_{F1}$, line $5 - x_D$ at $D/F = 0$, line $6 - x_B$ at $D/F = 1$.

In the case of a finite number of stages, we have nonsharp separation; in that of an infinite number, we have semisharp (Fig. 3.2a) or sharp separation with a distributed component (Fig. 3.2c), or sharp separation without distributed components (Fig. 3.2b).

The set of product points at all values of parameters N and D/F is shown in Fig. 3.3 (Petlyuk & Avet'yan, 1971; Stichlmair, Fair, & Bravo, 1989). For each feasible point of the top product, there is some corresponding point of the bottom product lying at the intersection of the c-line, passing through the point of top product, and the material balance line, passing through the points of the top product and the feed point.

The set of product points is restricted by the limit values of parameters $N = \infty$, $D/F = 0$, and $D/F = 1$. At $D/F = 0$, only the bottom product is being withdrawn from the column, and at $D/F = 1$, only the top product. Therefore, at $D/F = 0$, the composition of the bottom product coincides with that of the feeding and, at $D/F = 1$, composition of the distillate coincides with that of the feeding. As far as in such mode the point of one of the products coincides with the point of feeding, the distillation trajectory lies in the c-line passing through the feed point (lines 5 and 6 at Fig. 3.3). With the increase of N, the point of the product, the withdrawing of which is zero, is moving away along this c-line from the feed point to the corresponding node.

The set of product points under the infinite reflux ($R = \infty$) and at the infinite number of stages ($N = \infty$) is a subset of the total set of product points at infinite reflux ($R = \infty$). As far as the mentioned *subset ($R = \infty$ and $N = \infty$) depends on one parameter (the only parameter is D/F), it is the line in the concentration triangle and, in general, in the concentration simplex of any dimensionality*. In Fig. 3.3, this subset for points x_D consists of lines 1 and 3 and for points x_B consists of lines 2 and 4.

3.3.3.   Product Composition Regions for Ideal Four-Component Mixtures

Let's examine a set of product points at $R = \infty$ and its subset at $R = \infty$ and $N = \infty$ for a four-component ideal mixture (Fig. 3.4). Some point of the bottom product belonging to the *possible bottom product region at set feed composition* (dark shaded region to the right of point F) corresponds to the top product point

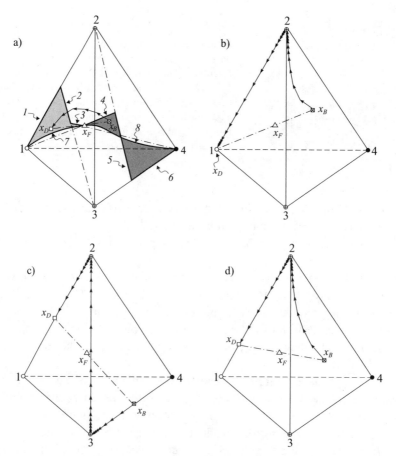

Figure 3.4. (a) Product region under infinite reflux for given x_F and different number of trays and D/F. Ideal four-component mixture ($K_1 > K_2 > K_3 > K_4$), line $1 - x_D$ at $N = \infty$ and $x_{F1} < D/F < (x_{F1} + x_{F2})$, line $2 - x_D$ at $N = \infty$ and $(x_{F1} + x_{F2}) < D/F < (x_{F1} + x_{F2} + x_{F3})$, line $3 - x_D$ at $N = \infty$ and $(x_{F1} + x_{F2} + x_{F3}) < D/F < 1$, line $4 - x_B$ at $N = \infty$ and $0 < D/F < x_{F1}$, line $5 - x_B$ at $N = \infty$ and $x_{F1} < D/F < (x_{F1} + x_{F2})$, line $6 - x_B$ at $N = \infty$ and $(x_{F1} + x_{F2}) < D/F < (x_{F1} + x_{F2} + x_{F3})$, line $7 - x_D$ at $D/F = 0$, line $8 - x_B$ at $D/F = 1$; (b) distillation trajectory for split $1 : 2,3,4$; (c) distillation trajectory for split $1,2 : 3,4$; (d) distillation trajectory for split $1,2 : 2,3,4$. Product regions are shaded, bottom region is darker shaded. Short segments with arrows, conjugated tie-lines liquid–vapor.

belonging to the *possible top product region at set feed composition* (light shaded region to the left of point F), if these points lie at the same c-line and material balance line. Feasible points of distillate at $R = \infty$ and $N = \infty$ lie on lines *1*, *2*, and *3*; feasible points of the bottom product lie on lines *4*, *5*, and *6*.

As can be seen in Figs. 3.3 and 3.4, at $R = \infty$, $N = \infty$, and with the increase of the parameter D/F, the top product point shifts from the unstable node (vertex 1) to the feed point and the bottom product point shifts from the feed point to the stable node (to vertex 3 in Fig. 3.3 or vertex 4 in Fig. 3.4). For this, the number of distillate components increases and the number of the bottom product components decreases.

3.3.4. Feasible Splits for Ideal Mixtures

At some boundary values of the parameter D/F, at which it is equal to the concentration of the lightest component or to the sum of concentrations of a few light components in the feeding, we have sharp separation without distributed component and at other values of the parameter D/F we have sharp separation with one distributed component. These are sharp splits without distributed components: $1:2,3,4; 1,2:3,4; 1,2,3:4$ (here and further the components of the top product are shown before the colon and those of the bottom product follow the colon).

In the case of sharp separation of ideal mixture without distributed components, the initial mixture is separated into two different groups of components: the *top product components* and the *bottom product components*. The heaviest component among the top product components is called the *light key component* and the lightest component among bottom product components is called the *heavy key component*. The light and the heavy key components neighbour in volatility.

The following splits in Fig. 3.4 belong to the splits with one distributed component: $1,2:2,3,4$, lines *1* and *5*; $1,2,3:3,4$, lines *2* and *6*.

Splits with the number of distributed components bigger than one at $R = \infty$ and $N = \infty$ are impossible (e.g., for four-component mixture, the split $1,2,3 : 2,3,4$ with two distributed components is impossible).

Another important property of the mode of $R = \infty$ and $N = \infty$ consists in the following: feasible splits do not depend on the form of c-lines inside the concentration simplex and on the availability of α-lines. For example, for the ideal mixture in Fig. 3.5a and for the zeotropic mixture in Fig. 3.5b, the set of feasible splits is one and the same: $1:2,3; 1,2:3$ and $1,2:2,3$.

At an arbitrary location of the point x_F, any point of edges 1-2 and 3-4 and any point of faces 1-2-3 and 2-3-4 (Fig. 3.4) can be top x_D or bottom x_B product point. Further, we call the set of possible product points in each of the boundary

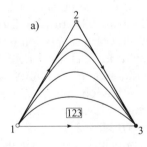

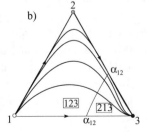

Figure 3.5. C-lines for ternary zeotropic mixtures: (a) ideal mixture, and (b) mixture with α-line. *123, 213,* component order regions.

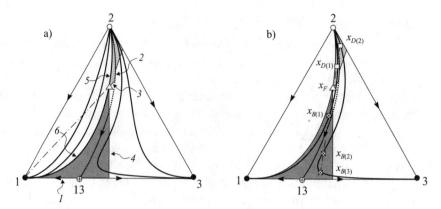

Figure 3.6. (a) Product regions (shaded) under infinite reflux for given x_F and different N and D/F for ternary azeotropic mixture: line $1 - x_B$ at $N = \infty$ and $x_{F2} < D/F$, line $2 - x_D$ at $N = \infty$ and $x_{F2} < D/F$, line $3 - x_D$ at $D/F < x_{F2}$, line $4 - x_B$ at $D/F < x_{F2} + x_{F3}$. (b) Some product points for given x_F ($x_{D(1)}$ and $x_{B(1)}$, $x_{D(2)}$ and $x_{B(2)}$, $x_{D(2)}$ and $x_{B(3)}$).

elements of concentration simplex at an arbitrary location of feed point the *region of possible product composition at sharp distillation* Reg_D and Reg_B. This term is widely used in sequencing.

3.3.5. Product Composition Regions for Azeotropic Three-Component Mixtures

Let's examine three-component azeotropic mixtures with one binary azeotrope and with two regions of distillation at infinite reflux Reg^∞ (Fig. 3.6a). There is some region (triangle to the right of separatrix) where two points of the bottom product corresponding to one top product point exist. This fact is explained by the S-shape of c-lines in this region (Fig. 3.6b, points $x_{B(2)}$ and $x_{B(3)}$).

The main difference between the azeotropic mixtures (and also nonideal zeotropic mixtures) and the ideal ones are that, to determine possible splits of an azeotropic mixture, special analysis is required. The availability of a few distillation regions under the infinite reflux Reg^∞ can result in sharp separation becoming completely impossible or in a decrease in sharp splits number. Let's note that for ideal mixtures the line of possible products compositions at $R = \infty$ and $N = \infty$ and set feed composition goes partially inside the concentration simplex and partially along its boundary elements. For azeotropic mixtures, this line can go along the boundary elements of the distillation region (Fig. 3.6a, line 2).

The question about feasible splits is one of the principal questions in the distillation theory. The understanding of this question was gradually transformed and became more precise.

The original oversimplified view on feasible azeotropic mixtures splits consists of the following: the feed point and product points have to belong to one distillation region ($x_D \in \text{Reg}^\infty$ and $x_B \in \text{Reg}^\infty$ if $x_F \in \text{Reg}^\infty$). This view is quite accurate if the separatrix of distillation regions is linear. In a general case, at curvilinear separatrixes, the feed point can lie in one distillation region at infinite reflux and

a)

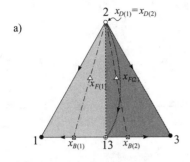

b)

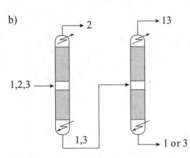

Figure 3.7. (a) Product simplexes Reg_{simp} for ternary azeotropic mixture (shaded), and (b) two-column sequence (product points – 2,13,1 for feed point $x_{F(1)}$ or 2,13,3 for feed point $x_{F(2)}$).

both product points in the other one ($x_{D(2)} \in \text{Reg}^{\infty}$ and $x_{B(2)} \in \text{Reg}^{\infty}$, but $x_F \notin \text{Reg}^{\infty}$; Fig. 3.6b). This property was noted in the works (Balashov, Grishunin, & Serafimov, 1970; Balashov, Grishunin, & Serafimov, 1984; Balashov & Serafimov, 1984). For example, in Fig. 3.6a there is shaded triangle to the right from separatrix 2-13 filled with possible bottom points x_B, while the feed point x_F lies to the left of this separatrix.

This property allowed to propose sequences of columns with recycles (Balashov et al., 1970; Balashov & Serafimov, 1984; Balashov et al., 1984). Recently, much attention is devoted to such sequences (Laroche et al., 1992).

Figure 3.6a shows that at $R = \infty$ and $N = \infty$ for the type of azeotropic mixtures under consideration, there is only one sharp split $2 : 1, 3$ regardless of the feed point location. However, if the point x_F lies to the left of straight line 2-13, then the bottom product point appears at the segment 1-13, otherwise, at the segment 13-3 (Fig. 3.7a). Correspondingly, in the second column, the bottom product will be component 1 or 3. Thus, at sharp separation of such azeotropic mixture in each column, the set of column sequence products depends only on the feed point location relative to the straight line 2-13.

Further, we call triangles 1-2-13 and 3-2-13 *product simplexes* Reg_{simp}. This notion has great significance for separation flowsheets synthesis, because for a feed point x_F located inside the product simplex one can get all the components and azeotropes that are vertexes of this simplex in a sequence of $(n - 1)$ columns.

In Fig. 3.8, bundles of c-lines for some types of azeotrope mixtures are shown and, in Fig. 3.9, possible products compositions regions at $R = \infty$ and at the given feed compositions x_F.

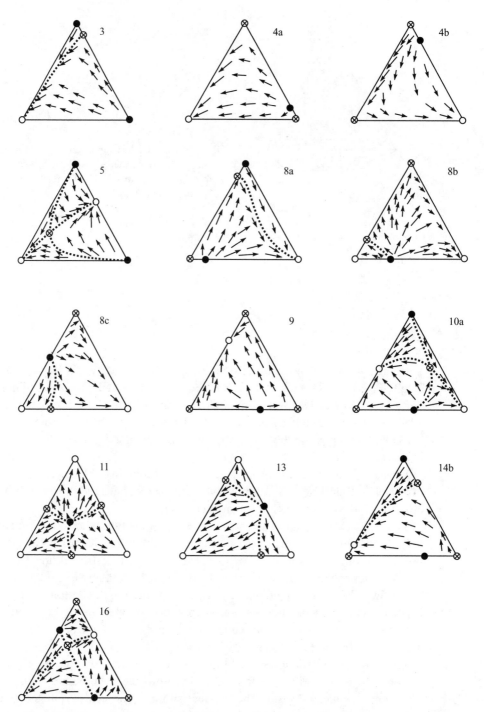

Figure 3.8. Computer simulation-derived tie-lines liquid–vapor for some structures of three-component mixtures. 3,4a,4b..., classification according to Gurikov (1958). Short lines with arrows, tie-lines liquid–vapor.

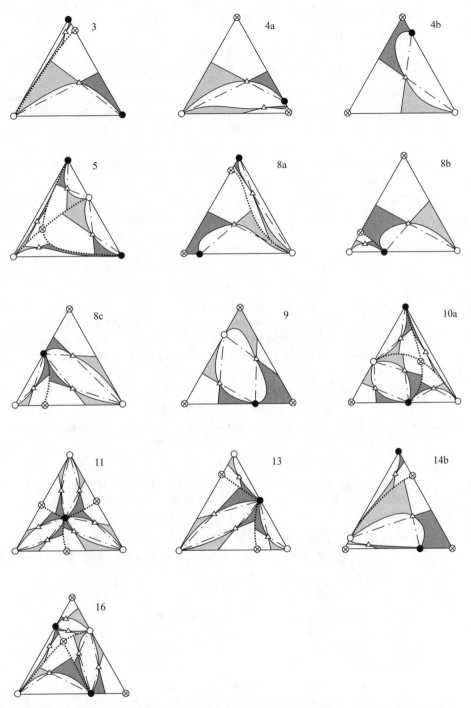

Figure 3.9. Subregions of distillation under infinite reflux Reg$_{sub}$ for some structures of three-component mixtures and product regions for given x_F, 3,4a,4b ... – classification according to Gurikov (1958). Product regions are shaded; bottom regions are darker shaded.

3.4. Rule for the Checkup of Azeotropic Mixtures Separability at $R = \infty$ and $N = \infty$

3.4.1. Distillation Trajectories Location at $R = \infty$ and $N = \infty$

To deduct the general rule for the checkup of possibility one or another sharp split, let's examine peculiarities of sharp distillation trajectories' location at $R = \infty$ and $N = \infty$ (Petlyuk, Avet'yan, & Inyaeva, 1977; Petlyuk, 1979; Petlyuk & Serafimov, 1983). In Figs. 3.2b, c, the distillation trajectories location for splits with one distributed component and without distributed component is shown. In Fig. 3.2c, distillation trajectory from the top product point, lying on side 1-2 of the concentration triangle, goes along this side (boundary element of the concentration triangle $\text{Reg}_D \equiv \text{Reg}^\infty_{bound, D}$) to the *stable node* N^+_D of this side (i.e., to vertex 2). Vertex 2 is, at the same time, a stable node for side 1-2, an *unstable node* N^-_B for side 2-3 ($\text{Reg}_B \equiv \text{Reg}^\infty_{bound, B}$), and a saddle point for the concentration triangle. From the vertex 2, distillation trajectory goes along side 2-3 to the bottom product point. We can briefly describe the distillation trajectory as follows: $x_D \to N^+_D \equiv N^-_B \to x_B$.
$_{1,2}_2_{2,3}$

Thus, the distillation trajectory in this case consists of two parts. The first part is located in the boundary element $\text{Reg}^\infty_{bound, D}$ the top product point belongs to; it joins the top product point x_D with the *stable node* N^+_D of this boundary element. The second part is located in the boundary element $\text{Reg}^\infty_{bound, B}$ the bottom product point belongs to – it joins the *unstable node* N^-_B of this boundary element with the bottom product point. In Fig. 3.2b, the top product point coincides with a boundary element of zero dimensionality – vertex 1. In this case, trajectory consists of the same two parts – the whole side 1-2 and part of the side 2-3 ($x_D \equiv N^+_D \to N^-_B \to x_B$).
$_1_2_{2,3}$

Let us examine the case of four-component mixture (Fig. 3.4). Let us consider the split 1 : 2,3,4. The distillation trajectory goes from vertex $1 \equiv \text{Reg}_D$ at edge 1-2, to vertex 2 and further inside face $2\text{-}3\text{-}4 \equiv \text{Reg}_B$ by c-line to the bottom point $x_B \in 2\text{-}3\text{-}4$ ($x_D \equiv N^+_D \to N^-_B \to x_B$; Fig. 3.4b). Let us also consider the split 1,2 : 3,4.
$_1_2_{2,3,4}$
The distillation trajectory goes from point x_D on edge 1-2 ($x_D \in 1\text{-}2 \equiv \text{Reg}_D$) along it to vertex 2, then along the edge 2-3, and further along edge 3-4 ($3\text{-}4 \equiv \text{Reg}_B$) to bottom product point $x_B \in 3\text{-}4$ ($x_D \to N^+_D \to N^-_B \to x_B$; Fig. 3.4c).
$_{1,2}_2_3_{3,4}$

In Figs. 3.10a, b, distillation trajectories at $R = \infty$ and $N = \infty$ for two types of three-component azeotrope mixtures are shown.

In the splits mentioned, the common rule is valid – the trajectory consists of three parts located in boundary elements of distillation regions for the top and bottom products points and in bond or in a few bonds, connecting *stable node* N^+_D *of the boundary element of top product* Reg_D and *unstable node* N^-_B *of the boundary element of the bottom product* Reg_B. In split 13 : 1,2 (Fig. 3.10b), there are two bonds: ($x_D \equiv N^+_{D(1)} \to N^+_{D(2)} \to N^-_{B(1)} \to x_B$). Let us note that split 2,3 :
$_{13}_3_2_{1,2}$
1,2 at big R and N numbers for this type of mixture was proposed for separation of binary azeotropic mixture 1,3 with small amount of entrainer 2 in the work (Laroche et al., 1992).

a)

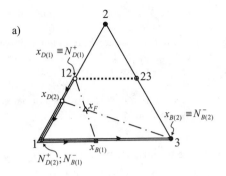

b)

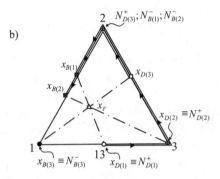

Figure 3.10. Condition of connectedness for two structures of three-component mixtures: (a) 8c, (b) 4b. 8c and 4b — classification according to Gurikov (1958). Thick lines with arrows, bonds and c-lines; N_D^+, stable node of distillate boundary element $\mathrm{Reg}_{bound,D}^{\infty}$; N_B^-, unstable node of bottom boundary element $\mathrm{Reg}_{bound,B}^{\infty}$; $x_{D(1)} : x_{B(1)}$, $x_{D(2)} : x_{B(2)}$, $x_{D(3)} : x_{B(3)}$, possible splits.

One of three parts of trajectory can transform into nodal or saddle point of concentration simplex.

3.4.2. Application of the Rule of Connectedness

Relatively simple examined examples allow formulation of the *general rule* (Petlyuk et al., 1977), with which the product points should comply at $R = \infty$ and $N = \infty$ at sharp separation: (1) *the stable node N_D^+ of the top product boundary element* Reg_D *and the unstable node N_B^- of the bottom product boundary element* Reg_D *should coincide* ($N_D^+ \equiv N_B^-$) *or should be connected with each other by the bond* ($N_D^+ \rightarrow N_B^-$) *or chain of bonds in direction to the bottom product*; and (2) the product points and the feed point should meet the conditions of material balance.

For the sake of briefness, we call the first of these conditions the term of *connectedness*. It has general nature – it can be applied to mixtures with any number of components and azeotropes. Moreover, the term of connectedness embraces not only sharp splits, when the product points lie in the boundary elements of the concentration simplex, but also the semisharp and nonsharp splits, when the product points lie in the boundary elements of the distillation region.

In Fig. 3.11, three constituent parts ($x_D \rightarrow N_D^+$, $N_D^+ \rightarrow N_B^-$, and $N_B^- \rightarrow x_B$) of n-component mixture distillation trajectory at $R = \infty$ and $N = \infty$ are shown. The term of connectedness establishes mutual location of distillation products feasible points at $R = \infty$ and $N = \infty$. Together with conditions of material balance, the

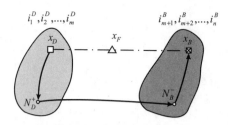

$i_1^D, i_2^D, ..., i_m^D$ $i_{m+1}^B, i_{m+2}^B, ..., i_n^B$

Figure 3.11. Condition of connectedness. Thick lines with arrows – bond, c-lines, indices: D and B, distillate and bottom boundary elements $\text{Reg}_{bound,D}^\infty$ and $\text{Reg}_{bound,B}^\infty$ of concentration symplex correspondingly; $i_1^D \div i_m^D$, components of distillate; $i_{m+1}^B \div i_n^B$, components of bottom. Product regions are shaded; bottom region is darker shaded.

term of connectedness determines feasible variants of product points' location in concentration symplex (i.e., separation products feasible compositions at any feed composition). This property of connectedness rule can be used for the solution sequencing tasks. The rule of connectedness follows from the fact that at $N = \infty$ and $R = \infty$ distillation trajectory should go at boundary elements of trajectory bundle at the infinite reflux (i.e., at the boundaries of distillation region $\text{Reg}_{bound,D}^\infty$ and $\text{Reg}_{bound,B}^\infty$). At sharp split, these boundary elements should be located at the boundary elements of concentration symplex. But, in the general case, $\text{Reg}_{bound,D}^\infty$ and/or $\text{Reg}_{bound,B}^\infty$ may be located within the concentration symplex on separatrixes, separating one distillation region Reg^∞ from the other. We examine such examples later (see Fig. 3.14). The rule of connectedness holds good in these cases, too.

Let's examine the application of the rule of connectedness in a few more cases. At the beginning, the trivial case of impossible separation of the ideal three-component mixture split $2:1,3$ (Fig. 3.3) does not meet the rule of connectedness. Really, stable node N_D^+ of top product region $\text{Reg}_D \equiv \text{Reg}_{bound,D}^\infty$ is vertex 2 and unstable node N_B^- of the bottom product region $\text{Reg}_B \equiv \text{Reg}_{bound,B}^\infty$ is vertex 1. Bond 1-2 is directed to the top but not to the bottom product.

Let's examine four-component azeotropic mixture (Fig. 3.12) with one region

$$\begin{array}{c} \uparrow \rightarrow \rightarrow 2 \rightarrow \rightarrow \downarrow \\ \text{of distillation: } 12 \Rightarrow 24 \text{ or } 12 \rightarrow 1 \rightarrow 3 \rightarrow 23 \rightarrow 24. \text{ Split } 1,3:2,4 \, (x_{D(1)}:x_{B(1)}), \\ \downarrow \rightarrow 4 \rightarrow \uparrow \end{array}$$

according to the rule of connectedness, is possible because the stable node N_D^+ of top product boundary element $\text{Reg}_{bound,D}^\infty$ (vertex 3) is connected with the unstable node N_B^- of the bottom product boundary element $\text{Reg}_{bound,B}^\infty$ (vertex 4):

$$\begin{array}{cccc} x_{D(1)} & \rightarrow N_{D(1)}^+ & \rightarrow N_{B(1)}^- & \rightarrow x_{B(1)} \\ \text{Reg}_{bound,D}^\infty & \text{Reg}_{bound,D}^\infty & \text{Reg}_{bound,B}^\infty & \text{Reg}_{bound,B}^\infty \end{array}$$. Here, the points through

which the distillation trajectory goes are given in the upper line, and the boundary elements they belong to are given in the lower one. However, split $x_{D(2)}:x_{B(2)}$ is impossible because between points $N_{D(2)}^+$ and $N_{B(2)}^-$ there is azeotrope 23 and the rule of connectedness is not kept.

To determine whether one or another sharp occurs split occurs in the case of $R = \infty$ and $N = \infty$, it is necessary (1) to define products compositions x_D and x_B; (2) to find the stable node N_D^+ of the boundary element Reg_D to which the point x_D belongs; (3) to find the unstable node N_B^- of the boundary element Reg_B, to which point x_B belongs; and (4) to establish whether points N_D^+ and N_B^- coincide with each other ($N_D^+ \equiv N_B^-$), or there is a bond or chain of bonds $N_D^+ \rightarrow N_B^-$, or this condition is not kept.

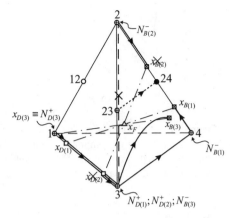

Figure 3.12. Examples of possible $(x_{D(1)} : x_{B(1)}, x_{D(3)} : x_{B(3)})$ and impossible $(x_{D(2)} : x_{B(2)})$ splits of four-component azeotropic mixture. Thick lines with arrows, bond and c-lines; dotted crossed line, absence of bond; dotty line, separatrix; impossible product points are striked out.

As we already saw, for three- and four-component mixtures, all the operations enumerated can be graphically implemented.

3.4.3. *n*-Component Mixture

At $n > 4$, it is necessary to use special algorithms. Determining products x_D and x_B compositions for distillation without distributed components $1, 2, \ldots k : k + 1$, $k + 2, \ldots n$ is implemented according to formulas:

$$x_{Di} = z_{Fi} \bigg/ \sum_{1}^{k} z_{Fi} \qquad (i = 1, 2, \ldots k) \tag{3.5}$$

$$x_{Bi} = z_{Fi} \bigg/ \sum_{k+1}^{n} z_{Fi} \qquad (i = k + 1, k + 2, \ldots n) \tag{3.6}$$

Nodes N_D^+ and N_B^- are defined by means of calculation of line of conjugated liquid–vapor tie-lines from point x_D or line of conjugated vapor–liquid tie-lines from point x_B. In Fig. 3.12, line of conjugated vapor–liquid tie-lines from point $x_{B(3)}$ is shown for four-component azeotropic mixture at separation according to split 1 : 2,3,4. In this case, N_B^- coincides with vertex 3.

For an *n*-component mixture, the rule of connectedness in some cases can be used without graphic interpretation of the concentration simplex and without application of structural matrix (Petlyuk et al., 1977; Knight & Doherty, 1990). Such elementary cases are the following: (1) $N_D^+ \equiv N_B^-$, and (2) N_D^+ and N_B^- are two vertexes of concentration simplex and at the edge (k), $(k + 1)$ between them, there is no binary azeotrope and temperature in the point $N_B^- = (k + 1)$ is higher than at the point $N_D^+ = (k)$. In these cases, there exists the link $N_D^+ \to N_B^-$, which makes such a split possible. In more complicated cases with a large number of components and azeotropes, it is necessary to use structural matrix. The structural matrix describes the structure of the whole concentration space and checks the possibility of separation for any feed composition of distillation column. It is

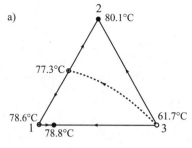

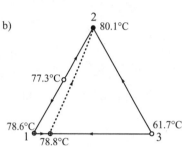

Figure 3.13. An example of bonds ambiguity at the same azeotropes and components boiling temperatures for methyl-ethyl ketone(1)-benzene(2)-chloroform(3) mixture. Arrows, directions of residium curves.

necessary, in particular, at the examination of sequences with recycles when feed composition depends on the recycle flow rate.

To build a structural matrix, it is necessary to obtain information about compositions and boiling temperatures of all the pure components and azeotropes. It is possible to get this information in various reference books on azeotropy (Gmehling et al., 1994a, 1994b) and/or by calculation using the known models of phase equilibrium. In Fidkowski, Malone, & Doherty (1993), there is a general algorithm based on the method of homotopy that allows all azeotropes of n-component mixture to be found simultaneously.

In the majority of cases, the information about azeotropes' and components' boiling temperatures is sufficient for the unique determination of connections between them, but sometimes it is not sufficient. The example of such ambiguity is shown in Fig. 3.13 for the mixture methyl-ethyl ketone(1)-benzene(2)-chloroform(3), the boiling temperatures for which are the following:

$$T_1 = 79.6°C, \ T_2 = 80.2°C, \ T_3 = 61.2°C, \ T_{12} = 78.1°C, \ T_{13} = 79.9°C.$$

The special algorithms of structural matrix synthesis were developed. In Petlyuk, Kievskii, & Serafimov (1975a, 1975b) and Petlyuk et al. (1977), the algorithm is based only on the information about azeotropes' and components' boiling temperatures. This algorithm includes the organized sorting out of the stationary points pairs and the checking of the possibility of connection between them. First, the binary constituents of n-component mixtures are examined, then the three-component constituents, four-component constituents, etc.

This algorithm was tested on a number of industrial polyazeotropic mixtures: fractions of oxidate of naphtha (14 components, 23 binary and 6 ternary azeotropes), fractions of coal tar (20 components, 38 binary and 16 ternary azeotropes), and a mixture processed in the resin industry (9 components, 15 binary and 3 ternary azeotropes).

In Safrit & Westerberg (1997), a heuristic algorithm is based on the information about local characteristics of stationary points, and checked by the authors at large amounts of three-component mixtures and at some four-component mixtures. This algorithm takes into consideration azeotropes formed by any number of components.

In industry, it is necessary to deal with very complicated mixtures for which structural matrices can serve as an instrument of separation flowsheets synthesis. In Wahnschafft (1997), the example of plant for separation of coal tar in South Africa (20 components, more than 200 azeotropes) consisting of 40 columns is given.

Having a structural matrix and knowing compositions of products x_D and x_B, it is easy to find the nodal points N_D^+ and N_B^- in the boundary elements of the concentration space and to determine the availability of a bond or chain of bonds $N_D^+ \rightarrow N_B^-$.

3.5. Feasible Splits at $R = \infty$ and $N = \infty$

When dealing with practical tasks, the designer of separation flowsheet should have on hand the set of feasible splits in the first column. Of course, this set of splits will hardly allow the separation of the mixture in the system of columns into pure components without the use of recycles or special methods. But frequently it is sufficient to separate only some product components. Sometimes it is reasonable to separate the mixture into several fractions that can be the subject of separation by more complicated methods, for example, using entrainers. In any case, at first stage the designer has to determine the set of the splits.

For mixtures of any number of components, it is the easiest to determine two splits: the direct and the indirect one. To do that, it is enough to calculate from x_F point the line of conjugated liquid–vapor tie-lines and the line of conjugated vapor–liquid tie-lines. The first of them will lead to the unstable node N^- of the distillation region Reg^∞, to which point x_F belongs, and the other will lead to the stable node N^+. Full or partial separability of these components or pseudo-components (azeotropes) from the mixture is always possible and corresponds to the direct or indirect split. At direct split $x_D \equiv N^-$, point x_B belongs to the distillation region boundary (i.e., $x_B \in \text{Reg}^\infty_{bound} \equiv \text{Reg}_B$). At indirect split $x_B \equiv N^+$, x_D belongs to the distillation region boundary (i.e., $x_D \in \text{Reg}^\infty_{bound} \equiv \text{Reg}_D$). The distillation structural matrix allows determination of stationary points of the boundaries between the distillation region and other regions, and states by calculation to which of these boundaries the second product point belongs at direct and indirect split. For three- and four-component mixtures, even for structurally complicated ones, these operations may be rendered clear and demonstrable if the necessary software is at hand. Figure 3.14 shows examples of which distillation

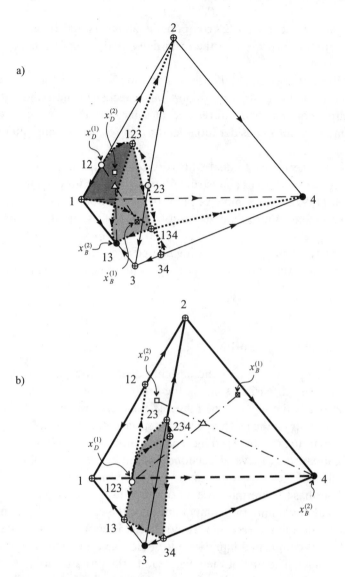

Figure 3.14. Examples of determination whether the feed points x_F belong to distillation regions Reg$^\infty$ and what are the compositions of x_D and x_B products at direct $x_{D(1)} : x_{B(1)}$ and indirect $x_{D(2)} : x_{B(2)}$ splits of the mixture of (a) acetone(1)-methanol(2)-chloroform(3)-ethanol(4), the region $12 \Rightarrow 13$; (b) isopropanol(1)-benzene(2)-cyclohexane(3)-n-butanol(4), the region $123 \Rightarrow 4$. The distillation regions boundaries are shaded; the thick lines stand for distillation region frames.

region the given feed compositions belong to for four-component mixtures with several binary and ternary azeotropes and the boundaries of the distillation regions found with the adjacent regions Reg$^\infty_{bound}$: (1) for the mixture acetone(1)-methanol(2)-chloroform(3)-ethanol(4) of composition $x_F(0.65, 0.15, 0.15, 0.05)$, and (2) for the mixture i-propanol(1)-benzene(2)-cyclohexane(3)-n-butanol(4) of composition $x_F(0.15, 0.40, 0.15, 0.30)$. In addition, the location of points x_D and x_B at direct and indirect split is shown [for mixture 1 at direct split $x_D \equiv 12(0.785, 0.215, 0.0, 0.0), x_B(0.414, 0.036, 0.412, 0.138)$, at indirect split $x_D(0.687, 0.168, 0.089, 0.056), x_B \equiv 13(0.344, 0.0, 0.656, 0.0)$; for mixture 2 at direct split $x_D \equiv 123(0.376, 0.169, 0.455, 0.0), x_B(0.047, 0.510, 0.0, 0.443)$, at indirect split $x_D(0.215, 0.570, 0.215, 0.0), x_B \equiv 4(0.0, 0.0, 0.0, 1.0)$].

However, the above-described method is unfit for splits other than the direct and indirect ones, and the number of such splits grows dramatically with the increase of n. The general methods of *splits set determination* are based on the usage of *structural matrix* and of *method of product simplex for distillation subregions* (Petlyuk, Kievskii, & Serafimov, 1979).

3.5.1. Method of Product Simplex for Distillation Subregions ($m = n$)

Let's return to the notion of *distillation subregion* $\text{Reg}_{sub}(\text{Reg}_{sub} \in \text{Reg}^\infty)$. It is a polygon, a polyhedron, or a hyperpolyhedron, the vertexes (components and azeotropes) of one bonds chain connecting nodes of distillation region Reg^∞ and including all the components of a mixture. The types of boundary elements of distillation subregion are the following: (1) parts of boundary elements of concentration simplex coincident with boundary elements of distillation region $\text{Reg}^\infty_{bound}$; (2) boundary elements of distillation region $\text{Reg}^\infty_{bound}$, separating it from other distillation regions; (3) straight lines, planes, or hyperplanes connecting nodes of distillation region and separating the subregion under consideration from the other subregions inside one distillation region.

The term *distillation subregion* was introduced in a number of works (Petlyuk et al., 1975a; Petlyuk et al., 1977; Petlyuk & Serafimov, 1983). In contrast to that of *distillation region*, the notion of distillation subregion includes not only location of c-lines bundles, but also definite conditions of material balance. In Safrit & Westerberg (1997), for distillation subregion the terminology *the region of continuous distillation* in contrast to the terminology *region of batch distillation* was used.

Let's show that *if product points belong to the first and second types boundary elements of distillation subregion* $\text{Reg}_{sub}(x_D \in \text{Reg}_{sub}$ and $x_B \in \text{Reg}_{sub})$, *then these product points meet the conditions of connectedness* (e.g., all these splits are feasible; at $R = \infty$ and $N = \infty$, product points should lie on these boundary elements).

Let distillation subregion correspond to the following *chain of bonds*: $A_1 \rightarrow A_2 \rightarrow A_3 \rightarrow \cdots \rightarrow A_{m-1} \rightarrow A_m$, where $A_1, A_2, \ldots A_m$ are stationary points, the set of which includes all the n components of the mixture, $m \geq n$. (One should bear in mind that we call bonds chain the sequence of bonds for which the end of one is the beginning of the next one.)

Let's examine the case when $m = n$. In this case the distillation subregion is a simplex ($\text{Reg}^\infty_{sub} \equiv \text{Reg}_{simp}$), the amount of vertexes of which is equal to the amount of components. From the practical point of view, it is frequently convenient to substitute such simplex for the simplex with linear boundary elements, which we call the *product simplex* Reg_{simp}. Here are the examples of such distillation subregions:

1. $2 \to 13 \to 1$ and $2 \to 13 \to 3$ ($2 \Rightarrow 1$ and $2 \Rightarrow 3$) at Fig. 3.6
2. $12 \to 23 \to 2$, $12 \to 1 \to 3$ and $12 \to 23 \to 3$ ($12 \Rightarrow 2$ and $12 \Rightarrow 3$) at Fig. 3.10a
3. $12 \to 2 \to 23 \to 24$ ($12 \Rightarrow 24$) at Fig. 3.12

Such simplex is analogous to an ideal mixture, components of which correspond to the stationary points of the simplex (in both cases, we have only one bonds chain). That is why at sharp separation, when product points lie in boundary elements of distillation subregion, the set of stationary points (pseudocomponents) is divided into two subsets of stationary points of the top and bottom products, just as at separation of ideal mixture, a set of components is divided into two subsets.

At sharp separation of azeotropic mixture, the *key stationary points* (*key pseudocomponents*), that is, stationary points that are adjacent in the bonds chain, play the key components. At separation without distributed pseudocomponents, the set of stationary points $A_1, A_2 \ldots A_m$ is divided into two subsets: $A_1, A_2 \ldots A_k$ and $A_{k+1}, A_{k+2} \ldots A_m$. These subsets are the boundary elements of distillation subregion $\text{Reg}_{bound}^{\infty}$, which the top and bottom product points belong to (Reg_D and Reg_B). Dimensionality of these boundary elements is $k - 1$ and $m - (k + 1)$, correspondingly. The summary dimensionality of these boundary elements is equal to $m + 2$.

As far as the *stable node* of boundary element $A_1, A_2 \ldots A_k$ (Reg_D) is stationary point $A_k(A_k \equiv N_D^+)$ and *unstable node* of boundary element $A_{k+1}, A_{k+2} \ldots A_m$ (Reg_B) is stationary point A_{k+1} ($A_{k+1} \equiv N_B^-$) and as far as there is bond $A_k \to A_{k+1}$ (A_k and A_{k+1} are adjacent stationary points of one bonds chain), separation into considered subsets of stationary points meets the rule of connectedness (i.e., it is feasible). In exactly the same way, it is possible to show that splits with one distributed pseudocomponent are feasible. It is noteworthy that the boundary elements $A_1, A_2 \ldots A_k$ and $A_{k+1}, A_{k+2} \ldots A_m$ are curvilinear, and three constituent parts of the distillation trajectory $x_D \to N_D^+ \to N_B^- \to x_B$ are also curvilinear.

Thus, *if the feed point lies inside some distillation subregion* ($x_F \in \text{Reg}_{sub}$) *to which the chain of bonds $A_1 \to A_2 \to \cdots \to A_m$ (where $m = n$), corresponds, then at $R = \infty$ and $N = \infty$ the following splits without distributed pseudocomponents are feasible:* (1) $A_1 : A_2, A_3, \ldots A_m$; (2) $A_1, A_2 : A_3, \ldots A_m$; $\ldots$; $(m-1) A_1, A_2, \ldots A_{m-1} : A_m$, *and also the following splits with one distributed pseudocomponent:* (1) $A_1 : A_1, A_2 \ldots A_m$; (2) $A_1, A_2 : A_2, \ldots A_m$; $\ldots (m-1) A_1, A_2, \ldots A_{m-1} : A_{m-1}, A_m$; (m) $A_1, A_2, \ldots A_{m-1}, A_m : A_m$.

Let's call the above-stated method of determination of the set of feasible splits at $R = \infty$ and $N = \infty$ the *method of product simplex*.

3.5.2. Method of Product Simplex for Distillation Subregions ($m > n$)

Let's examine the case of $m > n$. Here are the examples of such distillation sub-regions:

1. $13 \to 3 \to 2 \to 1$ ($13 \Rightarrow 1$) at Fig. 3.10b
2. $12 \to 1 \to 3 \to 4 \to 24$ and $12 \to 1 \to 3 \to 23 \to 24$ ($12 \Rightarrow 24$) at Fig. 3.12

If we choose among m stationary points any n ones, then we return to the previous case. *Let's call part of distillation subregion* Reg_{sub}, *containing n stationary points and having linear boundary elements, a product simplex* Reg_{simp} ($\text{Reg}_{simp} \in \text{Reg}_{sub}$). It is noteworthy that here the linearity is assumed solely in order to make it easier to determine if the point x_F belongs to this or that product simplex. *For a product simplex, the separation in one column of a feed pseudocomponents (stationary points) set into two product subsets is feasible if it meets the rule of connectedness (if* $x_F \in \text{Reg}_{simp}$, *then* $x_D \in \text{Reg}_{simp}$ *and* $x_B \in \text{Reg}_{simp}$). Product simplex is an analog of distillation subregion under condition when distillation subregion has n stationary points because the stationary points of the product simplex are connected with one bonds chain and a number of stationary points is also equal to n.

Product simplex for three-component mixtures is a triangle; for four-component mixtures, it is a tetrahedron; for five-component mixtures, it is a pentahedron; etc. Inside one distillation subregion at $m > n$, product simplexes cross each other (i.e., one and the same feed point can simultaneously enter several product simplexes).

Thus, the product simplex is an elementary cell in the general structure of concentration space at $R = \infty$ and $N = \infty$. In the example shown in Fig. 3.15 (the

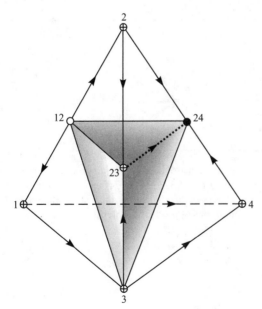

Figure 3.15. An example of product simplex Reg_{simp} of four-component azeotropic mixture (shaded).

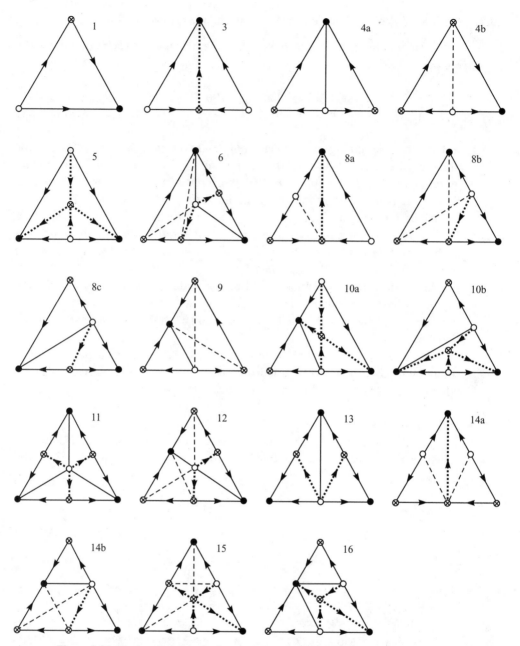

Figure 3.16. Regions Reg$^\infty$ and subregions Reg$_{sub}$ of distillation and product simplexes Reg$_{simp}$ of some structures of three-component mixtures. 1,3,4a,..., classification according to Gurikov (1958). Dotty lines with arrows, boundaries of distillation regions; thin lines, boundaries of distillation subregions; dotted lines, boundaries of product simplexes.

structure is as in Fig. 3.12), the vertexes of the simplex enter in the link chain $12 \rightarrow 1 \rightarrow 3 \rightarrow 23 \rightarrow 24$. Let's consider this example in more detail because it has some peculiarities. If the feed point belongs to the product simplex shown in this figure, then the feasible top product compositions x_D, according to the rule of connectedness, can be (1) point 12, which is the unstable node N^- of the whole distillation region Reg^∞, (2) any point at the segment 12,3, and (3) any point in the triangle 12, 3, 23, for which the stable node N_D^+ of the distillation region boundary elements $\text{Reg}_D = \text{Reg}^\infty_{bound,D}$ is point 23. The feasible bottom product compositions can be (1) point 24, (2) any point at the segment 23,24 for which the unstable node N_B^- of the distillation region boundary element $\text{Reg}^\infty_{bound,B}$ is point 23 (the boundary element of distillation region $\text{Reg}^\infty_{bound,B}$ is separatrix $23 \rightarrow 24$), and (3) any point in the triangle 3, 23, 24, for which the unstable node N_B^- of the distillation region boundary element $\text{Reg}^\infty_{bound,B}$ is point 3. Thus, the following splits without distributed components are possible: 12 : 3,23,24 (trajectory $12 \rightarrow 1 \rightarrow 3 \rightarrow 3,23,24$), 12,3 : 23,24 (trajectory $12,3 \rightarrow 23 \rightarrow 23,24$), 12,3,23 : 24 (trajectory $12,3,23 \rightarrow 23 \rightarrow 24$), with one distributed component: 12,3,23 : 23,24 (trajectory $12,3,23 \rightarrow 23 \rightarrow 23,24$) is also possible, but the split 12,3 : 3,23,24 is impossible because at this split $N_D^+ \equiv 23$ and $N_B^- \equiv 3$; that is, the link $N_D^+ \rightarrow N_B^-$ does not exist. This is why at $m > n$ it is necessary to check the rule of connectedness for the supposed possible splits inside the product simplex. Besides that, it is necessary to find out if the boundary element to which the supposed product point belongs is one of the first two types of boundary elements of distillation subregion $\text{Reg}^\infty_{bound,D}$. Thus, if the feed point belongs to a product simplex $x_F \in \text{Reg}_{simp}(m > n)$, the components and pseudocomponents of this product simplex can be separated at $R = \infty$ and $N = \infty$, just as components of ideal mixture if $x_D \in \text{Reg}^\infty_{bound,D}$ and $x_B \in \text{Reg}^\infty_{bound,B}$, and if the rule of connectedness is not broken.

The notion of product simplex coincides with the notion of region of batch distillation that was used in Safrit & Westerberg (1997).

A method similar to that of product simplex and considering azeotropes as pseudocomponents was proposed for synthesis of separation flowsheets in Sargent (1994).

Division of main types of three-component mixtures phase diagrams according to classification of Gurikov (1958) into product triangles is given in Fig. 3.16. Each type pertains to two different antipodal phase diagrams, which differ from each other by direction of all the bonds and replacement of stable nodes by unstable ones and vice versa. This figure can serve as the basis for determination of feasible sharp and semisharp splits of various azeotropic mixtures at any feed composition.

3.5.3. Algorithm of Product Simplex for n-Component Mixtures

For n-component mixtures, the method of product simplex is based on application of structural matrix. It includes the following steps:

1. Determination from structural matrix by means of sorting out the bonds chains, set of stationary points of which includes in itself in total all the

a)

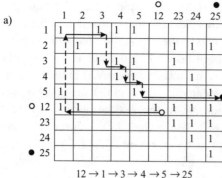

$$12 \to 1 \to 3 \to 4 \to 5 \to 25$$

b)

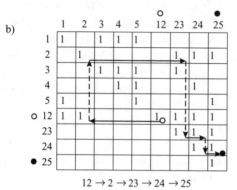

$$12 \to 2 \to 23 \to 24 \to 25$$

Figure 3.17. An example of identification of connection chains by means of structural matrix. Thick line with arrow, bond; dotted line, transfer to next bond: (a) first chain $12 \to 1 \to 3 \to 4 \to 5 \to 25$, and (b) second chain $12 \to 2 \to 23 \to 24 \to 25$.

components of the mixture being separated (i.e., determination of the distillation subregions Reg_{sub})

2. Determination of the product simplexes Reg_{simp} from each bonds chain, for which $m > n$
3. Checkup of which product simplexes feed point belongs to ($x_F \in Reg_{simp}$)

To illustrate the first step, determination of the first and second bonds chain from structural matrix is shown in Figs. 3.17a, b, respectively. Let's note that the attempt to use bond $1 \to 4$ at creation of the second bonds chain does not lead to positive result (bonds chain $12 \to 1 \to 4 \to 5 \to 25$ does not include component 3). To isolate the product simplexes Reg_{simp} from bonds chain with $m > n$ (second step), the combinations of n stationary points from m are being examined. For example, for the chain $12 \to 1 \to 3 \to 4 \to 5 \to 25(12 \Rightarrow 25)$, it is possible to get the following product simplexes Reg_{simp}:

1. $12 \to 1 \to 3 \to 4 \to 5$
2. $12 \to 3 \to 4 \to 5 \to 25$
3. $1 \to 3 \to 4 \to 5 \to 25$

The other combinations of six stationary points 12, 1, 3, 4, 5, and 25, five at a time, do not form product simplexes because they do not contain all the components. The checkup of belonging of the feed point to one or another product simplex (third step) should be performed for all the product simplexes. This checkup is being carried out by means of solution of the following linear equation system for each product simplex:

$$x_{F1} = x_1^1 a_1 + x_1^2 a_2 + \cdots + x_1^n a_n$$
$$x_{F2} = x_2^1 a_1 + x_2^2 a_2 + \cdots + x_2^n a_n \tag{3.8}$$
$$\ldots\ldots\ldots\ldots\ldots\ldots\ldots$$
$$x_{Fn} = x_n^1 a_1 + x_n^2 a_2 + \cdots + x_n^n a_n$$

where x_i^j is concentration of component i in stationary point A_j of the product simplex Reg_{simp}.

The system [Eq. (3.8)] is an expression for center of gravity of the product simplex, when gravity is applied only to its vertexes (stationary points). For all this, the center of gravity coincides with feed point x_F and relative distances of vertexes of the simplex A_j from feed point are equal to the corresponding coefficients a_j.

If the feed point belongs to product simplex Reg_{simp} being examined, then solution of the system [Eq. (3.8)] relative to coefficients a_j should answer inequalities $0 < a_j < 1$. If part of coefficients a_j does not answer these inequalities, then feed point is located out of the product simplex being examined. This method of determination of whether the feed point belongs to one or to another product simplex was proposed in Petlyuk et al. (1979).

If it is ascertained by means of such analyses that feed point belongs to several product simplexes, then the mixture at $R = \infty$ and $N = \infty$ can be separated into pseudocomponents at any border between two adjacent *key stationary points* of each product simplex if the rule connectedness is not broken. The splits with one pseudocomponent being distributed between products is also feasible.

The described approach allows ascertainment of a complete set of splits of fixed initial mixture x_F at $R = \infty$ and $N = \infty$ independently on the number of components and azeotropes. Thus, this method can be applied for synthesis of separation flowsheets of polyazeotropic mixtures. Let's show an application of product simplex method at obvious case of four-component mixture separation acetone(1)-benzene(2)-chloroform(3)-toluene(4) of composition $x_{F1} = 0.25, x_{F2} = 0.30, x_{F3} = 0.20$, and $x_{F4} = 0.25$. The mixture has one binary azeotrope 13 ($x_{az,1} = 0.66; x_{az,2} = 0.34$). The boiling temperatures of the components and of the azeotrope are $T_1 = 56.5°C$, $T_2 = 80.1°C$, $T_3 = 61.2°C$, and $T_{13} = 66°C$. Bonds between stationary points are shown in the concentration tetrahedron (Fig. 3.18a) and in the structural matrix (Figs. 3.18b,c). It is seen from the structural matrix that there are only two bonds chains: $1 \rightarrow 13 \rightarrow 2 \rightarrow 4$ ($1 \Rightarrow 4, m = n$) and $3 \rightarrow 13 \rightarrow 2 \rightarrow 4$ ($3 \Rightarrow 4, m = n$). These two bonds chains corresponds to two distillation regions Reg^∞, which differ from each other by their unstable nodes (points 1 and 3). Each distillation region contains one product simplex Reg_{simp} and feed point gets into product simplex $\text{Reg}_{simp} \equiv 1 \rightarrow 13 \rightarrow 2 \rightarrow 4$ ($x_F \in \text{Reg}_{simp}$ – it is seen even without solution of the system [Eq. (3.8)]). Therefore, feasible splits at $R = \infty$ and $N = \infty$ without distributed components or pseudocomponents: (1) 1 : 13,2,4; (2) 1,13 : 2,4; (3)1,13,2 : 4.

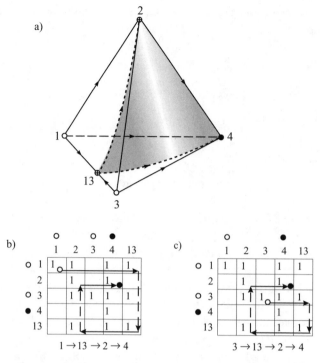

Figure 3.18. (a) A acetone(1)-benzene(2)-chloroform(3)-toluene (4) concentration tetrahedron, and (b) a structural matrix of this mixture and connection chains $1 \to 13 \to 2 \to 4$ and $3 \to 13 \to 2 \to 4$. Separatrix surface is shaded.

An example of determination if the feed composition belongs to this or that product simplex, which was formerly depicted in Fig. 3.14a, is shown in Fig. 3.19. Besides the direct and the indirect splits that were shown in Fig. 3.14a, we have three other possible splits: an intermediate split and two splits with one distributed component. In Fig. 3.19, lines of material balance are shown for all possible splits of the feed composition under consideration at $R = \infty$ and $N = \infty$.

As an example, let's examine the industrial polyazeotropic mixture, which is a by-product of wood pyrolysis (Petlyuk, Kievskii, & Serafimov, 1979). Approximate composition of this mixture and components boiling temperatures are given in Table 3.1. Boiling temperatures and compositions of azeotropes are given in Table 3.2. The structural matrix shown at Fig. 3.20 was synthesized for this mixture.

It follows from the structural matrix that concentration simplex contains three distillation regions $\text{Reg}^\infty (1 \Rightarrow 6, 1 \Rightarrow 8, \text{ and } 1 \Rightarrow 9)$, with common unstable node corresponding to the point of acetaldehyde-1. The points corresponding to ethanol-6, water-8, and diethylketone-9 are stable nodes.

The product simplex Reg_{simp} the feed point belongs to is $1 \to 234 \to 23 \to 38 \to 568 \to 58 \to 78 \to 89 \to 8$ (distillation region is $1 \Rightarrow 8$) (at Fig. 3.20 this bonds chain is marked). Here are feasible splits in one column without distributed pseudocomponents:

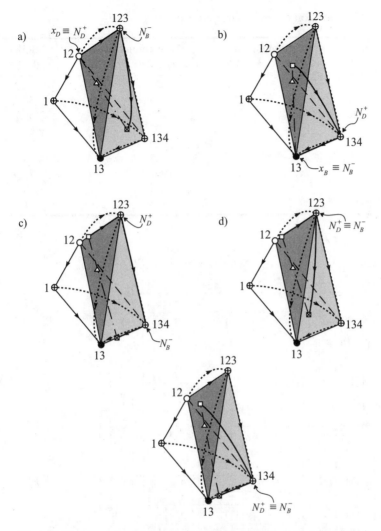

Figure 3.19. Example of determination whether the feed point (Fig. 3.14a) belongs to the product simplex Reg_{simp} (only the distillation region $\text{Reg}^{\infty} \equiv 12 \Rightarrow 13$ is shown). Material balance lines and distillation trajectories at $R = \infty$, $N = \infty$ (shown schematically on linear boundary elements of the distillation region $\text{Reg}_{bound}^{simp}$, not on curvilinear ones of the distillation region $\text{Reg}_{bound}^{\infty}$). The visible faces of the simplex are shaded (one darker, one lighter).

1. 1 : 234,23,38,568,58,78,89,8
2. 1,234 : 23,38,568,58,78,89,8
3. 1,234,23 : 38,568,58,78,89,8
4. 1,234,23,38 : 568,58,78,89,8
5. 1,234,23,38,568 : 58,78,89,8
6. 1,234,23,38,568,58 : 78,89,8
7. 1,234,23,38,568,58,78 : 89,8
8. 1,234,23,38,568,58,78,89 : 8

Table 3.1. Composition of wood pyrolysis product

Component name	Component no.	Feed composition (x_{Fi}, mass. %)	Boiling point (°C)
Acetaldehyde	1	0.997	20.2
Acetone	2	0.443	56.2
Methyl acetate	3	2.215	56.7
Methanol	4	1.107	64.5
Ethyl acetate	5	4.429	77.1
Ethanol	6	0.443	78.3
MEK	7	0.997	79.6
Water	8	88.594	100.0
Diethyl ketone	9	0.775	101.7

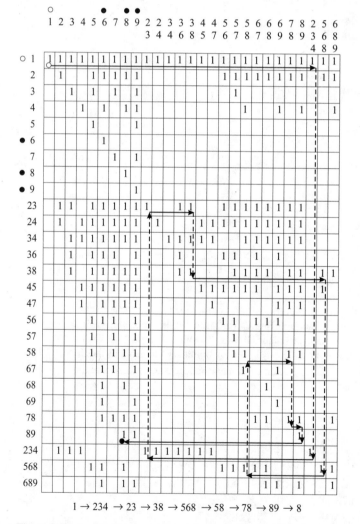

$$1 \to 234 \to 23 \to 38 \to 568 \to 58 \to 78 \to 89 \to 8$$

Figure 3.20. A structural matrix and one connection chain $1 \to 234 \to 23 \to 38 \to 568 \to 58 \to 78 \to 89 \to 8$ of wood pyrolysis product.

Table 3.2. Azeotropes for wood pyrolysis product

Azeotrope	Component no.	Boiling point (°C)	Composition of azeotrope (mass. %)
Acetone	23	55.4	64.7
Methyl acetate			35.3
Acetone	24	55.5	88.0
Methanol			12.0
Methyl acetate	34	53.5	81.3
Methanol			18.7
Methyl acetate	36	56.6	97.0
Ethanol			3.0
Methyl acetate	38	55.6	90.4
Water			9.6
Methanol	45	62.3	55.0
Ethyl acetate			45.0
Methanol	47	64.3	30.0
MEK			70.0
Ethyl acetate	56	71.8	69.1
Ethanol			30.9
Ethyl acetate	57	76.5	78.0
MEK			22.0
Ethyl acetate	58	70.4	91.5
Water			8.5
Ethanol	67	74.3	60.6
MEK			39.4
Ethanol	68	78.1	95.6
Water			4.4
Ethanol	69	78.2	91.2
Diethyl ketone			8.8
MEK	78	73.4	89.0
Water			11.0
Water	89	82.9	86.0
Diethyl ketone			14.0
Acetone	234	53.3	5.8
Methyl acetate			76.8
Methanol			17.4
Ethyl acetate	568	70.3	83.2
Ethanol			9.0
Water			7.8
Ethanol	689	77.4	71.7
Water			9.1
Diethyl ketone			19.2

In the case of direct split, acetaldehyde is being isolated as top product, and at indirect split, water is being isolated as bottom product. Among eight splits, only the first one is a sharp split without distributed components.

Splits with one distributed component shown below in brackets are the third one (component 3), the fourth one (component 8), the sixth one (component 8), the seventh one (component 8), and the eighth one (component 8).

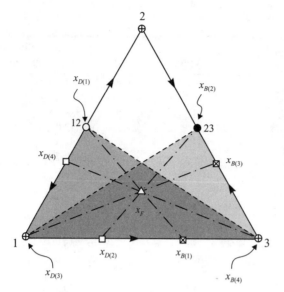

Figure 3.21. Some possible splits $(x_{D(1)} : x_{B(1)}, x_{D(2)} : x_{B(2)}, x_{D(3)} : x_{B(3)}, x_{D(4)} : x_{B(4)})$ of ternary azeotropic mixture at $m > n$. Product simplexes Reg_{simp} are shaded.

The *rule of product simplex* gives us the instrument that allows not only to determine feasible splits in the first column, but also gives an opportunity to determine at once the compositions of the products in the sequence of $(n - 1)$ columns. Compositions corresponding to the vertexes of the product simplex the feed point belongs to can be obtained as columns' system products.

If $m > n$, then feed point belongs to several product simplex. Therefore, several sets of products, corresponding to vertexes of each product simplex, can be got from such feeding. Let's examine a few examples.

For the product simplex $Reg_{simp} \equiv 2 \to 13 \to 1 \ (2 \Rightarrow 1)$, the set of products in the system of two columns is 2; 13; 1 (Fig. 3.6). For the product simplex $Reg_{simp} \equiv 12 \to 1 \to 3 \ (12 \Rightarrow 3)$, the set of products is 12; 1; 3 (Fig. 3.10a). For the product simplex $Reg_{simp} \equiv 1 \to 13 \to 2 \to 4 \ (1 \Rightarrow 4)$, the set of products is 1; 13; 2; 4 (Fig. 3.18a).

Therefore, the mixture acetone(1)-benzene(2)-chloroform(3)-toluene(4) of the composition (0.25; 0.30; 0.20; 0.25) can be separated into three columns without recycles into acetone, benzene, toluene, and the azeotrope of acetone and chloroform.

Feeding x_F at Fig. 3.21 ($12 \Rightarrow 23$) gets into two product simplexes $Reg_{simp} \equiv 12 \to 1 \to 3$ and $Reg_{simp} \equiv 1 \to 3 \to 23$. Therefore, this mixture can be separated into two columns and into products 12; 1 and 3 or products 1; 3 and 23.

Feeding at Fig. 3.15 ($12 \Rightarrow 24$) can get into several product simplex, for example, into simplexes $Reg_{simp} \equiv 1 \to 3 \to 23 \to 24$ and $Reg_{simp} \equiv 12 \to 3 \to 23 \to 24$ (bonds chain $12 \to 1 \to 3 \to 23 \to 24$). In this case, sets of products 1; 3; 23 and 24 or 12; 3; 23 and 24 are feasible (simplex $Reg_{simp} \equiv 12 \to 3 \to 23 \to 24$ shown at Fig. 3.15). At the other composition, feeding can get into simplexes $Reg_{simp} \equiv 12 \to 1 \to 3 \to 4$; $Reg_{simp} \equiv 12 \to 3 \to 4 \to 24$; and $Reg_{simp} \equiv 1 \to 3 \to 4 \to 24$

(bonds chain $12 \rightarrow 1 \rightarrow 3 \rightarrow 4 \rightarrow 24$). Here are feasible sets of products at such feeding: 12; 1; 3 and 4 or 12; 3; 4 and 24 or 1; 3; 4 and 24.

Let's return to the example of industrial mixture that is a by-product of wood pyrolysis. It is seen from the examination of product simplex $\text{Reg}_{simp} \equiv 1 \rightarrow 234 \rightarrow 23 \rightarrow 38 \rightarrow 568 \rightarrow 58 \rightarrow 78 \rightarrow 89 \rightarrow 8$ that in a column sequence consisting of eight columns this mixture can be separated into nine products. However, only component 1(acetaldehyde) and component 8(water) can be isolated purely. For all this, water can be isolated only partially because it is a constituent of azeotropes 38, 568, 58, 78, and 89. The rest of the seven components – 2, 3, 4, 5, 6, 7, 9 – cannot be isolated at all by means of distillation without recycles at $R = \infty$.

3.6. Separation of Azeotropic Mixtures in Sequence of Columns with Recycles at $R = \infty$ and $N = \infty$

Azeotropic mixtures can almost never be separated completely into pure components in the sequence of columns without recycles at $R = \infty$ and $N = \infty$. The set of products of such a system of columns almost always contains not only pure components, but also azeotropes (pseudocomponents). Mixtures, for which concentration simplex contains only one distillation region, are an exception. For three-component azeotropic mixtures, the only phase diagrams of such type are the diagram shown at Fig. 3.10b and antipodal it. Such a mixture can be separated into two columns and into pure components. Two variants of flowsheet with direct $1 : 2,3$ or indirect $1,2 : 3$ split in the first column are feasible.

Other types of azeotropic mixtures can be separated into pure components only in the sequence of columns with recycles using mode of $R = \infty$ and $N = \infty$. Such possibility was for the first time shown in Balashov et al. (1970), Balashov & Serafimov (1984) and Balashov et al. (1984) at the example of the mixture shown in Fig. 3.6. This possibility is caused by curvature of separatrix 2-13 and by location of feed point x_F at concave side from this separatrix. Usage in the first column of the best semisharp indirect separation: $2,13 : 1$ (point x_{D1} lies on separatrix 2-13), usage in the second column of direct separation $2 : 1,13$, and usage in the third column of separation $13 : 3$ with recycles of azeotrope 13 in feeding of the first column (Figs. 3.22a, b) were proposed in the above-mentioned works. Later, it was shown (Wahnschfft, Le Redulier, & Westerberg, 1993) that in order to completely separate the mixture it is necessary to use the recycle of component 2 and/or of component 1. If we limit ourselves only by recycle of azeotrope 13, then point of total feeding of the first column will move to the point 13 (Fig. 3.22a) that in its turn leads to the necessity of increasing of the recycle, etc. As a result, complete separation of the mixture at recycle of only azeotrope 13 cannot be achieved.

The smaller the curvature of the separatrix 2-13 and the smaller the concentration of component 2 in the point x_F, the bigger is the necessary value of recycles.

Usually, the curvature of the separatrix between distillation regions is not big for azeotropic mixtures. It leads to the necessity of bigger recycles and correspondingly of bigger energetic and capital expenses. This way for separation of

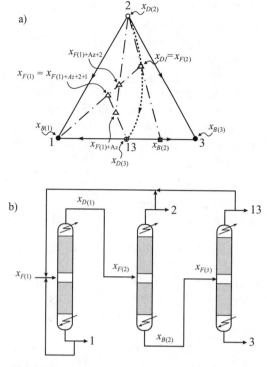

Figure 3.22. (a) A concentration triangle of three-component azeotropic mixture with one binary azeotrope and curvilinear separatrix, and (b) a column sequence with recycles. (1), (2), (3), columns; $x_{F(1)}$, initial feed; $x_{F(1)+Az}$, initial feed and recycle of column (3) overhead product; $x_{F(1)+Az+2}$, initial feed and recycles of columns (3) and (2) overhead products; $x_{F(1)+Az+2+1}$, initial feed and recycles of columns (3) and (2) overhead products and recycle of column (1) bottom product.

a number of binary azeotropic mixtures using various entrainers (butanol-water with methanol; methanol-methylacetate with hexane) is examined in the work (Laroche et al., 1992). In spite of the fact that in the last example ternary mixture with three binary and one ternary azeotropes appears, it can be separated into two columns with recycle of ternary azeotrope.

3.7.   Nonsingularity of Separation Products Compositions at $R = \infty$ and $N = \infty$

For ideal mixtures at $R = \infty$ and $N = \infty$ and at fixed feed composition, unique products compositions (point x_D and x_B at Figs. 3.3 and 3.4) correspond to each value of parameter D/F at the interval $[0,1]$. That is, monotonous increase of parameter D/F at the interval $[0,1]$ corresponds to movement of point x_D at the segments 1 and 3 at Fig. 3.3 and at the segments 1, 2 and 3 at Fig. 3.4 and to movement of point x_B at the segments 4 and 2 at Fig. 3.3 and the segments 4, 5 and 6 at Fig. 3.4.

For azeotropic mixtures, there is no such monotony. That is, at movement of points x_D and x_B along the line of products feasible compositions for fixed feed composition x_F, the value of parameter D/F goes through points of extremums if the number of stationary points of distillation subregion exceeds that of components ($m > n$). Such nonmonotony leads to existence of several products feasible compositions at fixed feed composition and fixed value of parameter D/F. It was shown in the work (Petlyuk & Avet'yan, 1971) for the mixture of type 9 at Fig. 3.16. The phenomenon of plurality of products composition for three-component mixtures is elaborated in the work (Bekiaris et al., 1993), and it is shown that for the mixture presented at Fig. 3.10b, in some interval of values of parameters D/F, there are three feasible sets of products compositions. In Petlyuk & Serafimov (1983), it is shown that for $n > 3$, sometimes infinite number of sets of products compositions is feasible at one and the same value of parameter D/F. Prognosis of plurality of products compositions is of great importance for the proper designing of separation units because this phenomenon can lead to the obtaining of undesired product.

3.8. Conclusion

At $R = \infty$ and $N = \infty$, distillation trajectories bundles fill up distillation regions Reg^∞ in concentration simplex limited by node and saddle stationary points (points of components and azeotropes) and by boundary elements of various dimensionality, part of which are located at boundary elements of concentration simplex and part of which are located inside it.

Product points at fixed feed composition and at various values of parameter D/F fill up some line in concentration simplex connecting nodes of distillation region Reg^∞ with the feed point.

Feasible splits are only those without distributed components or pseudocomponents (azeotropes) at boundary values of parameter D/F or with one distributed component or pseudocomponent at all the intermediate values of parameter D/F.

Split is feasible if points N_D^+ and N_B^- coincide with each other or if there is bond or bonds chain $N_D^+ \to N_B^-$ (the rule of connectedness). This rule is fulfilled if the feed point is located inside the concentration simplex, the vertexes of which are stationary points entering in one link chain. The stationary points of the link chain must include all the components, and the number of links must be no less than that of components. *Split is feasible if points x_D and x_B belong to boundary elements of the same product simplex (the rule of product simpex) and if the number of stationary points of the links chain equals the number of components. If the links number exceeds the number of components, the rule of connectedness must be checked for each split.*

An azeotropic mixture can be separated in a sequence of columns with recycles if the point x_F is located at the concave side of separatrix between the distillation regions Reg^∞.

3.9.   Questions

1. What is the difference between distillation subregion Reg_{sub} and distillation region Reg^{∞}?

2. What is the difference between product simplex Reg_{simp} and distillation subregion Reg_{sub}?

3. What parameters determine separation mode in the finite column with infinite reflux?

4. What parameter determines separation mode in the infinite column with infinite reflux?

5. What is the set of feasible points of products of the finite column with infinite reflux in concentration triangle and tetrahedron?

6. What is the set of feasible points of products of the infinite column with infinite reflux in concentration triangle and tetrahedron?

7. Where within the concentration simplex can the product points of distillation at $R = \infty$ and $N = \infty$ be located?

8. Formulate the rule of connectedness.

9. How many parts a distillation trajectory at $R = \infty$ and $N = \infty$ can consist of?

10. Which boundary elements of distillation region Reg^{∞} or subregion Reg_{sub} do the product points at $R = \infty$ and $N = \infty$ belong to?

11. How many types of boundary elements can a distillation subregion Reg_{sub} have?

12. Can two boundary elements of distillation subregion to which product points at $R = \infty$ and $N = \infty$ belong have common stationary points? If the answer is yes, then how many common stationary points?

13. How many stationary points a distillation trajectory at $R = \infty$ and $N = \infty$ can go through?

14. How many sharp splits at $R = \infty$ and $N = \infty$ can be?

15. Formulate the rule of product simplex.

16. In which way can the rule of connectedness be checked without structural matrix?

17. Fill up structural matrix for four-component azeotropic mixture the structure of concentration space of which is shown at Fig. 3.15.

18. Single out all the bonds chains from the structural matrix of question 17.

19. Feed point belongs to product simplex $Reg_{simp} \equiv 12 \rightarrow 1 \rightarrow 3 \rightarrow 4$ at Fig. 3.15. Enumerate all the boundary elements of dimensionality 0, 1, and 2 of this product simplex. Which splits are feasible for this feeding?

3.10.   Exercises with Software

1. For a mixture of acetone(1)-benzene(2)-chloroform(3)-toluene(4), determine the location of the boundary between distillation regions.

2. For a given feed point x_F in the product simplex 1-13-2-4 for the mixture of exercise 1, find the bottom point x_B for the best direct separation.

3. For a mixture of acetone(1)-methanol(2)-chloroform(3)-ethanol(4), determine the location of the boundary between distillation regions. List all the distillation subregions and product simplexes for this mixture. For each product simplex, state all feasible splits without distributed components or pseudocomponents.

4. For the mixture of exercise 3 of composition $x_F(0.2, 0.4, 0.3, 0.1)$, determine to which distillation region and to which product simplex (or simplexes) x_F belongs, what splits are possible, and what products may be obtained at $R = \infty$ and $N = \infty$.

References

Balashov, M. I., Grishunin, V. A., & Serafimov, L. A. (1970). The Rules of Configuration of Boundaries of Regions of Continuous Distillation in Ternary Systems. *Transactions of Moscow Institute of Fine Chemical Technology*, 2, 121–6 (Rus.).

Balashov, M. I., Grishunin, V. A., & Serafimov, L. A. (1984). Regions of Continuous Rectification in Systems Divided into Distillation Regions. *Theor. Found. Chem. Eng.*, 18, 427–33.

Balashov, M. I., & Serafimov, L. A. (1984). Investigation of the Rules Governing the Formation of Regions of Continuous Rectification. *Theor. Found. Chem. Eng.*, 18, 360–366.

Bekiaris, N., Meski, G. A., Radu, C. M., & Morari, M. (1993). Multiple Steady States in Homogeneous Azeotropic Distillation. *Ind. Eng. Chem. Res.*, 32, 2023–38.

Doherty, M. F. (1985). Presynthesis Problem for Homogeneous Azeotropic Distillations Has Unique Explicit Solution. *Chem. Eng. Sci.*, 40, 1885–9.

Doherty, M. F., & Caldarola, G. A. (1985). Design and Synthesis of Homogeneous Aseotropic Distillations. 3. The Sequencing of Columns for Azeotropic and Extractive Distillations. *Ind. Eng. Chem. Fundam.*, 24, 474–85.

Fidkowski, Z. T., Malone, M. F., & Doherty, M. F. (1993). Computing Azeotropes in Multicomponent Mixtures. *Comput. Chem. Eng.*, 17, 1141–4.

Gmehling, J., Menke, J., Fischer, K., & Krafczyk, J. (1994a). *Azeotropic Date. Part 1.* New York: VCH.

Gmehling, J., Menke, J., Fischer, K., & Krafczyk, J. (1994b). *Azeotropic Date. Part 2.* New York: VCH.

Gurikov, Yu. V. (1958). Some Questions Concerning the Structure of Two-Phase Liquid-Vapor Equilibrium Diagrams of Ternary Homogeneous Solutions. *J. Phys. Chem.*, 32, 1980–96 (Rus.).

Knight, J. R., & Doherty, M. F. (1990). Systematic Approaches to the Synthesis of Separation Schemes for Azeotropic Distillation. In *Foundation of Computer-Aided Process Design*. Sirola, J. J., Grossmann, I. E., & Stephanopoulos, G., editors. New York: Elsevier.

Laroche, L., Bekiaris, N., Andersen, H. W., & Morari, M. (1992). Homogeneous Azeotropic Distillation: Separability and Flowsheet Synthesis. *Ind. Eng. Chem. Res.*, 31, 2190–209.

Ostwald, W. (1900). Dampfdrucke ternarer Gemische, Abhandlungen der Mathematisch-Physischen Classe der Konige Sachsischen. *Gesellschaft der Wissenschaften*, 25, 413–53 (Germ.).

Petlyuk, F. B. (1979). Structure of Concentration Space and Synthesis of Schemes for Separating Azeotropic Mixtures. *Theor. Found. Chem. Eng.*, 683–9.

Petlyuk, F. B., & Avet'yan, V. S. (1971). Investigation of Three-Component Distillation at Infinite Reflux. *Theor. Found. Chem. Eng.*, 5, 499–510.

Petlyuk, F. B., Avet'yan, V. S., & Inyaeva, G. V. (1977). Possible Product Compositions for Distillation of Polyazeotropic Mixtures. *Theor. Found. Chem. Eng.*, 11, 177–83.

Petlyuk, F. B., Kievskii, V. Ya., & Serafimov, L. A. (1975a). Thermodynamic and Topological Analysis of Phase Diagrams of Polyazeotropic Mixtures. 1. Determination of Distillation Regions Using a Computer. *J. Phys. Chem.*, 49, 1834–5 (Rus.).

Petlyuk, F. B., Kievskii, V. Ya., & Serafimov, L.A. (1975b). Thermodynamic and Topological Analysis of Phase Diagrams of Polyazeotropic Mixtures. 2. Algorithm for Construction of Structural Graphs for Azeotropic Ternary Mixtures. *J. Phys. Chem.*, 49, 1836–7 (Rus.).

Petlyuk, F. B., Kievskii, V. Ya., & Serafimov, L. A. (1977). Method for Isolation of Regions of Rectification Polyazeotropic Mixtures Using an Electronic Computer. *Theor. Found. Chem. Eng.*, 11, 1–7.

Petlyuk, F. B., Kievskii, V. Ya., & Serafimov, L. A. (1979). Determination of Product Compositions for Polyazeotropic Mixtures Distillation. *Theor. Found. Chem. Eng.*, 13, 643–9.

Petlyuk, F. B., & Serafimov, L. A. (1983). Multicomponent Distillation. Theory and Calculation. Moscow: Khimiya (Rus.).

Rooks, R. E., Julka, V., Doherty, M. F., & Malone, M. F. (1998). Structure of Distillation Regions for Multicomponent Azeotropic Mixtures. *AIChE J.*, 44, 1382–91.

Safrit, B. T., & Westerberg, A. W. (1997). Algorithm for Generating the Distillation Regions for Azeotropic Multicomponent Mixtures. *Ind. Eng. Chem. Res.*, 36, 1827–40.

Sargent, R. W. S. H. (1994). *A Functional Approach to Process Synthesis and Its Application to Distillation Systems*. Tech. Rep. Centre for Process Systems Engineering. London: Imperial College.

Schreinemakers, F. A. H. (1901). Dampfdrucke ternarer Gemische. *Z. Phys. Chem.*, 36, 413–49 (Germ.).

Stichlmair, J. G., Fair, J. R., & Bravo, J. L. (1989). Separation of Azeotropic Mixtures via Enhanced Distillation. *Chem. Eng. Prog.*, 85, 63–6.

Thormann, K. (1928). *Destillieren and Rektifizieren*. Leipzig: Verlag von Otto Spamer (Germ.).

Wahnschafft, O. M. (1997). *Advanced Distillation Synthesis Techniques for Nonideal Mixtures Are Making Headway in Industrial Applications*. Paper presented at Distillation and Absorption Conference, Maastricht, pp. 613–23.

Wahnschafft, O. M., Le Redulier, J. P., & Westerberg A. W. (1993). A Problem Decomposition Approach for the Synthesis of Complex Separation Processes with Recycles. *Ind. Eng. Chem. Res.*, 32, 1121–40.

Zharov, V. T. (1968). Phase Representations and Rectification of Multicomponent Solutions. *J. Appl. Chem.*, 41, 2530–41 (Rus.).

Zharov, V. T., & Serafimov, L. A. (1975). *Physico-Chemical Foundations of Bath Open Distillation and Distillation*. Leningrad: Khimiya (Rus.).

4

Trajectories of Thermodynamically Reversible Distillation

4.1. Introduction

Although the thermodynamically reversible process of distillation is unrealizable, it is of great practical interest for the following reasons: (1) it shows in which direction real processes should be developed in order to achieve the greatest economy, and (2) the analysis of this mode is the important stage in the creation of a general theory of multicomponent azeotropic mixtures distillation.

First investigations of thermodynamically reversible process concerned binary distillation of ideal mixtures (Hausen, 1932; Benedict, 1947). Later works concerned multicomponent ideal mixtures (Grunberg, 1960; Scofield, 1960; Petlyuk & Platonov, 1964; Petlyuk, Platonov, & Girsanov, 1964).

The analysis of the thermodynamically reversible process of distillation for multicomponent azeotropic mixtures was made considerably later. Restrictions at sharp reversible distillation were revealed (Petlyuk, 1978), and trajectory bundles at sharp and nonsharp reversible distillation of three-component azeotropic mixtures were investigated (Petlyuk, Serafimov, Avet'yan, & Vinogradova, 1981a, 1981b).

Restrictions at nonsharp reversible distillation of three-component azeotropic mixtures were studied by Poellmann and Blass (1994).

Trajectories of adiabatic distillation at finite reflux for given product points should be located in concentration space in the region limited by trajectories at infinite reflux and by trajectories of reversible distillation (Petlyuk, 1979; Petlyuk & Serafimov, 1983).

The algorithm, based on this principle and checking for three-component mixtures whether it is possible to get either products of given composition at distillation, was developed by Wahnschafft et al. (1992).

As far as stationary points of trajectory bundles of distillation at finite reflux lay on trajectories of reversible distillation, these trajectories were also called the lines of stationarity (pinch lines, lines of fixed points) (Serafimov, Timofeev, & Balashov, 1973a, 1973b). These lines were used to deal with important applied tasks connected with ordinary and extractive distillation under the condition of finite reflux: to determine approximately minimum reflux number (Koehler, Aguirre, & Blass, 1991) and minimum entrainer rate (Knapp & Doherty, 1994).

Significance of reversible distillation theory consists in its application for analysis of evolution of trajectory bundles of real adiabatic distillation at any splits. Numerous practical applications of this theory concern creation of optimum separation flowsheets; determination of optimum separation modes, which are close to the mode of minimum reflux; and thermodynamic improvement of distillation processes by means of optimum intermediate input and output of heat.

4.2.   Essence of Reversible Distillation Process and Its Peculiarities

4.2.1.   Essence of Reversible Distillation Process

The process, in which transformation of the system in direct and indirect directions is being accomplished through continuous series of equilibrium states, is understood as reversible one. At the reversible process,

$$\Delta S = \int dQ/T \tag{4.1}$$

Equation (4.1) concerns not only reversible distillation process, but also any thermodynamically reversible process. For the distillation,

$$\Delta S_{dist} = S_F - S_D - S_B \tag{4.2}$$

dQ is equal to input or output of heat at temperature T in the reboiler, condenser, and in intermediate relatively to column height reboilers and condensers. At the reversible process of distillation,

$$\int dQ/T = S_F - S_D - S_B \tag{4.3}$$

Decrease of entropy of distillation products compared with entropy of feed is written in the right side of Eq. (4.3), and increase of entropy of heat sources and receivers is written to the left. The entropy of separation products is always below that of feed, and the entropy of heat sources and receivers always increases during the process of distillation because there is transmission of heat from the sources with a higher temperature to the receivers with a lower temperature.

Total change of entropy in the incoming and outgoing flows of the column and in the sources and receivers of heat should be equal to zero [Eq. (4.3)] in the case of the thermodynamically reversible process of distillation.

In real processes of distillation, total change of entropy is always above zero because of thermodynamic losses, and here lies the reason of nonreversibility:

$$\int dQ/T - \Delta S_{dist} + \Delta S_{ir} > 0 \tag{4.4}$$

Growth of the entropy in real processes of distillation in view of nonreversibility has a number of reasons: (1) nonequilibrium flows of liquid and vapor meet each other at the plates of the column (it becomes especially apparent in the point of feed input and in the ends of the column where the flows from condenser and reboiler are brought in); (2) there is a loss of pressure because of hydraulic resistance of contact devices at the plates of the column; and (3) heat is brought and removed in reboilers and condensers at nonzero temperature differences.

The real process thermodynamic efficiency of distillations equals to

$$\eta = \Delta S_{dist}/(\Delta S_{dist} + \Delta S_{ir}) \tag{4.5}$$

In the case of real process, the thermodynamic efficiency is quite low; in air separation units, it is equal to 18%; in crude units, it is equal to 12%; and in units for ethylene and propylene production, it is equal to 5% (Haselden, 1958).

4.2.2. Location of Reversible Distillation Trajectories

The main peculiarity of thermodynamically reversible distillation process consists of the fact that flows of two different phases (vapor and liquid) found in any cross-section are in equilibrium, and flows found in the feed cross-section are of the same composition as feed flows.

Using the equation of material balance and of phase equilibrium for an arbitrary cross-section of the upper section, we get

$$V y_i = L x_i + D y_{Di} \tag{4.6}$$

$$y_i = K_i x_i \tag{4.7}$$

Similarly, for the lower section,

$$L x_i = V y_i + B x_{Bi} \tag{4.8}$$

It follows from Eq. (4.6) that the points of the upper section x_i, y_i, y_{Di} lie in one straight line in the concentration simplex.

Similarly, it follows from Eq. (4.8) that points x_i, y_i, x_{Bi} of the lower section are also colinear. Equations (4.6) and (4.8) are simultaneously valid for the feed cross-section. Therefore, points x_{Fi}, y_{Fi}, y_{Di}, and x_{Bi} should belong to the same straight line ($x_{fi} = x_{Fi}$, $y_{fi} = y_{Fi}$, where x_{fi}, y_{fi} are concentrations of the component i in the feed cross-section and x_{Fi}, y_{Fi} are concentrations in the feed). The validity of Eqs. (4.6) and (4.8) leads to the conclusion: the product points in the concentration simplex should lie on the prolongation of liquid–vapor tie-lines in each cross-section of the column (Fig. 4.1). *The reversible distillation trajectory is the locus of the points where the straight lines passing through product points are tangent to the residue curves because liquid–vapor tie-lines are tangent to these curves.*

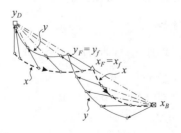

Figure 4.1. Location of reversible section trajectories and liquid–vapor tie-lines in arbitrary tray cross-section illustrating that extended tie-lines pass through product points x_B and y_D. x and y, composition in arbitrary tray cross-section (little circles).

In addition, the product points should lie on the straight line passing through the liquid–vapor tie-line of feeding. Hence, it follows that the maximum length of reversible distillation trajectory is achieved at the intersection of this straight line with the hyperfaces of concentration simplex that (hyperfaces) correspond to $(n-1)$-component constituents C_{n-1} of the initial mixture (sharp separation).

4.2.3. Sharp and Nonsharp Reversible Distillation of Ideal Mixtures

For the ideal mixture, one and the same order of increase and decrease of phase equilibrium coefficients is consistent throughout the whole concentration simplex:

$$K_1^F > K_2^F > \cdots > K_{n-1}^F > K_n^F \tag{4.9}$$

At the sharp separation, the top product contains all the components except the heaviest component n and the bottom product contains all the components except the lightest component 1.

Feasible sharp reversible distillation split of ideal mixtures can be presented as follows: $1, 2, \ldots (n-1) : 2, 3 \ldots n$. Therefore, at the reversible distillation, components $2, 3, \ldots (n-1)$ are distributed among the top and the bottom products. At nonsharp and semisharp reversible distillation, both products contain all the components or one of the products does not contain the lightest or the heaviest component. At nonsharp reversible distillation, product points lie in the same straight line as at sharp distillation but at some distance from the hyperfaces of the concentration simplex.

The mode of sharp reversible distillation is the most interesting. As far as $x_{Dn} = 0$, Eq. (4.6) for this mode for component n looks as follows:

$$V K_n x_n = L x_n \tag{4.10}$$
$$\text{i.e.,} \quad L/V = K_n \tag{4.11}$$

Similarly, for the lower section and for component 1,

$$L x_1 = V K_1 x_1 \tag{4.12}$$
$$\text{i.e.,} \quad L/V = K_1 \tag{4.13}$$

Therefore, in an arbitrary cross-section of upper (lower) section at sharp reversible distillation the ratio of liquid and vapor flows is equal to the phase equilibrium coefficient of the heaviest (lightest) component (i.e., the component absent in the product of the section).

If we take an ideal mixture with constant relative volatilities ($\alpha_i = const$),

$$K_i = \alpha_i K_r \tag{4.14}$$

then we'll get from Eqs. (4.6) and (4.8) for an arbitrary cross-section of the sharp reversible distillation column:

$$x_i/x_j = x_{Fi}/x_{Fj} \quad (i = 1, 2, \ldots n-1) \text{ (for the top section)} \tag{4.15}$$
$$x_i/x_j = x_{Fi}/x_{Fj} \quad (i = 2, 3, \ldots n) \text{ (for the bottom section)} \tag{4.16}$$

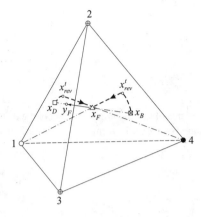

Figure 4.2. Location of reversible section trajectories of an ideal four-component mixture at sharp split and liquid–vapor tie-line of the feed point ($x_F \rightarrow y_F$).

Therefore, *at $\alpha_i = const$, the sharp reversible trajectory of the upper (lower) section goes from the feed point to hyperface of concentration simplex along secant, passing through the vertex of the simplex corresponding to the heaviest (the lightest) component* (Fig. 4.2) (i.e., to the components absent in the section products).

4.2.4. Column Sequence of Ideal Mixtures Reversible Distillation

Figure 4.3 shows the change of the liquid flow rate at the height of a binary reversible distillation column (the column height is characterized by the concentration of the light component) for sharp and nonsharp separation. It is typical of sharp separation that the input of heat and of cold, which is not equal to zero, is required at the ends of the column and, for nonsharp separation, this input makes an infinitesimal quantity.

As far as only one component can be exhausted in each section of the reversible distillation column (i.e., this component is absent in the product of this section), the system of columns shown in Fig. 4.4 (for $n = 3$) or in Fig. 4.5 (for $n = 4$) will be required to perform the complete separation of a multicomponent mixture into pure components.

Figure 4.6 shows the change of the liquid flow rate and of components concentrations for $n = 3$ at the height of all the columns (the height of the columns is characterized by the inverse value of phase equilibrium coefficient of the reference component (the third one) $1/K_3$). As can be seen in Figs. 4.4 and 4.5, the

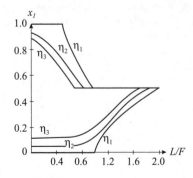

Figure 4.3. Liquid flow rate profiles L/F of a binary reversible distillation under different product purities ($\eta_1 = 1.0, \eta_2 = 0.95, \eta_3 = 0.9; \alpha = 2.0, x_F = 0.5, L_F/F = 1.0$).

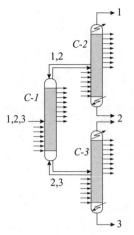

Figure 4.4. A sequence of three-component reversible distillation. *C-1*, *C-2*, *C-3*, columns; arrows, heat input and output.

total number of columns of reversible distillation exceeds the number of columns of ordinary sharp distillation. This difference rapidly augments with the increase of components number. One cannot install reboilers and condensers at the ends of intermediate columns of reversible distillation because it will lead to nonequilibrium of the liquid and vapor flows. Therefore, the inner vapor and liquid flows of the intermediate columns are formed by those removed from feed cross-sections of the columns that follow.

4.2.5. Main Peculiarities of Reversible Distillation Column

The following main peculiarities of the columns of reversible distillation for separation of multicomponent mixtures into pure components arise from the aforesaid:

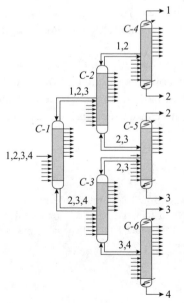

Figure 4.5. A sequence of four-component reversible distillation. *C-1*, *C-2*, *C-3*, *C-4*, *C-5*, *C-6*, columns; arrows, heat input and output.

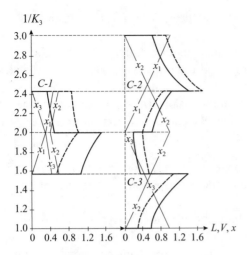

Figure 4.6. Liquid and vapor flow rate and composition profiles of a three-component reversible distillation. C-1, C-2, C-3, columns; x_1, x_2, x_3, concentrations of component 1, 2, 3 on columns trays; K_3, vapor–liquid phase equilibrium coefficient of component 3; thick line, liquid flow rate profile; dotted line, vapor flow rate profile; thin line, concentration profile.

(1) all the columns are infinite; (2) one component is exhausted in each section of each column; (3) infinitesimal amount of heat is brought to or drained from each intermediate cross-section of each column; (4) liquid and vapor flows are created at the ends of the intermediate (nonproduct) columns by removing these flows from feed cross-sections of the columns that follow; and (5) liquid and vapor flows are created at the ends of the product columns with the help of condensers and reboilers with a finite value of output and input of heat.

The analysis of the reversible distillation of ideal mixtures has led to important practical results. The use of flowsheets of the same type as those shown in Fig. 4.4 but consisting of real adiabatic columns (i.e., columns with finite numbers of separation stages), without intermediate input of heat and cold and with product columns joined into one complex column (Petlyuk, Platonov, & Slavinskii, 1965), allows energy consumption for separation to be reduced by thirty to forty percent.

4.3. Trajectory Bundles of Sharp Reversible Distillation

4.3.1. Bundles and Regions of Sharp Reversible Distillation

Such a set of trajectories for which not more than one trajectory passes through each nonsingular point is convenient for understanding as a distillation trajectory bundle. This notion is different for various modes of distillation because the number of parameters influencing the location of the trajectories is different.

It is clear from Fig. 4.1 that the location of a reversible distillation section trajectory is determined only by the location of its product point (i.e., it does not depend on any parameter). One component is absent at sharp reversible distillation [i.e., the product point of the section is located at some $(n - 1)$ component edge, face, or hyperface of the concentration simplex]. That is why, in all the points of a section trajectory at sharp reversible distillation, liquid–vapor tie-lines should be directed to this edge, face, or hyperface or from it [i.e., one and the same component should be the heaviest (for the top section) and the lightest (for the bottom section) along this trajectory].

Therefore, one or several trajectory bundles, filling up that region inside the concentration simplex, where one and the same component is the heaviest (for the top section) or the lightest (for the bottom section), will correspond to *all the product points* located at one and the same $(n-1)$ component edge, face, or hyperface of concentration simplex. We call this region the *region of reversible distillation* $\mathrm{Reg}_{rev,r}^{(n-1)\,j=h}$ or $\mathrm{Reg}_{rev,s}^{(n-1)\,j=l}$ (Petlyuk, 1978). It is the unification of several regions of equal order of components $\mathrm{Reg}_{ord}^{i\,ijk}$, and it is separated from other regions of reversible distillation by the corresponding α-lines, surfaces, or hypersurfaces.

As one can see in Fig. 4.2, the trajectory of each section at sharp reversible distillation consists of two parts: the part, located inside the $(n-1)$ component boundary element C_{n-1} of concentration simplex, lying between the product point x_D or x_B and the *tear-off point of the trajectory from this boundary element* x_{rev}^t, and the part located inside concentration simplex C_n, lying between the tear-off point of the trajectory and the feed point x_F. Only the second part should be located inside a region of reversible distillation $\mathrm{Reg}_{rev,r}^h$ or $\mathrm{Reg}_{rev,s}^l$, and product point x_D or x_B can lie outside this region.

Unlike trajectories of distillation at infinite reflux, which may come off the boundary elements of the concentration simplex in the saddle points S only, reversible distillation trajectories come off in ordinary points x_{rev}^t.

Let's illustrate the location of trajectory bundles of reversible distillation by the example of three-component acetone(1)-benzene(2)-chloroform(3) mixture with one binary saddle azeotrope with a maximum boiling temperature (Fig. 4.7)

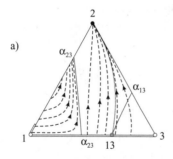

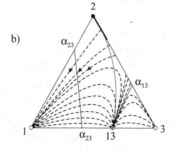

Figure 4.7. Sharp reversible section trajectories of acetone(1)-benzene(2)-chloroform(3) mixture: (a) rectifying section, (b) bottom section. Double line, possible overhead product $\mathrm{Reg}_{rev,D}$; thick solid line, possible bottom product $\mathrm{Reg}_{rev,B}$; dotted lines with arrows, reversible section trajectories; dotted lines without arrows, lines of stationarity.

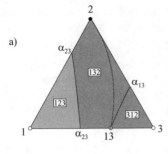

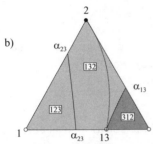

Figure 4.8. Regions of reversible distillation Reg_{rev} (shaded) and regions of component order $\text{Reg}_{ord}^{1,2,3}$, $\text{Reg}_{ord}^{1,3,2}$, and $\text{Reg}_{ord}^{3,1,2}$: (a) rectifying section, and (b) bottom section. Dotty line, separatrix; thin lines, α-lines; $\text{Reg}_{rev,r}^{2}$ (region in which component 2 is heaviest) and $\text{Reg}_{rev,s}^{3}$ (region in which component 3 is lightest), darker shaded.

for which antipodal mixture the location of points in the trajectories was obtained by means of calculation (Petlyuk et al., 1981a).

Figure 4.8 shows the *regions of components order* Reg_{ord}^{ijk} *and of reversible distillation* $\text{Reg}_{rev,r}^{h}$ or $\text{Reg}_{rev,s}^{l}$ for this example. In all points of one trajectory in Fig. 4.7b, liquid–vapor tie-lines are directed from one and the same point of the product x_B located at side 2-3 (for the regions of components order $\text{Reg}_{ord}^{1,2,3}$ and $\text{Reg}_{ord}^{1,3,2}$ forming common region of reversible distillation $\text{Reg}_{rev,s}^{1}$ of the bottom section in which the lightest component is component 1− lighter shaded in Fig. 4.8b) or at side 1-2 (for the region of components order $\text{Reg}_{ord}^{3,1,2} \equiv \text{Reg}_{rev,s}^{3}$ − darker shaded in Fig. 4.8b). In all the points of one trajectory in Fig. 4.7a, liquid–vapor tie-lines are directed to one and the same point of the product x_D located at side 1-2 (for the region of components order $\text{Reg}_{ord}^{1,2,3} \equiv \text{Reg}_{rev,r}^{3}$ − lighter shaded in Fig. 4.8a) or at side 1-3 (for the regions of components order $\text{Reg}_{ord}^{1,3,2}$ and $\text{Reg}_{ord}^{3,1,2}$ forming common region of reversible distillation $\text{Reg}_{rev,r}^{2}$ − darker shaded in Fig. 4.8a). Therefore, each trajectory in Figs. 4.7a,b is a *line of stationarity* for the point of the product of sharp distillation x_D and x_B, located at the corresponding side of the concentration triangle. Moreover, in Fig 4.7a, all lines of stationarity consist of two parts, one of which goes from point x_D at the side of the concentration triangle $\text{Reg}_D^{(2)} \equiv$ 1-3 (i.e., they are *true trajectories of reversible distillation*). At Fig. 4.7b, only a number of stationarity lines go from the points of the products x_B located at some segment of side 2-3 adjacent to vertex 2. Only these stationarity lines are true trajectories of reversible distillation, and this segment of side 2-3 is a segment of feasible points of the bottom product $\text{Reg}_B^{(2)}$. The rest of stationarity lines do not reach that side of the concentration triangle where the corresponding product points are located. Therefore, these stationarity lines are *fictitious trajectories of reversible distillation*. They all begin and end in the points of components and azeotrope (points 1, 3, and 13) that are node points of stationarity lines bundles ($N_{rev} \Leftrightarrow N_{rev}$). True trajectories of reversible distillation, in contrast to the fictitious ones, always start in the product point and end in the node point of the bundle (point of the component or azeotrope) $Reg_D^{(2)} \Rightarrow N_{rev}$ or $Reg_B^{(2)} \Rightarrow N_{rev}$ or at the same boundary element of the concentration simplex where the product point is located $Reg_D^{(2)} \Leftrightarrow Reg_D^{(2)}$ or $Reg_B^{(2)} \Leftrightarrow Reg_B^{(2)}$. In the last case, the trajectory bundle does not have node points (bundle in Fig. 4.7a to the left of line α_{23} can serve as an example of such bundle).

Therefore, the concentration simplex can contain one or several *regions of reversible distillation* $Reg_{rev,r}^h$ or $Reg_{rev,s}^l$ for each section.

The region of reversible distillation can contain one or several reversible distillation trajectory bundles (lines of stationarity). Some of these bundles can be true; some of them can be fictitious. Fictitious bundles always have two node points and true ones have one node point or no node point.

4.3.2.   Condition in Tear-Off Points of the Reversible Distillation Trajectories

Let's examine the tear-off points of the trajectories of reversible distillation from the boundary elements of the concentration simplex (Fig. 4.9). These points are points of branching: one branch of the trajectory is being torn off from the boundary element and goes inside the concentration simplex, and the second branch stays inside the boundary element. Conditions [Eqs. (4.11) or (4.13)] should be

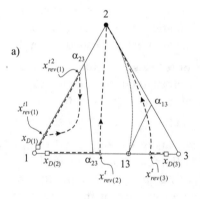

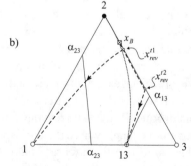

Figure 4.9. Reversible section trajectories of acetone (1)-benzene(2)-chloroform(3) mixture for given product points: (a) rectifying section, and (b) bottom section $x_{D(1)}, x_{D(2)}, x_{D(3)}, x_B$, product points; x_{rev}^t, tear-off points.

valid in the tear-off point of the trajectory for the component that is absent in the product and conditions [Eqs. (4.6) or (4.8)] should be valid for the rest of components. After simple transformations of these equations for the tear-off point x_{rev}^t, we get (Petlyuk & Serafimov, 1983)

$$x_{Di} = x_{rev,i}^t (K_i^t - K_n^t)/(1 - K_n^t) \quad \text{(for top section)} \tag{4.17}$$

$$x_{Bi} = x_{rev,i}^t (K_1^t - K_i^t)/(K_1^t - 1) \quad \text{(for bottom section)} \tag{4.18}$$

In Eq. (4.17), the component n is the heaviest one and, in Eq. (4.18), the component 1 is the lightest one.

Equalities [Eqs. (4.17) and (4.18)] can be written as follows:

$$x_{rev,i}^t = x_{Di}(1 - K_n^t)/(K_i^t - K_n^t) \quad (i = 1, 2, \ldots n - 1) \tag{4.19}$$

$$x_{rev,i}^t = x_{Bi}(K_i^t - 1)/(K_1^t - K_i^t) \quad (i = 2, 3, \ldots n) \tag{4.20}$$

The location of reversible distillation trajectories in the concentration simplex at sharp separation may be presented in the following brief form:

$$x_D \to x_{rev,r}^t \to x_F \text{ and } x_B \to x_{rev,s}^t \to x_F (x_F \in C_n, x_D \in C_{n-1}, x_B \in C_{n-1})$$

4.3.3. Possible Product Composition Regions

If product point x_D or x_B belongs to the *possible product point region* Reg_D or Reg_B, the condition [Eq. (4.19) or (4.20)] is valid in one or two points $x_i^{t(n-1)}$ along the trajectory of reversible distillation located at $(n - 1)$ component boundary element C_{n-1} of the concentration simplex (i.e., there is one tear-off point x_{rev}^t of the trajectory or there are two x_{rev}^t). In the last case, right side of the expression [Eq. (4.19) or (4.20)] should have an extremum.

If the product point x_D or x_B does not belong to the possible product point region ($x_D \notin \text{Reg}_D$ or $x_B \notin \text{Reg}_B$), then the condition [Eq. (4.19) or (4.20)] is not valid anywhere (i.e., tear-off points of the trajectory are absent).

Therefore, Eqs. (4.19) and (4.20) allow determination of the boundaries of the possible product composition region Reg_D or Reg_B at sharp reversible distillation in $(n - 1)$-component boundary elements C_{n-1} of the concentration simplex.

In Fig. 4.9b, there are two tear-off points x_{rev}^t of the reversible distillation trajectory for any point x_B of the possible product composition segment $\text{Reg}_B^{(2)}$ at side 2-3. The trajectory goes from one of these points x_{rev}^{t1} to $N_{rev}^1 \equiv 1$ and from the other one x_{rev}^{t2} the trajectory goes to $N_{rev}^2 \equiv 13$ (

$$
\begin{array}{ccccc}
 & & \uparrow & \to t_{rev}^{t2} & \to N_{rev}^2 \\
x_B & \to & t_{rev}^{t1} & \to & N_{rev}^1 \\
\text{Reg}_B & \text{Reg}_{rev,s}^t & 1 & & 13
\end{array}
$$

). For the point x_B located at the end of the possible product composition segment, both these points coincide with each other ($x_{rev}^{t1} = x_{rev}^{t2}$) (i.e., two trajectories go inside from this common tear-off point x_{rev}^t). The set of trajectories tear-off points for any points x_B at Fig. 4.9b fills up the tear-off segment $\text{Reg}_{rev,s}^{t(2)}$ of side 2-3 from vertex 2 up to point α_{13}.

In Fig. 4.9a, there are two trajectories tear-off points x_{rev}^t of the reversible distillation trajectory for any points x_D of the possible product segment $\mathrm{Reg}_D^{(2)}$ at side 1-2. For point x_D located at the end of the possible product composition segment, both these trajectories tear-off points x_{rev}^t coincide with each other, but the length of the trajectory inside the concentration triangle becomes equal to zero. The segment from vertex 1 to point α_{23} is *trajectory tear-off segment* $\mathrm{Reg}_{rev,r}^{t(2)}$. At the same figure at side 1-3 from point α_{23} to vertex 3, there is a segment of trajectories tear-off $\mathrm{Reg}_{rev,r}^{t(2)}$ and possible product composition segment $\mathrm{Reg}_D^{(2)}$ coincides with side 1-3. Along with that, for any point x_D there is one trajectory tear-off point x_{rev}^t.

4.3.4. Necessary Condition of Sharp Reversible Distillation

It follows from the aforesaid that sharp separation in a reversible distillation column is feasible only if the liquid–vapor tie-line of feeding is directed to the possible product composition region $\mathrm{Reg}_D^{(n-1)}$ at the boundary element C_{n-1} of the concentration simplex and from region $\mathrm{Reg}_B^{(n-1)}$ at other boundary element C_{n-1}.

In other words, for sharp separation in a two-section column it is necessary that the feed point belongs to the intersection of true bundles of reversible distillation trajectories of the two sections $x_F \in (\mathrm{Reg}_{rev,r} \bullet \mathrm{Reg}_{rev,s})$ (Fig. 4.10). The reversible

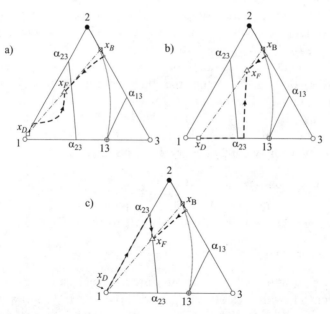

Figure 4.10. Sharp reversible section trajectories of acetone(1)-benzene(2)-chloroform(3) mixture for two-section columns: (a) split $1,2:2,3$, $x_F \in (\mathrm{Reg}_{rev,r}^3 \bullet \mathrm{Reg}_{rev,s}^1)$; (b) split $1,3:2,3$, $x_F \in (\mathrm{Reg}_{rev,r}^2 \bullet \mathrm{Reg}_{rev,s}^1)$; and (c) split $1:2,3$, $x_F \in (\mathrm{Reg}_{rev,r}^{2=3} \bullet \mathrm{Reg}_{rev,s}^1)$ and $x_F \in \alpha_{23}$.

distillation column trajectory may be briefly described as follows:

$$
\underset{\mathrm{Reg}_D}{x_D^{(2)}} \;\to\; \underset{\mathrm{Reg}_{rev,r}^t}{x_{rev,r}^{t(2)}} \;\to\; \underset{\mathrm{Reg}_{rev,rs}}{x_F^{(3)}} \;\leftarrow\; \underset{\mathrm{Reg}_{rev,s}^t}{x_{rev,s}^{t(2)}} \;\leftarrow\; \underset{\mathrm{Reg}_B}{x_B^{(2)}}.
$$

If the feed point lies on the α-line, α-surface, or α-hypersurface, then the liquid–vapor tie-line of feeding is directed to some $(n-2)$-component boundary element or from some $(n-2)$-component boundary element. If, along with that, the liquid–vapor tie-line is directed to the possible product composition region at this boundary element or from this region, then the product of reversible distillation section can contain $n-2$ components. For example, if the feed point in Fig. 4.10c lies on the α_{23}-line within the true bundle of bottom section trajectories then the liquid–vapor tie-line of feeding is directed to vertex 1 [i.e., the component $1 = \mathrm{Reg}_D$ can be a product of the section (the product contains $n-2$ components)].

The general rule: if the phase equilibrium coefficients of k light or heavy components are equal to each other in the feed point, then the section product can contain $(n-k)$ components.

We previously examined the process of reversible distillation for a given feed point. Below we examine trajectories of reversible distillation sections for given product points located at any k-component boundary elements C_k of the concentration simplex ($x_D \in C_k$ or $x_B \in C_k$). If $k < (n-1)$, then in the general case such trajectories should consist of two parts: the part located in the same k-component boundary element where the product point lies and the part located at some $(k+1)$-component boundary element C_{k+1} adjacent to it. Along with that, the product point should belong to the possible product composition region $\mathrm{Reg}_D^{(k)}$ or $\mathrm{Reg}_B^{(k)}$ for the examined (k)-component boundary element, and the boundaries of this region can be defined with the help of Eqs. (4.19) and (4.20).

Such an approach on the basis of product points will be necessary at the analysis of the location of adiabatic sections trajectories bundles (at finite reflux) which products consist less $(n-1)$ components (see Chapter 5).

4.3.5. Liquid and Vapor Flow Rates Changing along the Reversible Distillation Trajectories

Besides the location of reversible distillation trajectories in the concentration simplex, the character of the liquid and vapor flow rates changing is of great importance. In accordance with the formulas [Eqs. (4.11) and (4.13)], the ratio of liquid and vapor flow rates in each cross-section in the top section should be equal to the phase equilibrium coefficient of the heaviest component and in the bottom section to that of the lightest component. For ideal mixtures, these phase equilibrium coefficients should change monotonously along the sections trajectories, which leads to maximum liquid and vapor flow rates in the feed cross-section (see Figs. 4.3 and 4.6).

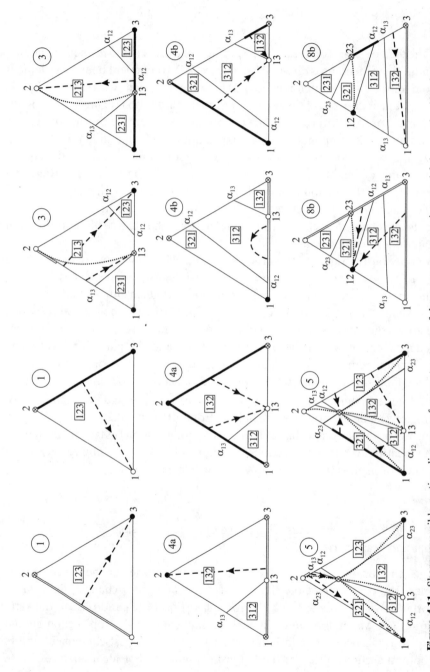

Figure 4.11. Sharp reversible section diagrams of some structures of three-component mixtures. 1,3,4a,..., classification according to Gurikov (1958). Double line, possible composition of overhead product Reg_D; thick solid line, possible composition of bottom product Reg_B; dotted lines with arrows, reversible section trajectories; *123, 132, 312*..., regions of component order Reg_ord; dotty lines, separatrixes; thin lines, α-lines.

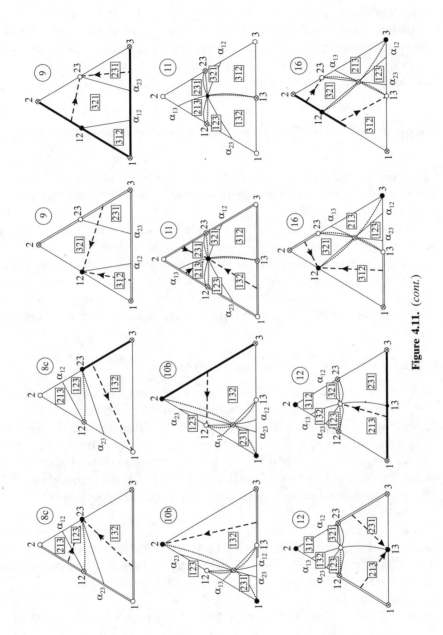

Figure 4.11. (*cont.*)

Therefore, an infinitesimal amount of heat should be drawn off in each cross-section of the top section and should be brought in in each cross-section of the bottom section. For azeotropic mixtures, the phase equilibrium coefficients field is of complicated character, which leads to nonmonotony of the liquid and vapor flow rates changing along the sections trajectories (i.e., to the necessity of input or output of heat in various cross-sections of the section). Such character of the flow rates changing at reversible distillation influences on the conditions of minimum reflux mode in adiabatic columns, which results in a number of cases in the phenomenon of "tangential pinch" (see Chapter 5).

4.4.　Diagrams of Three-Component Mixture Reversible Distillation

Locations of trajectories bundles Reg_{rev}, of node points of these bundles N_{rev}, and of possible product segments $Reg_D^{(2)}$ and $Reg_B^{(2)}$ can be shown in diagrams of three-component azeotropic mixtures sharp reversible distillation for various types of such mixtures (Fig. 4.11).

Along with the diagrams of open evaporation (see Chapter 3), these diagrams contain a great deal of information necessary to design separation units.

The location of trajectory bundles and possible product composition segments at reversible distillation of three-component mixtures determines the location of trajectory bundles, and of possible product composition regions of multicomponent mixtures and the locations of trajectory bundles of real adiabatic columns.

4.4.1.　Calculation of Reversible Distillation Trajectories

Diagrams of reversible distillation of various types of three-component mixtures can be obtained in various ways with the help of the model of phase equilibrium describing these types of mixtures. It is possible to calculate the trajectory consequently for each chosen product point using Eq. (4.6) ÷ (4.13) and increasing step by step the concentration of the component that is absent in the product, after the definition of trajectory tear-off point with the help of Eqs. (4.19) and (4.20). The iteration procedure for such calculation was proposed by Koehler et al. (1991) and further improved by Petlyuk & Danilov (2001).

Let's consider, for example, that some point x_i^0 on the trajectory of reversible distillation for given product point x_D. Further concentration of the heaviest component is increased $x_h = x_h^0 + \Delta$. Concentrations of the rest of components are defined by normalizing $x_i = x_i^0/(1 + \Delta)$. In a new point, $K_i(x_i)$ and $L/V = K_n(x_n)$ are calculated. New values x_i for $i = 1, 2 \ldots (n-1)$ and the corresponding new values K_i and L/V are defined with the help of Eq. (4.6). Such procedure goes on until concentrations x_i stop changing. To ensure convergence of the iteration process, the procedure illustrated graphically in Fig. 4.12 is used. After the final definition of point x_i at reversible distillation trajectory, the next increase of concentration of the heaviest component Δ is to be done. Such a method of calculation for reversible distillation trajectories is good for mixtures with any number

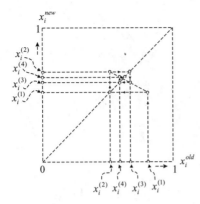

Figure 4.12. Iteration of component concentration at calculation reversible section trajectories. x_i^{old} and x_i^{new}, initial and calculated concentration of component; $x_i^{(1)}, x_i^{(2)}, x_i^{(3)}, x_i^{(4)}$, successive iteration concentration of component i.

of components. An easier way to get the general picture of the location of trajectory bundles is to scan the whole area of the concentration triangle, defining at each point the values K_i and – with the help of Eq. (4.6) ÷ (4.13) – values x_D and x_B (Petlyuk & Serafimov, 1983).

4.4.2.   Scanning the Sides of the Concentration Triangle

It was shown (Petlyuk, 1986) that the diagram of reversible distillation of any three-component mixture can be forecasted by scanning only the sides of the concentration triangle, defining at each point the values of phase equilibrium coefficients of all the components and using Eqs. (4.19) and (4.20). The latter way defines trajectory tear-off segments $Reg_{rev,r}^{t(2)}$ or $Reg_{rev,s}^{t(2)}$ and possible product segments $Reg_D^{(2)}$ or $Reg_B^{(2)}$. The node points N_{rev} of the trajectory bundles are determined hypothetically on the basis of the data on the location of azeotropes points and α-points.

The diagrams of reversible distillation were constructed for some types of three-component azeotropic mixtures. It is interesting that some types of mixtures with one binary azeotrope and with two distillation regions [types 3 and 5 according to classification (Gurikov, 1958)] permit sharp separation into component and binary zeotropic mixture at some feed compositions. The mixture acetone(1)-benzene(2)-chloroform(3) is an example of such mixture.

4.5.   Trajectories Bundles of Reversible Distillation for Multicomponent Mixtures

Let's examine the analysis of structure of reversible distillation trajectory bundles at the concrete example of four-component mixture acetone(1)-benzene(2)-chloroform(3)-toluene(4). At the beginning, the *segments of the components order* Reg_{ord}^{ijk} at the edges of the concentration tetrahedron are defined by means of scanning and calculation of the values K_i (Fig. 4.13a). The corresponding *regions of components order* Reg_{ord}^{ijk} in the tetrahedron are shown in Fig. 4.13b and in its faces – in Fig. 4.14. The whole face 1-2-3, where the component 4 that is absent

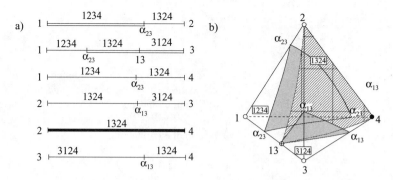

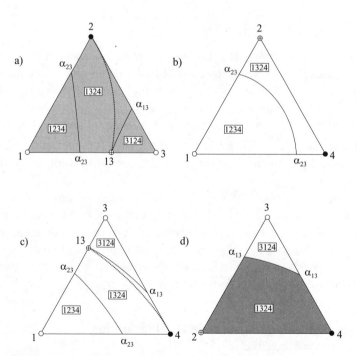

Figure 4.13. (a) Component-order Reg_{ord} and tear-off $Reg^t_{rev,r}$ and $Reg^t_{rev,s}$ (double line for the overhead product and thick line for the bottom product) segments on edges of the acetone(1)-benzene(2)-chloroform(3)-toluene(4) concentration tetrahedron. (b) Component-order regions $Reg^{1,2,3,4}_{ord}$, $Reg^{1,3,2,4}_{ord}$, and $Reg^{3,1,2,4}_{ord}$ inside the concentration tetrahedron. The boundary of distillation regions under infinite reflux hatched, α-surfaces are shaded.

Figure 4.14. Component-order Reg_{ord} and tear-off $Reg^t_{rev,r}$ and $Reg^t_{rev,s}$ regions on three-component faces of the concentration tetrahedron (shaded for the overhead product and darker shaded for the bottom product) for the acetone(1)-benzene(2)-chloroform(3)-toluene(4) mixture: (a) $Reg^4_{rev,r}$, (d) $Reg^1_{rev,s}$, *1234, 1324,* and *3124* — component-order regions $Reg^{1,2,3}_{ord}$, $Reg^{2,3,4}_{ord}$, $Reg^{1,3,2,4}_{ord}$, $Reg^{1,3,2,4}_{ord}$, and $Reg^{3,1,2,4}_{ord}$.

in the top product is the heaviest, is a *region of the trajectories tear-off of the top section* $\text{Reg}_{rev,r}^{t}$, and the *region of reversible distillation* $\text{Reg}_{rev,r}^{4}$ contains one trajectory bundle and fills up the whole concentration tetrahedron. In face 2-3-4, there is the *region of the trajectories tear-off of the bottom section* $\text{Reg}_{rev,s}^{t}$ between the α_{13}-line and side 2-4, where the component 1 that is absent in the bottom product is the lightest. Product points can be located only in faces where one can find the trajectories tear-off regions, (i.e., in face 1-2-3 for the top product and in face 2-3-4 for the bottom product). The contour of *possible product regions* Reg_D and Reg_B in these faces consists of the segments at their sides. These segments are *possible product composition segments* at separation of three-component mixtures formed by the components of the side under consideration and by the component, which is absent in the face under consideration (for face 2-3-4, it is component 1). These segments are shown in Fig. 4.15a ÷ c for face 2-3-4. In face 2-3-4, the whole side 2-4 and segments at edges 2-3 and 4-3 that are adjacent to vertexes 2 and 4 are such segments. Corresponding to it *possible bottom product composition region* $\text{Reg}_{B}^{1}{}_{2,3,4}$

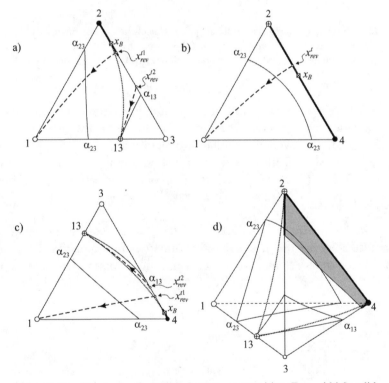

Figure 4.15. Segments of possible bottom composition Reg_B (thick solid lines) on the sides of face 2-3-4 of the acetone(1)-benzene(2)-chloroform(3)-toluene(4) concentration tetrahedron: (a) $\text{Reg}_{B}^{1}{}_{2,3}$ (side 2-3); (b) $\text{Reg}_{B}^{1}{}_{2,4}$ (side 2-4); (c) $\text{Reg}_{B}^{1}{}_{3,4}$ (side 3-4); and (d) region of possible bottom of reversible distillation $\text{Reg}_{B}^{1}{}_{2,3,4}$ on the face 1-2-3 (shaded).

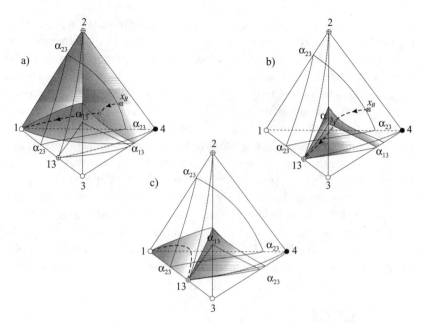

Figure 4.16. Bundles of sharp reversible stripping trajectories in region reversible distillation $\mathrm{Reg}^1_{rev,s}$ for the acetone(1)-benzene(2)-chloroform(3)-toluene(4) mixture: (a) node is component 1, (b) node is azeotrope 13, and (c) nodes are component 1 and azeotrope 13.

in face 2-3-4 is shown in Fig. 4.15d, and *trajectory bundles of reversible distillation* $\mathrm{Reg}^1_{rev,s}$ are shown in Fig. 4.16. If the feed point is located at the intersection of true trajectory bundles of reversible distillation of the sections $x_F \in (\mathrm{Reg}^4_{rev,r} \bullet \mathrm{Reg}^1_{rev,s})$, then sharp reversible distillation in a two-section column is feasible according to the split 1,2,3 : 2,3,4 (The reversible distillation column trajectory may be briefly described as follows:

$$\begin{array}{ccccccccc} x_D^{(3)} & \to & x_{rev,r}^{t(3)} & \to & x_F^{(4)} & \leftarrow & x_{rev,s}^{t(3)} & \leftarrow & x_B^{(3)} \\ \mathrm{Reg}_D & & \mathrm{Reg}^t_{rev,r} & & \mathrm{Reg}_{rev,rs} & & \mathrm{Reg}^t_{rev,s} & & \mathrm{Reg}_B \end{array} \; ; \text{Fig. 4.17).}$$

The above-described way of definition of the possible composition region contour in face 2-3-4 has the most general nature. It can be applied for any $(n-1)$-component boundary elements of the concentration simplex of n-component mix-

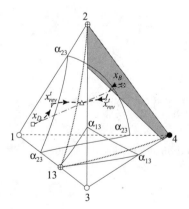

Figure 4.17. Reversible section trajectories of acetone (1)-benzene(2)-chloroform(3)-toluene(4) mixture for two-section column (split 1,2,3 : 2,3,4); region of possible bottom Reg^1_{B} is shaded.

tures. It is necessary to determine phase equilibrium coefficients of all the components in points of all the edges of the $(n - 1)$-component boundary element under consideration. After that, all possible product points x_D and x_B at each edge are defined according to Eq. (4.19) or (4.20) for all tear-off points x_{rev}^t from this edge into three-component boundary element, containing the component absent in the $(n - 1)$-component boundary element under consideration.

4.6. Diagrams of Extractive Reversible Distillation for Three-Component Mixtures

4.6.1. Condition in Tear-Off Points of the Extractive Reversible Distillation Trajectories

Let's examine a column of sharp reversible distillation with two feedings (Fig. 4.18a) for separation of a three-component mixture (Petlyuk & Danilov, 1999). For an intermediate section of such a column, Eq. (4.6) is as follows (F_2, upper feeding):

$$V y_i = L x_i + D y_{Di} - F_2 z_{F2,i} \tag{4.21}$$

Let's designate:
$$d_i' = D y_{Di} - F_2 z_{F2,i} \tag{4.22}$$

$$D' = \Sigma_i d_i' \tag{4.23}$$

$$x_{Di}' = d_i' / D' \tag{4.24}$$

For the intermediate section, point x_{Di}' plays the same role as the product point x_{Di} for the top section (i.e., at reversible distillation in each cross-section of the *intermediate section*, the continuation of the liquid–vapor tie-line goes through point x_{Di}'). Let's call point x_{Di}' the *pseudoproduct point*. It is seen from Eq. (4.22) that, in contrast to the product point, the pseudoproduct point can lie without the concentration triangle (i.e., the values x_{Di}' can be negative or greater than 1).

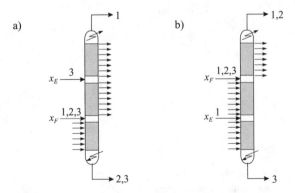

Figure 4.18. (a) Column of autoextractive reversible distillation of ideal ternary mixture ($K_1 > K_2 > K_3$). (b) Column of opposite autoextractive reversible distillation of ideal ternary mixture ($K_1 > K_2 > K_3$).

Let's show in the beginning that, for example, for an ideal mixture, for which $K_1 > K_2 > K_3$, in the intermediate section not the heaviest component as for the top section and not the lightest component as for the bottom section, but the *intermediate component* $m = 2$ can be exhausted.

Let's take a column, the main feeding of which contains all three components and additional upper feeding (entrainer, $E \equiv F_2$, $x_E \equiv z_{F2}$) contains only the heaviest component 3 ($x_{E3} = 1$). We call this the column of autoextractive distillation (Fig. 4.18a).

Let's accept that the top product point coincides with vertex 1 ($x_D = 1$). Then it follows from Eq. (4.22) that the pseudoproduct point should lie on the straight line passing though side 1 (top product)-3 (entrainer).

It follows from Eq. (4.22) that

$$d'_3 = -F_2 z_{F2,3} = -F_2 \tag{4.25}$$

Two cases are possible:

$$D' < 0 \tag{4.26}$$

$$D' > 0 \tag{4.27}$$

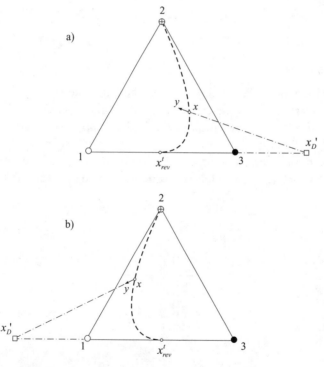

Figure 4.19. Reversible intermediate section trajectories of sharp auto-extractive distillation of ideal ternary mixture ($K_1 > K_2 > K_3$): (a) $D < E$, and (b) $D > E$. Component 1, overhead product; component 3, entrainer; x and y, composition in arbitrary cross-section; x'_D, composition of pseudoproduct.

In the first case, it follows from Eqs. (4.22) to (4.24), that $d_1' > 0$, $d_3' < 0$, $x_{D1}' < 0$, $x_{D3}' > 1$ (i.e., the pseudoproduct point is located without the concentration triangle at the continuation of side 1-3 beyond vertex 3) (Fig. 4.19a). In the second case, $d_1' > 0$, $d_3' < 0$, $x_{D1}' > 1$, $x_{D3}' < 0$ (i.e., the pseudoproduct point is located without the concentration triangle at the continuation of side 1-3 beyond vertex 1) (Fig. 4.19b).

In both cases, the trajectory tear-off point of sharp reversible distillation in the intermediate extractive section should lie at side 1-3 and the trajectory of intermediate section is a line, which is a geometric locus of points where the straight lines passing through a given point of pseudoproduct are tangent to residue curves. This trajectory reaches side 1-3 at the tear-off point x_{rev}^t, and vertex 2 is the node N_{rev} of the trajectory bundle at different pseudoproduct points. The location of point x_{rev}^t and of the whole trajectory of extractive reversible distillation depends on that of the pseudoproduct point x_D' (i.e., on the ratio E/F between the flow rates of the entrainer and the main feeding). Changing the parameter E/F, we get the trajectory bundle of extractive reversible distillation that, for an ideal mixture, fills up the whole concentration triangle.

We got an important result: *at reversible distillation in the intermediate section of the column of extractive or autoextractive distillation the component, which is intermediate in the value of phase equilibrium coefficient between the component separated as top product and the component brought in as an entrainer, is exhausted.*

4.6.2. Azeotropic Mixtures

This result also remains valid for azeotropic mixtures. A necessary condition for exhausting of the some component in the intermediate (extractive) section at reversible distillation consists of the fact that the whole trajectory of intermediate (extractive) section should be located in the region where this component is intermediate in phase equilibrium coefficient (in the region of reversible distillation of the intermediate section $\text{Reg}_{rev,e}^m$). The segment of the side containing only the component separated as top product and component brought in as an entrainer is a boundary element of this region $\text{Reg}_{D,E}^{(k)}$.

The same result we get if we use the process of reverse autoextractive (reextractive) distillation (Kiva et al., 1983; Petlyuk, 1984; Petlyuk & Danilov, 1999), that is, entering the lightest component in the form of vapor lower than the point of main feeding and withdrawing pure heaviest component as bottom product (Fig. 4.18b).

The application of extractive distillation is of great practical importance because it ensures the possibility of sharp separation of some types of azeotropic mixtures into zeotropic products, which is impossible in a column with one feeding. The mixture acetone(1)-water(2)-methanol(3) is an example of this type of mixture. Trajectories of reversible distillation of three sections of extractive distillation column, the feeding of which is binary azeotrope acetone-methanol, the extractive

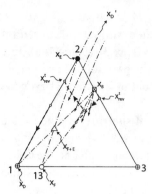

Figure 4.20. Reversible section trajectories for acetone(1)-water(2)-methanol(3) extractive distillation. Short segments with arrows, liquid–vapor tie-lines in arbitrary cross-sections of stripping and intermediate sections; little circles, composition in main and entrainer feed cross-section.

agent is water, and the top product is acetone, are shown in Fig. 4.20 and may be briefly described as follows:

$$\left(\begin{matrix} x_D^{(1)} & \leftarrow x_{e-1}^{(2)} \Downarrow \Rightarrow & x_{rev,e}^{t(2)} & \leftarrow & x_{f-1}^{(3)} & \Downarrow \Rightarrow & x_f^{(3)} & \leftarrow & x_{rev,s}^{t(2)} & \leftarrow & x_B^{(2)} \\ \text{Reg}_D & & \text{Reg}_{rev,e}^t & \text{Reg}_{rev,e} & & & \text{Reg}_{rev,s} & \text{Reg}_{rev,s}^t & & \text{Reg}_B \end{matrix} \right).$$

Trajectories are constructed for a fixed value of the parameter E/F. Let's note that, unless all sections of the column work reversibly, as a whole the process is irreversible because irreversibility arises in the points where the entrainer and the main feeding enter.

As we see, in contrast to simple column, we succeeded in obtaining pure acetone and binary zeotropic mixture methanol-water, which can be separated in the second column.

Let's examine now the structure of trajectory bundles of sharp reversible distillation for the intermediate (extractive) section of the column with two feedings at separation of different types of azeotropic mixtures, the way we did it for the top and the bottom sections (Fig. 4.21). While composing these diagrams, we used, just as we did before, the data on the phase equilibrium coefficients of present and absent components at the sides of the concentration triangle and the general regularities of the location of the trajectory bundles of sharp reversible distillation.

It is obvious that for the separation of three-component mixtures by means of extractive distillation, the mixtures of the type 4a (according to the classification given by Gurikov) that are widespread in practice are of the biggest interest. For these mixtures, according to Fig. 4.21, one can get pure component 1 as top product and zeotropic mixture 2,3 as bottom product. One can get the same result for the mixtures of type 4b.

4.7.   Trajectory Bundles of Extractive Reversible Distillation for Multicomponent Mixtures

Figures 4.22a, b show two different flowsheets of autoextractive distillation of a four-component ideal mixture ($K_1 > K_2 > K_3 > K_4$). Both flowsheets ensure sharp separation in intermediate extractive section, because for any inner point of

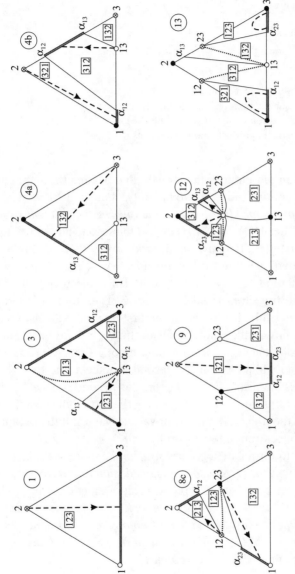

Figure 4.21. Reversible intermediate section diagrams of some structures of three-component mixtures. 1,3,4a,....; classification according to Gurikov (1958); gray segments, tear-off segments of intermediate section trajectories $\text{Reg}_{rev,e}^t$; dotted lines with arrows, reversible intermediate section trajectories.

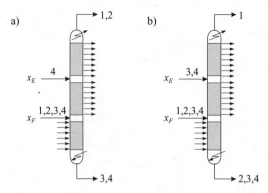

Figure 4.22. Columns for reversible extractive distillation of four-component mixtures: (a) mixture 1,2 is overhead product, and component 4 is entrainer; (b) component 1 is overhead product, and mixture 3,4 is entrainer.

the concentration tetrahedron the straight line passing through liquid–vapor tie line crosses the continuations of two faces: 1-2-4 and 1-3-4 (Fig. 4.23). At the first flowsheet, the *pseudoproduct point of the intermediate section* should lie in the continuation of face 1(D)-2(D)-4(E) and, at the second flowsheet, at continuation of face 1(D)-3(E)-4(E). For any pseudoproduct point located in the continuation of face 1-2-4, there is a point inside this face where the trajectory of extractive reversible distillation x_{rev}^t tears off from it. This trajectory joins point x_{rev}^t with node N_{rev} of the trajectory bundle – vertex 3. Therefore, the trajectory bundle of extractive reversible distillation at the first flowsheet at various values of the parameter E/F fills up the whole concentration tetrahedron and has vertex 3 as its node (the component 3 does not rank among the top product components and the entrainer). Similarly, at the second flowsheet, the tear-off point of the extractive reversible distillation trajectories is located in face 1-3-4, the node is vertex 2 (component 2 does not rank among the top product components and the entrainer), and the trajectory bundle at various values of the parameter E/F fills up the whole concentration tetrahedron.

It is possible to formulate a *general structural condition that should be valid in the tear-off point of the extractive reversible distillation trajectory* from a $(n-1)$-component face or hyperface of the concentration simplex of any multicomponent azeotropic mixture: *the phase equilibrium coefficient of the component that is absent in this face or hyperface and does not rank among the top product components and of the entrainer should be smaller than that of the top product components and bigger than that of the entrainer components.*

4.8.   Boundaries of Nonsharp Reversible Distillation

4.8.1.   Three-Component Azeotropic Mixtures

If, at reversible distillation of three-component azeotropic mixtures, the product segment constitutes part of a side of the concentration triangle (e.g. Fig. 4.24) and

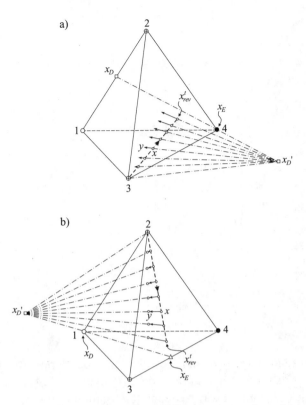

Figure 4.23. Reversible intermediate section trajectories for extractive distillation of four-component mixtures: (a) mixture 1,2 is overhead product, and component 4 is entrainer; (b) component 1 is overhead product, and mixture 3,4 is entrainer. Short segments with arrows, liquid–vapor tie-lines in arbitrary cross-sections.

the product point does not get into this segment, semisharp distillation, rather than sharp distillation, is feasible. In this example, the *boundary of semisharp reversible distillation* joining point x_{B3}^{max} at edge 2-3 with point of azeotrope 13 should exist. Reversible distillation is possible when both the bottom product point x_B and the feed point x_F are located to the left of this boundary, but it is impossible when the feed point is located to the left and the product point is located to the right of it. Therefore, the boundary of possible separation at distillation is not the boundary between the distillation regions at the infinite reflux joining vertex 2 with point of azeotrope 13, but the boundary of sharp and semisharp reversible distillation, including segment $[2, x_{B3}^{max}]$ at side 2-3 and boundary of semisharp distillation from point x_{B3}^{max} to point 13. Similarly, as for point x_{B3}^{max}, for any point x_B^{best} of the boundary of reversible distillation, the trajectory of reversible distillation goes from this point to point of branching x_{rev}^{branch}, from which three branches of the trajectory go to three bottom section nodes – points 1, 3, and 13

$$(x_B^{best} \rightarrow x_{rev}^{branch} \begin{array}{l} \uparrow \ \ \rightarrow N_{rev}^1 \\ \rightarrow N_{rev}^2 \ ). \\ \downarrow \ \ \rightarrow N_{rev}^{13} \end{array}$$

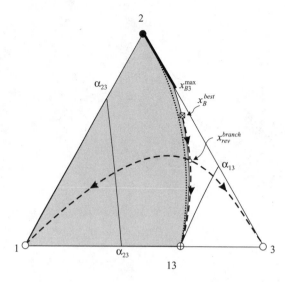

Figure 4.24. Boundary of semisharp reversible distillation region (shaded) of acetone(1)-benzene(2)-chloroform(3) mixture; $x_B^{best} \in \mathrm{Reg}_{rev,B}^{1,bound}$.

The location of semisharp reversible distillation boundaries for three-component mixtures was investigated in the works (Petlyuk et al., 1981b; Poellmann & Blass, 1994). The simplest way to define the location of point x_B^{best} for a given feed point x_F consists of the conduction of a series of trial calculations of reversible distillation trajectories for different points x_B at the straight line passing through the liquid–vapor tie-line of the feeding. For the points x_B to the left of boundary of semisharp reversible distillation, the node is vertex 1 and for the points to the right it is vertex 3.

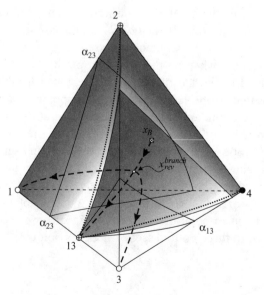

Figure 4.25. Boundary of semisharp reversible distillation region (shaded) of acetone(1)-benzene(2)-chloroform(3)-toluene mixture; $x_B^{best} \in \mathrm{Reg}_{rev,B}^{1,bound}$.

4.8.2. Four-Component Azeotropic Mixtures

Figure 4.25 shows the boundary of semisharp reversible distillation for the four-component mixture acetone(1)-benzene(2)-chloroform(3)-toluene(4). This boundary is a surface passing through point of azeotrope 13 and through the ends of possible product composition segments at edges 2-3 and 3-4 at separation of mixtures 1,2,3 and 1,3,4 correspondingly. For the points x_B^{best} lying on this surface, the trajectory of reversible distillation goes to point of branching x_{rev}^{branch}, from which three branches of the trajectory go to bottom section nodes 1, 3, and 13

$$(x_B^{best} \to x_{rev}^{branch} \to N_{rev}^3 \begin{matrix} \uparrow & \to N_{rev}^1 \\ & \\ \downarrow & \to N_{rev}^{13} \end{matrix}).$$

4.9. Conclusion

At sharp reversible distillation, the heaviest component is absent in the top product and the lightest one is absent in the bottom product. The product points of sharp reversible distillation should belong to possible product composition regions Reg_D or Reg_B on boundary elements of concentration simplex. The trajectory of the section at sharp reversible distillation goes from the product point x_D or x_B to the tear-off point $x_{rev,D}^t$ or $x_{rev,B}^t$ inside boundary element and then from this point to the feed point x_F inside the concentration simplex. In the points of the section trajectory, inside the concentration simplex including the tear-off point, the parameter L/V is equal to the phase equilibrium coefficient K_j of the component absent in the product of the section. In the points of reversible distillation trajectory of extractive section, the phase equilibrium coefficient K_j of the component that is absent in pseudoproduct should be smaller than that of the top product components $K_{D,i}$ and should be higher than that of the entrainer components $K_{E,i}$. In the tear-off points of reversible distillation trajectory from k-component boundary element into $(k+1)$-component boundary element, the equation of reversible distillation should be valid both for the components i present in the k-component boundary element and for the component j, which is absent there. It allows definition of the boundaries of possible product composition region Reg_D or Reg_B at sharp reversible distillation.

To define the frame of possible product composition regions Reg_D or Reg_B at the edges of concentration simplex, it is enough to determine values of phase equilibrium coefficients in the points of edges for the components present and for the component, which is absent in the product.

4.10. Questions

1. Define the main notions: (1) region of reversible distillation of top $Reg_{rev,r}^h$, bottom $Reg_{rev,s}^l$, and intermediate sections $Reg_{rev,e}^m$; (2) regions of trajectory tear-off $Reg_{rev,r}^{t(k)}$ and $Reg_{rev,s}^{t(k)}$ and $Reg_{rev,e}^{t(k)}$; (3) region of possible product points of sharp reversible distillation Reg_D, Reg_B, and $Reg_{D,E}$; and (4) node of trajectory bundle of reversible distillation N_{rev}.

2. Can trajectory bundle of reversible distillation go outside the boundaries of distillation region Reg^∞?

3. Let $K_2 > K_3 > K_1$ in the feed point. In which regions of components order Reg_{ord} can be located: (1) top reversible section trajectory, (2) bottom reversible section trajectory, (3) intermediate reversible section trajectory?

4. Answer the same question as in item 3 for a four-component mixture, in the feed point of which $K_3 > K_2 > K_4 > K_1$ (in extractive distillation column, the top product is component 3 and the entrainer is mixture 1, 4).

5. In face of concentration tetrahedron 2-3-4, there is a region of components order Reg_{ord}^{2413}. What should be the top product and the entrainer for this region to be the trajectory tear-off region $Reg_{rev,e}^{t(3)}$?

6. What minimum information is necessary to define the contour of possible product region Reg_D or Reg_B in face 1-3-4 of concentration tetrahedron 1-2-3-4?

4.11. Exercises with Software

1. Calculate α-lines for the faces of the concentration tetrahedron of a mixture of acetone(1)-benzene(2)-chloroform(3)-toluene(4).

2. Construct reversible distillation trajectories for points x_D: (1) 0.5, 0.0, 0.5, 0.0; (2) 0.05, 0.95, 0.0, 0.0; (3) 0.0, 0.5, 0.5, 0.0 of the above-mentioned mixture.

3. Do the same for points x_B: (1) 0.0, 0.5, 0.0, 0.5; (2) 0.0, 0.9, 0.1, 0.0; (3) 0.0, 0.0, 0.03, 0.97.

References

Benedict, W. (1947). Multistage Separation Processes. *Chem. Eng. Progr.*, 43(2), 46–60.

Grunberg, J. (1960). Reversible Separation of Multicomponent Mixtures. In *Advances in Cryogenic Engineering: Proceedings of the 1956 Cryogenic Engineering Conference*, Vol. 2, New York, pp. 27–38.

Gurikov, Yu. V. (1958). Some Questions Concerning the Structure of Two-Phase Liquid-Vapor Equilibrium Diagrams of Ternary Homogeneous Solutions. *J. Phys. Chem.*, 32, 1980–96 (Rus.).

Haselden, G. (1958). Approach to Minimum Power Consumption in Low-Temperature Gas Separation. *Trans. Inst. Chem. Engrs.*, 36, 123–32.

Hausen, H. (1932). Verlustfreie Zerlegung von Gasgemischen durch Umkehrbare Rektifikation. *Z. Techn. Physik.*, 13(6), 271–7 (Germ.).

Kiva, V. N., Timofeev, V. S., Vizhesinghe, A. D. M. C., & Chyue Vu Tam (1983). The Separation of Binary Azeotropic Mixtures with a Low-Boiling Entrainer. In *The Theses of 5th Distillation Conference in USSR*. Severodonezk (Rus.).

Knapp, J. P., & Doherty, M. F. (1994). Minimum Entrainer Flows for Extractive Distillation: A Bifurcation Theoretic Approach. *AIChE J.*, 40, 243–68.

Koehler, J., Aguirre, P., & Blass, E. (1991). Minimum Reflux Calculations for Nonideal Mixtures Using the Reversible Distillation Model. *Chem. Eng. Sci.*, 46, 3007–21.

Petlyuk, F. B. (1978). Thermodynamically Reversible Fractionation Process for Multicomponent Azeotropic Mixtures. *Theor. Found. Chem. Eng.*, 12, 270–6.

Petlyuk, F. B. (1979). Structure of Concentration Space and Synthesis of Schemes for Separating Azeotropic Mixtures. *Theor. Found. Chem. Eng.*, 683–9.

Petlyuk, F. B. (1984). Necessary Condition of Exhaustion of Components at Distillation of Azeotropic Mixtures in Simple and Complex Columns. In *The Calculation Researches of Separation for Refining and Chemical Industry*, 3–22. Moscow: Zniiteneftechim (Rus.).

Petlyuk, F. B. (1986). Rectification Diagrams for Ternary Azeotropic Mixtures. *Theor. Found. Chem. Eng.*, 20, 175–85.

Petlyuk, F. B., & Danilov, R. Yu. (1999). Sharp Distillation of Azeotropic Mixtures in a Two-Feed Column. *Theor. Found. Chem. Eng.*, 33, 233–42.

Petlyuk, F. B., & Danilov, R. Yu. (2001). Few-Step Iterative Methods for Distillation Process Design Using the Trajectory Bundle Theory: Algorithm Structure. *Theor. Found. Chem. Eng.*, 35, 229–236.

Petlyuk, F. B., & Platonov, V. M. (1964). The Thermodynamical Reversible Multicomponent Distillation. *Chem. Industry*, (10), 723–5 (Rus.).

Petlyuk, F. B., Platonov, V. M., & Girsanov, I. V. (1964). Calculation of Optimal Distillation Cascades. *Chem. Industry*, (6), 445–53 (Rus.).

Petlyuk, F. B., Platonov, V. M., & Slavinskii, D. M. (1965). Thermodynamical Optimal Method for Separating of Multicomponent Mixtures. *Int. Chem. Eng.*, 5(2), 309–17.

Petlyuk, F. B., & Serafimov, L. A. (1983). Multicomponent Distillation. Theory and Calculation. Moscow: Khimiya (Rus.).

Petlyuk, F. B., Serafimov, L. A., Avet'yan, V. S., & Vinogradova, E. I. (1981a). Trajectories of Reversible Distillation When One of the Components Completely Disappears in Every Section. *Theor. Found. Chem. Eng.*, 15, 185–92.

Petlyuk, F. B., Serafimov, L. A., Avet'yan, V. S., & Vinogradova, E. I. (1981b). Trajectories of Reversible Rectification When All Components Are Distributed. *Theor. Found. Chem. Eng.*, 15, 589–93.

Poellmann, P., & Blass, E. (1994). Best Products of Homogeneous Azeotropic Distillations. *Gas Separation and Purification*, 8, 194–228.

Scofield, H. (1960). The Reversible Separation of Multicomponent Mixtures. In *Advances in Cryogenic Engineering: Proceedings of the 1957 Cryogenic Engineering Conference*, Vol. 3, New York, pp. 47–57.

Serafimov, L. A., Timofeev, V. S., & Balashov, M. I. (1973a). Rectification of Multicomponent Mixtures. 3. Local Characteristics of the Trajectories Continuous Rectification Process at Finite Reflux Ratios. *Acta Chimica Academiae Scientiarum Hungarical*, 75, 235–54.

Serafimov, L. A. Timofeev, V. S., & Balashov, M. I. (1973b). Rectification of Multicomponent Mixtures. 4. Non-Local Characteristics of Continuous Rectification, Trajectories for Ternary Mixtures at Finite Reflux Ratios. *Acta Chimica Academiae Scientiarum Hungarical*, 75, 255–70.

Wahnschafft, O. M., Koehler, J. W., Blass, E., & Westerberg, A. W. (1992). The Product Composition Regions of Single-Feed Azeotropic Distillation Columns. *Ind. Eng. Chem. Res.*, 31, 2345–62.

5

Distillation Trajectories and Conditions
of Mixture Separability in Simple Infinite
Columns at Finite Reflux

5.1.   Introduction

This chapter is the central one of the book; all previous chapters being introductory
ones to it, and all posterior chapters arising from this one. Distillation process in
infinite column at finite reflux is the most similar to the real process in finite
columns. The difference in results of finite and infinite column distillation can be
made as small as one wants by increasing the number of plates. Therefore, the main
practical questions of distillation unit creation are those of separation flowsheet
synthesis and of optimal design parameters determination (i.e., the questions of
conceptual design) that can be solved only on the basis of theory of distillation in
infinite columns at finite reflux.

The significance of such theory and, in particular, the significance of development of minimum reflux number calculation methods has been clear for numerous
investigators all over the world since the beginning of distillation science development. A great number of publications have been devoted to these questions.
However, the general distillation theory at finite reflux was created only lately
on the basis of unification of several important ideas and theories of geometric
nature. One can refer to the latter the idea of examination of distillation trajectory bundles at finite reflux for fixed product composition, the conception of sharp
separation of multicomponent mixtures, the theory of location of reversible distillation trajectories in the concentration simplex, the theory of trajectory tear-off
from the boundary elements of concentration simplex at finite reflux, and the
theory of section trajectories joining.

The whole history of investigation of distillation in infinite columns at finite
reflux can be divided into three main stages: creation of distillation theory for
binary mixtures, creation of distillation theory for multicomponent ideal mixtures,
and creation of general distillation theory for all kinds of mixtures.

The development of the McCabe and Thiele diagram (McCabe & Thiele, 1925)
that allowed minimum reflux number and minimum number of theoretic plates for
binary mixtures was of decisive significance for the creation of distillation theory
for such mixtures. For many years, this diagram was the basis for investigation of

distillation process and the main instrument of approximate design of distillation unit.

Qualitative leap to the second stage (i.e., to the distillation theory of ideal multicomponent mixtures) was realized by Underwood (1945, 1946a, 1946b, 1948). Underwood succeeded in obtaining the analytical solution of the system of distillation equations for infinite columns at two important simplifying assumptions – at constant relative volatilities of the components (i.e., which depend neither on the temperature nor on mixture composition at distillation column plates) and at constant internal molar flow rates (i.e., at constant vapor and liquid flow rates at all plates of a column section). The solution of Underwood is remarkable due to the fact that it is absolutely rigorous and does not require any plate calculations within the limits of accepted assumptions.

The solution of Underwood gave impetus to numerous investigations based on this approach. Part of these works was directed to the creation of geometric interpretation of the results obtained from the solution of the Underwood equations system. It is impossible without such interpretation to form a true notion of the general regularities of the distillation process of ideal mixtures. For one-section columns, geometric analysis of trajectories, stationary points, and separatrixes of distillation was realized even before the works of Underwood by Hausen (1934, 1935, 1952) on the basis of calculations using the method "tray by tray."

The works (Franklin & Forsyth, 1953; White, 1953; Vogelpohl, 1964; Petlyuk, Avet'yan, & Platonov, 1968; Vogelpohl, 1970; Shafir, Petlyuk, & Serafimov, 1984; Franklin, 1986, 1988a, 1988b), in which the evolution of product and stationary points in concentration triangles and tetrahedrons for two-section columns at a given feed composition and variable reflux number was examined, appeared after the articles of Underwood. Underwood's method was generalized in the work (Acrivos & Amundson, 1955) for continuous mixtures (i.e., for mixtures consisting of components whose properties are changed continuously from one component to another).

Unfortunately, the method of Underwood cannot be applied to nonideal mixtures and even to ideal ones, relative volatilities of the components that depend on the temperature. Therefore, "tray by tray" method was used for the calculation of minimum reflux mode for such ideal mixtures (Shiras, Hanson, & Gibson, 1950; Erbar & Maddox, 1962; McDonough & Holland, 1962; Holland, 1963; Lee, 1974; Chien, 1978; Tavana & Hanson, 1979) and others.

In the mentioned works, it is suggested that "tray by tray" method should be used only for the part of the column located between zones of constant concentrations. The special equations, taking into account phase equilibrium between the meeting vapor and liquid flows, are applied to such zones. Approximations to the mode of minimum reflux are estimated by means of gradual increase of theoretical plates number in that part of the column for which "tray by tray" method is used.

The attempts to create a theory and to develop methods of minimum reflux number calculation for nonideal and azeotropic mixtures began later.

The notion of distillation trajectory bundles at finite reflux and at fixed product composition, which is important for the further development of the theory, was introduced in the works of Serafimov and coauthors (Serafimov, Timofeev, & Balashov, 1973a, 1973b).

Numerous works (Levy, Van Dongen, & Doherty, 1985; Levy & Doherty, 1986; Julka & Doherty, 1990) in which distillation trajectory bundles of three- and four-component mixtures for two sections of distillation column were used at fixed product compositions and at different values of reflux (vapor) number, are of great importance. They defined the conditions of two section trajectories joining in the feed cross-sections of the column in the mode of minimum reflux, and they developed the methods of this mode calculation for some splits.

However, numerous questions remained unsolved in these works: (1) the methods of prediction of possible product compositions for a given feed composition were absent, which does not allow to calculate minimum reflux mode; (2) the methods of calculation were good only for two special splits: direct and indirect ones, but these methods were not good for the intermediate splits; (3) the peculiarities arising in the case of availability of α-lines, surfaces, and hypersurfaces that are characteristic of nonideal and azeotropic mixtures were not taken into consideration; and (4) the sudden change of concentrations in the feed cross-section was not taken into consideration.

Calculation investigations (Petlyuk, 1978; Petlyuk & Vinogradova, 1980; Shafir et al., 1984) determined the conditions under which saddle and saddle-node stationary points of sections trajectory bundles at finite reflux arise inside the concentration simplex, but not only at its boundary elements, promoted the development of this trajectory bundles theory.

As far as stationary points of section trajectory bundles should be located at the trajectories of reversible distillation, the systematic analysis of these trajectories locations was of great importance (see Chapter 4).

The approximate calculation method of minimum reflux mode (Koehler, Aguirre, & Blass, 1991) – the method of "the smallest angle," which holds good for mixtures with any component numbers and for any splits, including frequently found at azeotropic mixtures separation cases of "tangential pinch," is based on the calculation of reversible distillation trajectories for the given product compositions.

The appearance of the "tangential pinch" in the mode of minimum reflux was investigated in the works (Levy & Doherty, 1986; Fidkowski, Malone, & Doherty, 1991).

The approach to calculation of the minimum reflux mode, based on eigenvalue theory, was introduced in the work (Poellmann, Glanz, & Blass, 1994). In contrast to the above-mentioned works of Doherty and his collaborators this method calculates the mode of minimum reflux not only for direct and indirect, but also for intermediate split of four-component mixtures.

However, this method is not effective for the mixtures with component numbers greater than four and, besides that, does not take into consideration the leap of concentration in feed cross-section.

The approximate method of calculation of the minimum reflux mode for three-component mixtures at the absence of "tangential pinch" was suggested in the work (Stichlmair, Offers, & Potthoff, 1993).

The previously enumerated methods of calculation of the minimum reflux mode for nonideal zeotropic and azeotropic mixtures have considerable defects: (1) they presuppose preliminary setting of possible separation product compositions, which is a complicated independent task for azeotropic mixtures; (2) they embrace only three- and four-component mixtures or only special splits; and (3) they do not take into consideration the leap of concentrations in feed cross-section.

In practice, the enumerated calculation methods are hardly used when designing distillation units because of these defects. Calculation of the minimum reflux mode is not conducted at all, and the working reflux number and number of plates in the sections are chosen, as a rule, arbitrarily, based on the designer's intuition and experience, which can lead to considerable overstating of separation costs.

To overcome these defects, it was necessary to apply the conception of sharp separation and to develop the theory of distillation trajectory tear-off from the boundary elements of concentration simplex at sharp separation (Petlyuk, Vinogradova, & Serafimov, 1984; Petlyuk, 1998) and also to develop the geometric theory of section trajectories joining in feed cross-section in the mode of minimum reflux that does not contain simplifications and embraces mixtures with any number of components and any splits (Petlyuk & Danilov, 1998; Petlyuk & Danilov, 1999b; Petlyuk & Danilov, 2001a; Petlyuk & Danilov, 2001b).

The significance of the methods, based on the geometric distillation theory, consists in their universality, rigor, and reliability. To obtain the result, it is not necessary to set any estimation parameters and possible separation product compositions; minimum reflux number for these compositions and distillation trajectory at this reflux number are defined in the process of this calculation.

The geometric distillation theory also allowed the development of the general methods of separation flowsheets synthesis for azeotropic mixtures and design calculation of simple and complex distillation columns, which is examined in the chapters to follow.

This chapter answers two fundamental questions of the conceptual designing of distillation units:

1. How to determine which sharp splits of any multicomponent mixture are feasible?
2. How to determine which minimum energy is necessary for the separation of any multicomponent mixture at any feasible split?

5.2. Calculation of Distillation at Minimum Reflux for Ideal Mixtures

A number of regularities of the minimum reflux mode are common for the ideal, nonideal, and even azeotropic mixtures. Among these regularities is the following: each section trajectory at minimum reflux and at sharp separation is partially

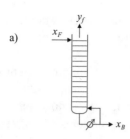

a)

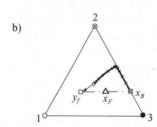

b)

Figure 5.1. (a) Stripping section. (b) Stripping trajectory (liquid compostion profile) for a ideal mixture; dashes, tray compositions.

located at the boundary elements of concentration simplex and partially located inside it, where they are joined with each other. Another common regularity is availability of constant concentration zones, to which stationary points at each section trajectory correspond. It is expedient to begin the analysis of distillation regularities at minimum reflux with the ideal mixtures, for which the whole concentration simplex is one component-order region $\text{Reg}_{ord}^{1,2...n}$ ($K_1 > K_2 > \cdots > K_n$). In a more particular case, we understand as the ideal mixtures those for which, besides that, relative volatilities of the components in all points of concentration simplex are the same (i.e., $\alpha_i = K_i/K_h = const$), and latent heats of evaporation of the components are the same (i.e., at distillations, molar vapor and liquid flow rates do not change along the section height $V_r = const$, $V_s = const$, $L_r = const$, $L_s = const$).

Availability of these conditions allowed Underwood (1948) to obtain general solution, connecting separation product compositions at minimum reflux with the mode parameters (e.g., with V_r and V_s). Even before (Hausen, 1934, 1935), distillation trajectories of the ideal mixtures in the one-section columns (Fig. 5.1a) were investigated by means of calculation, and it was shown that the part of distillation trajectory located inside the concentration triangle is rectilinear for the ideal mixture (Fig. 5.1b). Later, linearity of distillation trajectories of three-component ideal mixtures at sharp separation was rigorously proved (Levy et al., 1985).

5.2.1. Underwood System of Equations

The Underwood system of equations can be obtained from the conditions of componentwise material balance and of phase equilibrium in the cross-section of constant concentration zones of each section. For example, the following expression can be obtained from the equation of componentwise material balance at the contour, embracing the part of top section from the cross-section in the zone of constant concentration to the output of top product, taking into consideration the conditions of phase equilibrium between the incoming vapor flow and the outgoing liquid flow in this zone:

$$V_r y_i = \alpha_i d_i / (\alpha_i - L/V K_n)$$

Designating $\varphi = L/V K_n$ and summing up by all the components, we obtain one of the main equations of Underwood:

$$\sum_i \alpha_i d_i / (\alpha_i - \varphi) = V_r^{min} \tag{5.1}$$

The analogous equation for the bottom section is as follows:

$$\sum_i \alpha_i b_i / (\alpha_i - \psi) = -V_s^{min} \tag{5.2}$$

The main achievement of Underwood consists in the proof of equality of parameters ϕ and ψ in Eqs. (5.1) and (5.2) in the mode of minimum reflux. Sum up these equations is following main equation:

$$\sum_i \alpha_i f_i / (\alpha_i - \theta) = (1 - q)F \tag{5.3}$$

where q is a portion of liquid in the feed and θ is a common parameter (root) of Eqs. (5.1) and (5.2).

For the set composition of the top product and for the set reflux number, the number of roots of Eq. (5.1) equals that of the components in top product (k):

$$0 < \varphi_1 < \alpha_k < \varphi_2 < \alpha_{k-1} < \cdots < \alpha_2 < \varphi_k < \alpha_1$$

Similarly, the number of roots of Eq. (5.2) equals that of components in the bottom product (m):

$$\alpha_m < \psi_1 < \alpha_{m-1} < \cdots < \alpha_2 < \psi_{m-1} < \alpha_1 < \psi_m$$

Correspondingly, the number of roots of Eq. (5.3) is less by one than that of the components, present in the top and in the bottom products (i.e., the number of distributed components).

The Underwood equation system determines separation product compositions and internal liquid and vapor flows in the sections for the set values of two parameters, characterizing the separation process. The reflux number R and withdrawal of one of the products D/F or recoveries of some two components into the top product $\xi_i = d_i/f_i$ and $\xi_j = d_j/f_j$, etc., can be chosen as such two parameters. For example, at direct split of three-component ideal mixture 1(2) : (1)2,3 (here the top product contains component 1 and small admixture of component 2 and the bottom product contain components 2,3 and small admixture of component 1), Eq. (5.3) has only one common for both section root $\alpha_2 < \theta < \alpha_1$. If ξ_1 and ξ_2 are set, then d_1 and d_2 can be defined and V_r^{min} can be obtained from Eq. (5.1). The rest of internal flows in the column section can be defined with the help of the material balance equations.

In a more general case, when there are several distributed components, it is necessary to obtain from Eq. (5.3) the common roots for two sections. After the substitution of each of these roots into Eq. (5.1) or (5.2), we obtain the system of linear equations relatively to d_i and V_r^{min} or b_i and V_s^{min}, the solution of which determines separation product compositions and internal vapor and liquid flows in the column sections. In addition, one can find the compositions of equilibrium phases in the cross-sections of constant concentration zones (i.e., stationary points of sections trajectories bundles).

The main problem in solving the Underwood equation system, as it was shown in Shiras et al. (1950), is the correct determination of the list of distributed components at two specified parameters. At the wrong setting of this list, the solution of the equation system leads to unreal values of d_i and b_i for some components ($d_i > f_i$ or $d_i < 0$).

In this case, it is necessary to correct the list of distributed components, referring those components, for which unreal values of d_i or b_i were obtained, to undistributed ones.

5.2.2. Evolution of Separation Product Compositions of One-Section Columns at Set Feed Composition

The use of the Underwood equation system allows for examination of the evolution of separation product compositions of one- and two-section columns at set feed composition and at variable reflux number. Figure 5.2 shows such evolution for one-section rectifying column (Shafir et al., 1984). This figure also shows trajectory bundles $N^- \Rightarrow N^+$ ($N^- \rightarrow S \rightarrow N^+$) in accordance with the notion of the bundle introduced by Serafimov et al. (1973a). The trajectory of distillation section is a line in concentration space that connects the points in which the set of equations of distillation at given product point and reflux is satisfied. This line obtains by means of "tray by tray" method from any point of concentration space. *The trajectory bundle at given finite reflux R is a set of trajectories with the same initial and final stationary points (unstable N^- and stable N^+ nodes) at the same product point x_D or x_B.* At the small reflux numbers (the first class of fractioning), the stationary point of trajectory bundle (composition point in the zone of constant concentrations) coincides with point x_F (equilibrium to the feed point y_F) and the product point x_D, as at reversible distillation, lies at the continuation of the liquid–vapor tie-line of feeding (Fig. 5.2a). The stationary point x_F is a stable node N^+ of rectifying section bundle (region $\text{Reg}_{w,r}^R$). At the increase of reflux number, the top product point x_D moves along the straight line, passing through liquid–vapor tie-line of feeding, moving away from the feed point and coming nearer to side 1-2.

In conclusion, at the limit (boundary) value of reflux number R_{lim}^1, the product point x_D approaches side 1-2 (sharp separation, the second class of fractioning; Fig. 5.2b). At the same time, the saddle stationary point S (trajectory tear-off point x^t from side 1-2) appears at side 1-2. Therefore, at boundary reflux number in

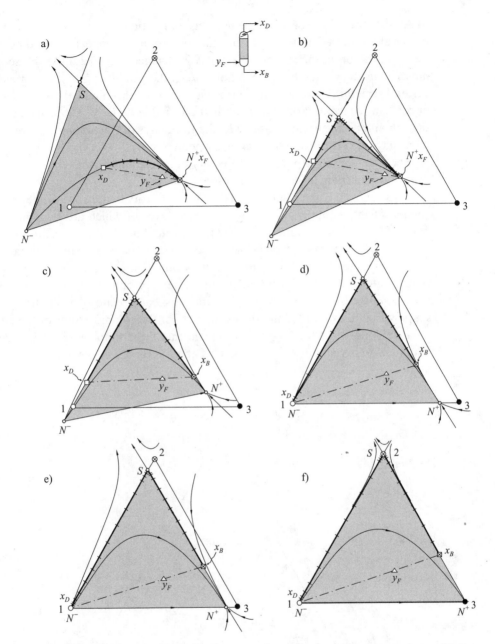

Figure 5.2. Evolution of a rectifying section region $\text{Reg}^R_{w,r}$ for a ternary ideal mixture: (a) $R < R^1_{\lim}$, (b) $R = R^1_{\lim}$, (c) $R^2_{\lim} > R > R^1_{\lim}$, (d) $R = R^2_{\lim}$, (e) $R > R^2_{\lim}$, (f) $R = \infty$.

one-section column, there are two zones of constant concentrations, the compositions of which coincide with the compositions in stationary points N^+ and S.

At $\alpha_i = const$ and $L/V = const$, the part of the trajectory located inside the concentration triangle $S \to N^+$ is rectilinear.

At further increase of R (the second class of fractioning), the product point x_D moves alongside 1-2 to vertex 1, the stationary point disappears in feeding, and the composition in the stationary point S is changed (Fig. 5.2c). At the second limit (boundary) value of reflux number R_{lim}^2, the product point x_D approaches vertex 1 (the third class of fractioning; Fig. 5.2d). At this mode, the second zone of constant concentrations (vertex 1 is the unstable node N^-) appears again in the column. At further increase of R (the third class of fractioning), the product point stays in vertex 1 and the stationary point S moves to vertex 2 (Fig. 5.2e). At $R = \infty$, the stationary points of trajectory bundle coincide with the vertexes of concentration triangle (Fig. 5.2f) and $\text{Reg}_{w,r}^R = \text{Reg}^\infty$.

It is seen from Fig. 5.2 that the distillation trajectory bundle of the one-section column fills up some triangle $\text{Reg}_{w,r}^R$, the vertexes of which are the stationary points. Some of these stationary points are located inside the concentration triangle $C^{(3)}$ and the rest of them outside it (i.e., they are of theoretical nature). The triangle $\text{Reg}_{w,r}^{R(3)}$ filled up with trajectory bundle is called *distillation triangle*. At a greater number of components, the trajectory bundle fills up some *distillation simplex* $\text{Reg}_{w,r}^{R(n)}$. In two-section columns, each section has its distillation simplex $\text{Reg}_{w,r}^R$ or $\text{Reg}_{w,s}^R$, and the availability of the common roots of the equations of Underwood for two sections means that these simplexes in the mode of minimum reflux adjoin to each other by their vertexes, edges, faces, or hyperfaces.

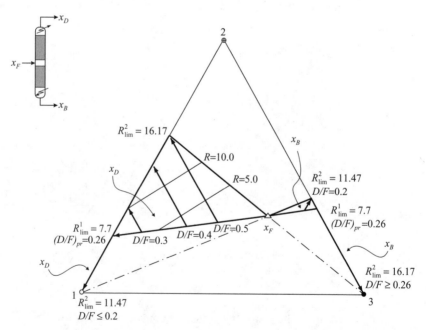

Figure 5.3. Product compositions x_D and x_B for an ideal mixture with $\alpha_1 = 1.5$, $\alpha_2 = 1.1$, $\alpha_3 = 1.0$, $x_{F1} = 0.2$, $x_{F2} = 0.3$, and $x_{F3} = 0.5$ at minimum reflux for different R_{min} and D/F. Thick lines with arrows $- D/F = $ const, thin lines $- R = $ const.

5.2.3. Evolution of Separation Product Compositions of Two-Section Columns at Set Feed Composition

The evolution of separation product compositions of two-section columns in contrast to one-section columns depends on two parameters: on the reflux number and on the withdrawal of top product D/F.

The Underwood equation system investigates the whole set of product compositions at the given feed composition and at the variable parameters R and D/F (Petlyuk et al., 1968).

Figures 5.3 and 5.4 show these sets of product compositions correspondingly for three- and four-component mixtures (calculations are made at $\alpha_1 = 1.5$, $\alpha_2 = 1.1$, $\alpha_3 = 1.0$, and $x_F = [0.2; 0.3; 0.5]$ for three-component mixture and at $\alpha_1 = 1.8$, $\alpha_2 = 1.5$, $\alpha_3 = 1.1$, $\alpha_4 = 1.0$, and $x_F = [0.1; 0.2; 0.3; 0.4]$ for four-component mixture).

Let's examine the evolution of product points of two-section column at the increase of R and at $D/F = const$. At small R (the first class of fractioning, nonsharp separation), both product points x_D and x_B lie at the straight line, passing though liquid–vapor tie-line of feeding x_F, and zones of constant concentrations of both

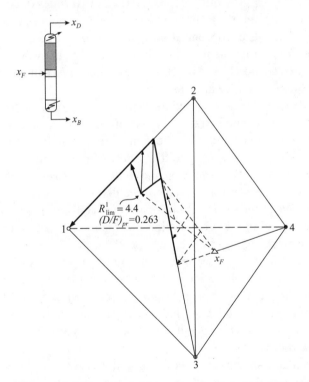

Figure 5.4. Top product compositions x_D for an ideal mixture with $\alpha_1 = 1.8$, $\alpha_2 = 1.5$, $\alpha_3 = 1.1$, $\alpha_4 = 1.0$, $x_{F1} = 0.1$, $x_{F2} = 0.2$, $x_{F3} = 0.3$, and $x_{F4} = 0.4$ at minimum reflux for different R_{min} and D/F. Lines with arrows – $D/F = const$.

sections adjoin to the feed cross-section (trajectories of both sections intersect each other in the feed point and section working regions have stable node N_r^+ and N_s^+ in the feed point x_F). At the increase of reflux number, the product points x_D and x_B come nearer to the boundary elements of concentration simplex (to the sides of concentration triangle in Fig. 5.3 or to the faces of concentration tetrahedron in Fig. 5.4).

Later, the process can develop in different ways, depending on the chosen value of D/F: (1) both product points x_D and x_B at $R = R_{lim}^1$ can simultaneously reach the boundary elements of concentration simplex (such a split was called a transitional one [Fidkowski, Doherty, & Malone, 1993] and a preferable one ([Stichlmair et al., 1993]); (2) the top product point x_D at $R = R_{lim}^1$ can reach the boundary element of concentration simplex, and the bottom product point x_B at the same time stays inside it (such a split was called a direct one); (3) the bottom product point x_B at $R = R_{lim}^1$ can reach the boundary element of concentration simplex, and the top product point x_D at the same time stays inside it (such a split was called an indirect one).

Designating withdrawal at preferable separation D_{pr}, at $D < D_{pr}$ there is a direct separation and at $D > D_{pr}$ there is an indirect separation.

At $D = D_{pr}$ and at $R = R_{lim}^1$ in both sections, there are two zones of constant concentrations – in the feed point x_F and in the trajectory tear-off points of sections x^t from the boundary elements of concentration simplex. For a three-component mixture there is a transition from the first class of fractioning right away into the third class, omitting the second class. At further increase of reflux number, the product compositions do not change any more.

At $D < D_{pr}$ and $R = R_{lim}^1$ in the top section, there are two zones of constant concentrations: in feed point x_F and in trajectory tear-off point from the boundary element of concentration simplex and in the bottom section there is one zone in feed point x_F. At $D > D_{pr}$ and $R = R_{lim}^1$, on the contrary, in the bottom section there are two zones of constant concentration and in the top the section there is one zone. In both cases there is a transition from the first class of fractioning to the second one (i.e., in one of the sections, zone of constant concentrations in feed cross-section disappears, and in the other section, the zone is preserved, but the composition in it starts to change with the change of R).

At further increase of R at direct separation, top product point x_D begins to move along side 1-2 to vertex 1 till component 1 will be completely in top product. After that, further movement of product points x_D and x_B is stopped (i.e., the third class of fractioning ensues). At indirect separation, bottom product point x_B moves to vertex 3 till component 3 will be entirely in bottom product. At the second class of fractioning, trajectory tear-off point x^t of one of the sections is not changed and, for mixtures with constant relative volatilities, part of trajectory of this section $x^t \equiv S \rightarrow N^+$ is also not changed (Stichlmair et al., 1993).

Depending on the parameter D/F for three-component mixtures at the transition to the third class of fractioning, the following splits are feasible: (1) 1 : 1,2,3; (2) 1 : 2,3; (3) 1,2 : 2,3; (4) 1,2 : 3; (5) 1,2,3 : 3. For four-component mixtures the

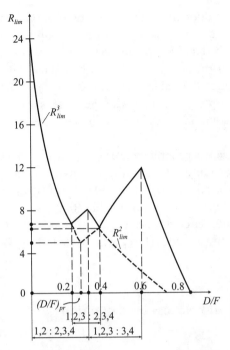

Figure 5.5. $R_{\lim}$ as function of D/F for the mixture described in Fig. 5.4. Segments with arrows, intervals of D/F value for different splits with distributed components. At $R_{\lim}$, the conversion take place from one split to another. At $R_{\lim}^3$, the conversion take place from second class of fractionation to third. Points on system axes, $D_{\lim}/F$ and $R_{\lim}$.

following splits are feasible: (1) 1 : 2,3,4; (2) 1 : 2,3,4; (3) 1,2 : 2,3,4; (4) 1,2,3 : 2,3,4; (5) 1,2 : 3,4; (6) 1,2,3 : 3,4; (7) 1,2,3 : 4; (8) 1,2,3,4 : 4.

We call such reflux number $R_{\lim}$, at which in one of the product one of the components disappears (i.e., at $R > R_{\lim}$ in one of the products, the components number is smaller than at $R < R_{\lim}$), a boundary one. We also call such value of withdrawal $D_{\lim}/F$, at which in both products one component disappears at some $R_{\lim}$ (i.e., at $D = D_{\lim}$ and $R > R_{\lim}$ in the top and bottom products, there are number of components smaller by one than at $R < R_{\lim}$), a boundary one. The sharp splits without distributed components appear at some boundary values of withdrawal. Besides that, for the splits with distributed components there are boundary values of withdrawal, at which reflux number is minimum. Figure 5.5 shows dependence of $R_{\lim}$ on D for the above-mentioned example of four-component mixture. It is well seen that at $D_{\lim}$ and $R_{\lim}$ for the separation modes with distributed components 2 and 3, the reflux number is minimum.

The conducted analysis of product points evolution, depending on R for ideal mixtures, determines a number of the important qualitative regularities of the minimum reflux mode: the existence of three classes of fractioning, the availability of one or two zones of constant concentrations in each section of the column, feasibility of various splits by means of a corresponding choice of two parameters of the mode – of R and D/F.

Besides that, Fig. 5.2 shows that at set value of R concentration simplex and surrounding it, space with unreal concentrations of the components are filled up with several trajectory bundles $N^- \Rightarrow N^+$. These bundles are separated from each other by separatrixes (at $n = 3$) or by dividing surfaces and hypersurfaces (at $n = 4$ and more).

At nonsharp separation, the stationary points of section working regions, except the stable node N^+, are located outside the concentration simplex (the direction of trajectory from the product is accepted). At sharp separation, other stationary points – trajectory tear-off points x^t from the boundary elements of concentration simplex – are added to the stable node. These are the saddle points S and, besides that, if the product point coincides with the vertex corresponding to the lightest or to the heaviest component, then this point becomes an unstable node N^-.

These qualitative regularities have a general nature and apply not only to ideal mixtures, but also to nonideal ones. Only the possibility of analytic solution for the minimum reflux mode (Underwood equation system) and linearity of separatrixes and of dividing surfaces and hypersurfaces of section regions $\text{Reg}_{w,r}^R$ and $\text{Reg}_{w,s}^R$ are specific for ideal mixtures. The latter circumstance was also extended to nonideal mixtures in a number of approximate methods (Levy et al., 1985; Julka & Doherty, 1990; Stichlmair et al., 1993).

5.3.   Trajectory Tear-Off Theory and Necessary Conditions of Mixture Separability

The task of determining distillation product compositions of ideal mixtures in infinite column at minimum reflux is discussed in the previous section. The Underwood equation system solves this task for set composition x_F and thermal state of feeding q at two set parameters (e.g., R and D/F or d_i and d_j).

For nonideal zeotropic and azeotropic mixtures, the solution of the task of minimum reflux mode calculation in such a statement run across the insurmountable calculating difficulties in the majority of cases.

The development of distillation trajectory bundles theory at finite reflux showed that the task of minimum reflux mode calculation for nonideal zeotropic and azeotropic mixtures can be solved in another statement: at set composition x_F and thermal state q of feeding, it is necessary to determine minimum reflux number $R_{\min}$ for the set product compositions x_D and x_B of sharp separation and set permissible concentrations of admixtures in the products.

If the problem is stated in this way, it is necessary to determine what product compositions x_D and x_B of sharp separation are feasible at distillation of nonideal zeotropic and azeotropic mixtures. The theory of distillation trajectory tear-off from the boundary elements of concentration simplex answers this question.

5.3.1. Conditions of Distillation Trajectory Tear-Off at Sharp Splits

Let's examine two constituent parts of section distillation trajectory at the example of sharp preferable split of three-component ideal mixture (Fig. 5.6a): the part located in the boundary element (the side of concentration triangle), and the part located inside concentration simplex (triangle). There is a trajectory tear-off point from the boundary element x^t between these two parts.

a)

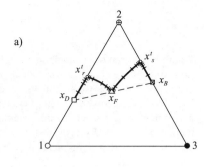

b)

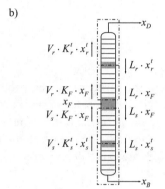

Figure 5.6. Preferrable split of a ideal mixture at minimum reflux: (a) the section trajectories, and (b) the pinches in column (shaded). Boundaries for material balances discussed in the text are indicated by dotted-dashed lines.

It is necessary for the distillation trajectory to be able to tear off from the boundary element that certain conditions be valid in the tear-off point x^t.

To obtain these conditions, let's examine material balance along the closed contour from the zone of constant concentrations to the end of the column and phase equilibrium in the cross-section of any zone of constant concentrations (see Fig. 5.6b):

$$L/V = \left(K_i^{st} x_i^{st} - x_i^D\right) / \left(x_i^{st} - x_i^D\right) \qquad \text{(for top section)} \qquad (5.4)$$

$$L/V = \left(K_i^{st} x_i^{st} - x_i^B\right) / \left(x_i^{st} - x_i^B\right) \qquad \text{(for bottom section)} \qquad (5.5)$$

If the component j is absent in the top product ($x_{Dj} = 0$) or in the bottom product ($x_{Bj} = 0$) (sharp separation), then it follows from Eqs. (5.4) and (5.5) for both sections that (Petlyuk et al., 1984; Petlyuk, 1998; Petlyuk & Danilov, 1998):

$$L/V = K_j^{st} \qquad (5.6)$$

Trajectory tear-off points x^t are a special kind of stationary points that can be called pseudostationary ones because, in the vicinity of these points, the concentrations of components, absent in the boundary element, along the distillation trajectory in the direction to the product decrease monotonously at infinite number of separation stages ($x_j^{(k+1)} > x_j^{(k)}$, where k is a separation stage closer to the product than $[k + 1]$).

Therefore, in the vicinity of tear-off points, x^t concentrations of components present and absent in the product behave differently at neighboring plates: concentrations of components present in the product are constant, and those of

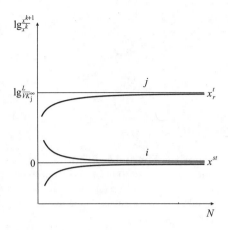

Figure 5.7. Variations in component concentration ratios at neighboring trays about stationary point (x^{st}) for any components i and about top section trajectory tear-off (pseudostationary) point (x_r^t) for absent in the boundary element components j. K_j^{∞} is the phase equilibrium coefficient of absent component j in the pseudostationary point.

components absent in the product decrease in the direction to the product. In contrast to it, in the vicinity of stationary points located inside concentration simplex, the concentrations of the components behave in one and the same way.

For the component j, absent in the product and in the boundary element from which the trajectory tears off, the conditions of materials balance for any stage of separation look like that:

$$VK_j^{(k+1)}x_j^{(k+1)} = Lx_j^{(k)} \qquad \text{(for top section)} \tag{5.7}$$

$$Lx_j^{(k+1)} = VK_j^{(k)}x_j^{(k)} \qquad \text{(for bottom section)} \tag{5.8}$$

Therefore, it follows from the inequality $x_j^{(k+1)} > x_j^{(k)}$ in trajectory tear-off points x^t from the boundary elements of concentration simplex, that:

$$L/V > K_j^t \qquad \text{(for top section)} \tag{5.9}$$

$$L/V < K_j^t \qquad \text{(for bottom section)} \tag{5.10}$$

We see that these conditions differ from those in other stationary points (Eq. [5.6]). The difference in change of concentrations ratio of the components at two neighboring plates in the stationary points and of the components absent in the boundary element in trajectory tear-off points x^t is shown for the top section in Fig. 5.7.

Inequalities (Eqs. [5.9] and [5.10]) for the components absent in the product and in the boundary element are valid inside concentration simplex not only in the vicinity of trajectory tear-off points x^t from the boundary elements, but also in other trajectory points that are not stationary.

Now it is necessary to examine the ratio between the parameter L/V and phase equilibrium coefficient K_i of the components present in the product in all stationary and pseudostationary points. It follows from Eqs. (5.4) and (5.5) that:

$$x_i^D = x_i^{st}(K_i^{st} - L/V)/(1 - L/V) \tag{5.11}$$

$$x_i^B = x_i^{st}(K_i^{st} - L/V)/(1 - L/V) \tag{5.12}$$

As far as $x_{Di} > 0$ and $x_{Bi} > 0$, $L/V < 1$ for the top section and $L/V > 1$ for the bottom section, it follows from Eqs. (5.11) and (5.12) that:

$$L/V < K_i^{st} \quad \text{(for top section)} \tag{5.13}$$

$$L/V > K_i^{st} \quad \text{(for bottom section)} \tag{5.14}$$

Comparison of these inequalities with equality (Eqs. [5.6]) for the stationary points and with inequalities (Eqs. [5.9]) and ([5.10]) for pseudostationary and other points of the trajectory leads to the important result: in all points of the trajectory and, in particular, in its tear-off points x^t from the boundary element, the following inequalities should be valid:

$$K_i^t > K_j^t \quad \text{(for top section)} \tag{5.15}$$

$$K_i^t < K_j^t \quad \text{(for bottom section)} \tag{5.16}$$

Equations (5.9), (5.10), (5.15), and (5.16) are necessary and sufficient conditions of trajectory tear-off from the boundary element of concentration simplex. *Equations (5.9) and (5.10) can be called operating ones because they depend on separation mode, and Eqs. (5.15) and (5.16) can be called structural ones because they depend only on the structure of the field of phase equilibrium coefficients.*

5.3.2.   Trajectory Tear-Off Regions and Sharp Distillation Regions

In trajectory tear-off points of the top section x_r^t phase equilibrium coefficients of the components present in the product K_i^t should be greater than those of the components absent in the product K_j^t, and vice versa in the bottom section. Therefore, tear-off of trajectories from the boundary elements of concentration simplex is feasible only if in the vicinity of this boundary elements there are component-order regions Reg_{ord}^{ijk} that meet these conditions of trajectory tear-off (Fig. 5.8). *We call the region where trajectory tear-off is feasible "trajectory tear-off region"* $\text{Reg}_r^{j,t(k)}$ or $\text{Reg}_s^{j,t(k)}$. Those α-manifolds, in points of which phase equilibrium coefficients of one of the present in the product component and of one of the absent in the product component are equal, are boundaries of trajectory tear-off regions.

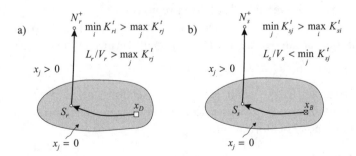

Figure 5.8. Tear-off conditions from boundary elements of the concentration simplex for the section trajectories: (a) rectifying section, and (b) stripping section.

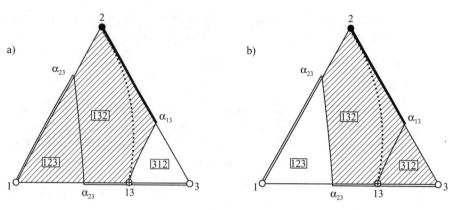

Figure 5.9. Component-order regions $\text{Reg}_{ord}^{i,j,k}$, trajectory tear-off regions $\text{Reg}_r^{t(2)}$ and $\text{Reg}_s^{t(2)}$, and sharp split regions $\text{Reg}_{sh}^{i:j}$ (hatched) for the acetone(1)-benzene(2)-chloroform(3) mixture for two splits: (a) 1 : 2,3 ($\text{Reg}_{sh}^{1:2,3}$) and (b) 1,3 : 2 ($\text{Reg}_{sh}^{1,3:2}$); *123, 132, 312*, component orders in which the phase equilibrium coefficients decrease in component-order region $\text{Reg}_{ord}^{i,j,k}$.

Now we can introduce the new important notion – *sharp split region for column section at set sharp split (i.e., at fixed set of the components present i and absent j in the product of column section).*

We call for the top section the region, in points of which phase equilibrium coefficients of the components present in the product ($x_{D,i} > 0$) are greater than those of the components absent ($x_{D,j} = 0$) "sharp split region" $\text{Reg}_{sh,r}^{i:j}(K_{Di} > K_{Dj})$, and vice versa for the bottom section $\text{Reg}_{sh,s}^{i:j}(K_{Bi} < K_{Bj})$.

As far as Eqs. (5.15) and (5.16) should be valid for the points of trajectory located inside concentration simplex, the whole trajectory of the section from tear-off point to point of junction with the trajectory of the second section should be located in one region of section sharp split $\text{Reg}_{sh,r}^{i:j}$ or $\text{Reg}_{sh,s}^{i:j}$. For each sharp split region, $\text{Reg}_{sh,r}^{i:j}$ or $\text{Reg}_{sh,s}^{i:j}$ consist of quite definite component-order regions Reg_{ord}^{ijk}. Therefore, for different splits section sharp regions, $\text{Reg}_{sh,r}^{i:j}$ and $\text{Reg}_{sh,s}^{i:j}$ are different. For splits without distributed components, sharp split regions of both sections coincide with each other $\text{Reg}_{sh,r}^{i:j} = \text{Reg}_{sh,s}^{i:j}$ (Fig. 5.9).

For splits with distributed components, sharp split regions of two sections are different (Fig. 5.10). Reversible distillation regions $\text{Reg}_{rev,r}^h$ and $\text{Reg}_{rev,s}^l$, which are discussed in Chapter 4, are a particular case of sharp split regions (in this case, component h is absent in overhead and component l is absent in bottom).

Figures 5.9 and 5.10 show trajectory tear-off regions $\text{Reg}_r^{t(2)}$ and $\text{Reg}_s^{t(2)}$ and sharp split regions $\text{Reg}_{sh,r}^{i:j}$ and $\text{Reg}_{sh,s}^{i:j}$ of sections at the examples of three-component azeotropic mixtures acetone(1)-benzene(2)-chloroform(3) and acetone(1)-methanol(2)-chloroform(3).

5.3.3. Necessary Condition of Mixture Separability for the Set Split

Let's formulate now the main necessary condition of mixtures separability for the set split: it is necessary that at the boundary element to which the top product

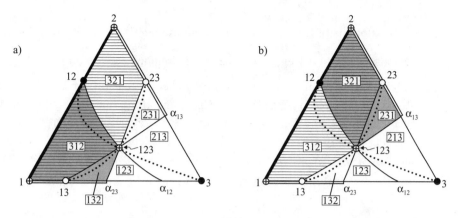

Figure 5.10. Component-order regions $\text{Reg}_{ord}^{i,j,k}$, trajectory tear-off regions $\text{Reg}_{r}^{t(2)}$ and $\text{Reg}_{s}^{t(2)}$, and sharp split regions $\text{Reg}_{sh,r}^{i:j}$ and $\text{Reg}_{sh,s}^{i:j}$ (hatched for bottom section and shaded for top section, hatched and shaded for two section) for the acetone(1)-chloroform(2)-methanol(3) mixture for splits: (a)1,3 : 1,2 ($\text{Reg}_{sh,r}^{1,3:1,2}$ and $\text{Reg}_{sh,s}^{1,3:1,2}$) and (b) 2,3 : 1,2 ($\text{Reg}_{sh,r}^{2,3:1,2}$ and $\text{Reg}_{sh,s}^{2,3:1,2}$).

point belongs (top product boundary element) there is a region where all the components absent in the product are the heaviest ones (trajectory tear-off region of the top section $\text{Reg}_{r}^{\underset{i}{j}\,t(k)}$), and at the boundary element to which the bottom product point belongs (bottom product boundary element) there is a region where all the components absent in the product are the lightest ones (trajectory tear-off region of the bottom section $\text{Reg}_{s}^{\underset{i}{j}\,t(k)}$).

If the split without distributed components is set, then the mentioned trajectory tear-off regions of sections should be boundary elements of one sharp split region $\text{Reg}_{sh,r}^{i:j} = \text{Reg}_{sh,s}^{i:j}$ (Fig. 5.9).

If the split with distributed components is set, then the mentioned trajectory tear-off regions of sections should be boundary elements of two different for the top and bottom sections, but partially overlapping regions of section sharp split ($\text{Reg}_{sh,r}^{i:j} \neq \text{Reg}_{sh,s}^{i:j}$, Fig. 5.10).

Product points can belong only to those boundary elements of concentration simplex that contain trajectory tear-off regions. Along with that, product points should be located at these boundary elements within the limits of some region, that we call *possible product region* $\text{Reg}_{D}^{\underset{i}{j}\,(k)}$ or $\text{Reg}_{B}^{\underset{i}{j}\,(k)}$.

The process of sharp distillation is feasible only if each product point belongs to possible product region in boundary element of concentration simplex ($x_D \in \text{Reg}_{D}^{\underset{i}{j}\,(k)}$ and $x_B \in \text{Reg}_{B}^{\underset{i}{j}\,(m)}$). The question of determination of boundaries of possible product regions at sharp distillation are discussed in the following sections of this Chapter and in Chapter 8.

The notions of trajectory tear-off regions $\operatorname{Reg}_r^{t(k)}{}_i^j$ or $\operatorname{Reg}_r^{t(k)}{}_i^j$, possible product regions $\operatorname{Reg}_D^{(k)}{}_i$ and $\operatorname{Reg}_B^{(k)}{}_i$ and sharp split regions $\operatorname{Reg}_{sh,r}^{i:j}$ and $\operatorname{Reg}_{sh,s}^{i:j}$ at finite reflux are analogous to the notions of trajectory tear-off regions $\operatorname{Reg}_{rev,r}^{t(n-1)}$ and $\operatorname{Reg}_{rev,s}^{t(n-1)}$, possible product regions Reg_D and Reg_B, and reversible distillation regions $\operatorname{Reg}_{rev,r}^h$ and $\operatorname{Reg}_{rev,s}^l$ for the process of reversible distillation. *The sharp distillation region of a section* $\operatorname{Reg}_{sh,r}^{i:j}$ *or* $\operatorname{Reg}_{sh,s}^{i:j}$ *includes the section's trajectories at the chosen split, at any reflux, at any product composition* (i.e., $\operatorname{Reg}_{w,r}^{R,i:j} \in \operatorname{Reg}_{sh,r}^{i:j}$ *and* $\operatorname{Reg}_{w,s}^{R,i:j} \in \operatorname{Reg}_{sh,s}^{i:j}$ *for any reflux R and any product point* x_D *or* x_B).

5.4.   Structure and Evolution of Section Trajectory Bundles for Three-Component Mixtures

To understand the structure of section trajectory bundles for multicomponent mixtures and their evolution with the increase of reflux number, let's examine first three-component mixtures, basing on the regularities of distillation trajectory tear-off at finite reflux and the regularities of location of reversible distillation trajectories.

We limit ourselves by examination, mostly, only of the top section in vies of symmetry of the distillation process and we use the parameter L/V instead of R (Petlyuk & Danilov, 1998).

5.4.1.   The Product Is a Pure Component ($k = 1$)

The pure component is a separation product of three-component mixture at direct and indirect splits ($1:2,3$ or $1,2:3$) if this component is the lightest or the heaviest one (i.e., if component point is node point of concentration triangle).

Let's examine, first, the ideal mixture ($K_1 > K_2 > K_3, x_{D1} = 1$; Fig. 5.11). We gradually increase the parameter L/V. At $L/V < K_3^1$ (K_3^1 is phase equilibrium coefficient of component 3 in vertex 1) Eq. (5.9) is not valid for sides 1-2 and 1-3 adjacent with vertex 1. Therefore, vertex 1 is the stable node N^+ (Fig. 5.11a) (i.e., it can not be distillation product point). At such values of the parameter L/V the process, opposite to distillation process, the process of distillation flows mixing is feasible (see Chapter 2).

At $L/V = K_3^1$, there is first bifurcation, Eq. (5.9) becomes valid for side 1-3 and not valid for side 1-2, vertex 1 turns into saddle S (Fig. 5.11b).

At $L/V = K_2^1$, there is second bifurcation, trajectory tear-off from vertex 1 along sides 1-2 and 1-3 becomes feasible (i.e., inside concentration triangle vertex turns into unstable node N^-) (Fig. 5.11c), distillation process for the product point under consideration becomes feasible, trajectory bundle with the saddle point S at side 1-2 and, with the stable node N^+ at side 1-3 in the vicinity of vertex 1, appears

$$\operatorname{Reg}_{w,r}^{R}{}_1^{2,3} \equiv N_r^- \overset{S_r^{(2)}}{\Rightarrow} N_r^{+(2)}(N^- \to S \to N^+)\,. \; L/V \text{ is equal to } K_2 \text{ in point } S \text{ and } K_3 \text{ in point } N^+ \text{ (see Eq. 5.6).}$$

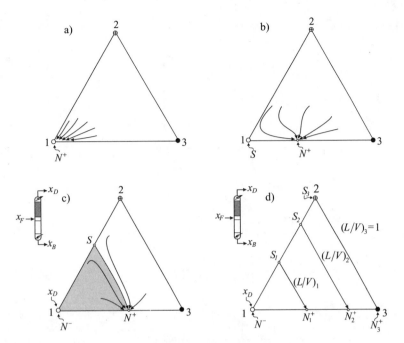

Figure 5.11. The evolution of trajectory (arrow-ended line) bundles of the rectifying section for an ideal mixture with $K_1 > K_2 > K_3$ (component 1 is the product): (a) $L/V < K_3^1$, (b) $K_3^1 < L/V < K_2^1$, (c) $L/V > K_2^1$ (the attraction region $\underset{1}{\overset{2,3}{\text{Reg}}}{}_{w,r}^R \equiv \text{Reg}_{att}^1$ shaded), (d) $(L/V)_3 = 1$, $(L/V)_2 > (L/V)_1 > K_2^1$. K_2^1 and K_3^1 are the phase equilibrium coefficients at vertex 1 for components 2 and 3, respectively.

At further increase of the parameter L/V, the points S and N^+ move away from vertex 1 (Fig. 5.11d) and reach vertexes 2 and 3 correspondingly at $L/V = 1$ (i.e., in the mode of infinite reflux).

Let's note that vertexes 2 and 3 cannot be top product points no matter what the value of parameter L/V because for these vertexes Eq. (5.15) is not valid (i.e., $x_D \neq [0,1,0]$ and $x_D \neq [0,0,1]$). Let's also note that section working region $\underset{1}{\overset{2,3}{\text{Reg}}}{}_{w,r}^R$ at $n = 3$ and $k = 1$ is two dimensional (i.e., at each value of $L/V > K_2^1$ in concentration triangle there is an attraction region $\text{Reg}_{att}^1 \equiv \underset{1}{\overset{2,3}{\text{Reg}}}{}_{w,r}^R$ of point x_D [1,0,0]). This region is called the attraction region because a calculated trajectory springing from any point of this region toward the product point is "attracted" to the product point.

The trajectories touch on the segment $N^- - S$ in the point N^-. One of these is the distillation working trajectory at minimum reflux. It goes through the saddle point S (see later in this Chapter).

Let's examine some deviations from the described evolution appearing at separation of nonideal zeotropic and azeotropic mixtures. One of such deviations can be caused by nonmonotonous dependence of the function $K_2(x_2)$ at side 1-2 and inside concentration triangle (Fig. 5.12a,b). Such nonmonotony leads in

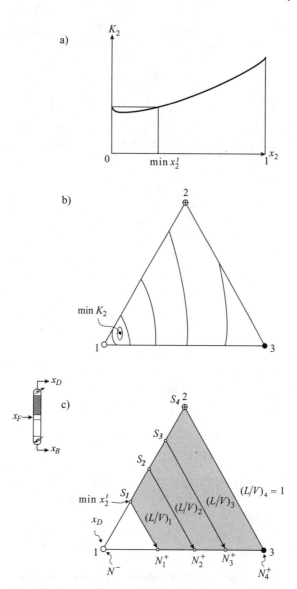

Figure 5.12. The evolution of trajectory bundle of the rectifying section ($\text{Reg}_{w,s}^{R} \overset{2,3}{\underset{1}{\equiv}} \text{Reg}_{att}^{1}$) for an mixture with $K_1 > K_2 > K_3$ (component 1 is the product) with the tangential pinch region Reg_{tang} (not shaded): (a) the function $K_2(x_2)$ with minimum on side 1-2, (b) the isolines $K_2(x_1, x_2, x_3)$ with minimum on triangle 1-2-3, (c) changes in the location of the separatrix as a function of $L/V((L/V)_4 = 1, (L/V)_4 > (L/V)_3 > (L/V)_2 > (L/V)_1)$, $S_1 = \min x_2^t$.

two-section columns to the phenomenon of "*tangential pinch*," which we examine in Section 5.6. In our case, when vertex 1 is separation product of three-component mixture, at $L/V = K_2^1$ the points S and N^+ appear at sides 1-2 and 1-3 not in the vicinity of vertex 1, as in the previous case, but at some distance from it (the tear-off point $s_1 \equiv \min x_2^t$). Therefore, the part of concentration triangle, adjacent to vertex 1 (pinch region Reg_{tang}), is not filled up with distillation trajectories at any values of the parameter L/V (Fig. 5.12c).

Another deviation appears in the case of *availability of α_{23}-line inside concentration triangle*. This line is one of the branches of reversible distillation trajectory for the product point $x_{D1} = 1$. At the increase of the parameter L/V first trajectory bundle appears $N^- \rightarrow S \rightarrow N^+_{(1)}$. Point S moves along side 1-2 to α_{23} – point, and point $N^+_{(1)}$ moves along side 1-3. Second trajectory bundle appears when point S coincides with α_{23} – point on side 1-2 $N^- \rightarrow S \rightarrow N^+_{(2)}$. Then point S moves along α_{23} – line, point $N^+_{(1)}$ moves along side 1-3 to α_{23} – point, and point $N^+_{(2)}$ moves along side 1-2 from α_{23} – point. First trajectory bundle disappears when point S coincides with α_{23} – point on side 1-3. Then point S moves along side 1-3 to vertex 3, and point $N^+_{(2)}$ moves along side 1-2 to vertex 2 (Fig. 5.13).

Along separatrix $S - N^+_{(1)}$ (Fig. 5.13b) the temperature decreases abnormally from point S on α_{23}-line to point $N^+_{(1)}$ (down along column). In contrast to that along separatrix $S - N^+_{(2)}$ the temperature increases normally from point S to point $N^+_{(2)}$. It is the result of inflection of residue curves on α_{23} – line. The first trajectory bundle is the working one, if composition point in feed crossection lies on separatrix $S - N^+_{(1)}$, and the second trajectory bundle is the working one, if composition point in feed crossection lies on separatrix $S - N^+_{(2)}$. Let's note that temperature cannot decrease down along column at the big enough value of parameter L/V, because temperature increases along residue curves. For the more the temperature can decrease only if the value of parameter L/V near enough to $(L/V)_{\min}$.

5.4.2. The Product Is a Binary Mixture ($k = 2$)

The binary mixture is the separation product of three-component mixture in the case of split with a distributed component (1,2 : 2,3), including the case of preferable split and in the case of the top section at indirect separation (1,2 : 3).

Let's start again with the ideal mixture ($K_1 > K_2 > K_3$, x_{D3}, $= 0$; Fig. 5.14). Reversible distillation trajectory tear-off from side 1-2 goes on in some point x_{rev}^t, defined by Eq. (4.17), at $(L/V)_{rev}^t = K_3(x_{rev}^t)$ (see Chapter 4). Therefore, at $L/V < (L/V)_{rev}^t$ point x_D at side 1-2 cannot be the distillation top product point. Along with that, at side 1-2 there is a stable node (Fig. 5.14a). At $(L/V) = (L/V)_{rev}^t$, the bifurcation goes on. Distillation trajectory tear-off from side 1-2 inside concentration triangle and the distillation process becomes feasible. The stable node at side 1-2 turns into a saddle in point x_{rev}^t. The trajectory bundle appears with saddle point S at side 1-2 and with the stable node N^+ inside concentration triangle at reversible distillation trajectory in the vicinity of point x_{rev}^t. At further increase

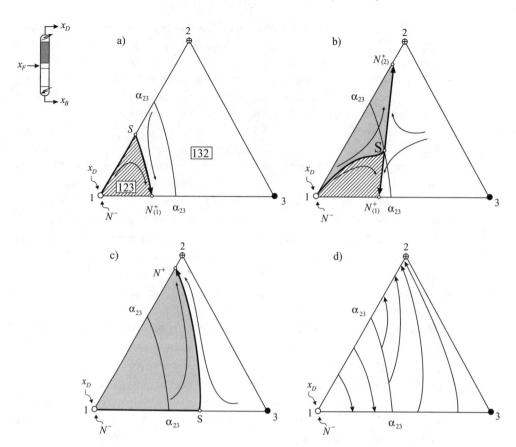

Figure 5.13. The evolution of trajectory bundles of the rectifying section ($\text{Reg}_{w,r}^{R} \equiv \text{Reg}_{att}^{1}$)
(shaded) for an mixture with $K_1 > K_2 > K_3$ and α_{23} line (component 1 is the product): (a)
$L/V > K_2^1 > K_3^1$, (b) $L/V > K_2$ at point α_{23} on side 1-2, (c) $L/V > K_2$ at point α_{23} on side 1-3,
(d) arrow-ended lines show the progressive change of the separatrix position as L/V increases.

of the parameter L/V, the points S and N^+ move away from point x_{rev}^t (Fig. 5.14b)
until $L/V = 1$, when point S reaches vertex 2, and point N^+ reaches vertex 3.
The section trajectory, the calculation direction taken into consideration, may be
presented in the following brief form (the calculation direction and characteristic
points given in the upper line [product components number in product points and
overall component number in other points in the upper index in parentheses], the
regions [including 0-dimensional] to which the points belong given in the lower
line): $x_D^{(2)} \rightarrow S_r^{(2)} \rightarrow N_r^{+(3)}$.
$\quad\quad \text{Reg}_D \quad \text{Reg}_r^t \quad \text{Reg}_{sep,r}^{sh,R}$

Let's note that the top product point cannot be located at sides 2-3 and 1-3
because Eq. (5.15) is not valid for these sides (i.e., $x_D \notin [2\text{-}3]$ and $x_D \notin [1\text{-}3]$).

Let's also note that the *separatrix sharp split region* of section trajectories bundle
$S \rightarrow N^+ \equiv \text{Reg}_{sep,r}^{sh,R}$ at $n = 3$ and $k = 2$ is one-dimensional (i.e., at each value of

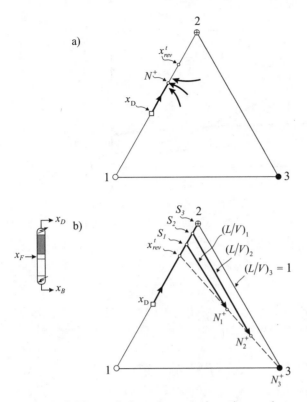

Figure 5.14. The evolution of separatrix trajectory bundle and separatrix sharp split region $S \to N^+ \equiv \mathrm{Reg}_{sep,r}^{sh,R}\ _{1,2}^{3}$ of the rectifying section for an ideal mixture with $K_1 > K_2 > K_3$ (mixture 1, 2 is the product): (a) $(L/V) < (L/V)_{rev}^t$, (b) $(L/V) > (L/V)_{rev}^t, (L/V)_3 = 1, (L/V)_3 > (L/V)_2 > (L/V)_1$.

$L/V > (L/V)_{rev}^t$ for top product point under consideration there is one distillation trajectory).

Deviations from the above evolution of distillation trajectory found at separation of nonideal zeotropic and, especially, of azeotropic mixtures are connected with the peculiarities of location of reversible distillation trajectories and with nonmonotony of change of the parameter L/V along these trajectories (see Chapter 4).

The most important peculiarities of location of reversible distillation trajectories of azeotropic mixtures, influencing the evolution of distillation trajectories at the change of the parameter L/V, are limitedness of trajectory tear-off segment $\mathrm{Reg}_{r}^{t(k)}\ _{i}^{j}$ or $\mathrm{Reg}_{s}^{t(k)}\ _{i}^{j}$ and of possible product composition segment $\mathrm{Reg}_{D}^{(k)}\ _{i}^{j}$ or $\mathrm{Reg}_{B}^{(k)}\ _{i}^{j}$ at the side of concentration triangle, availability of two node points or absence of node points of reversible distillation trajectories.

a)

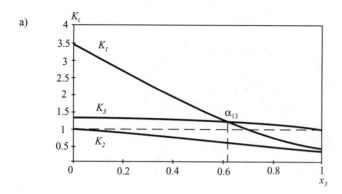

b)

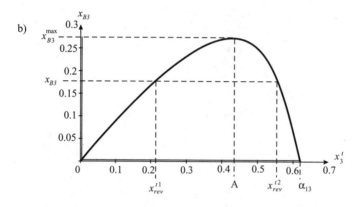

Figure 5.15. The determination of tear-off segment Reg^1_s and possible bottom product segment $\text{Reg}_B{}^1_{2,3}$ for bottom point x_B on the side 2-3 of the concentration triangle for the acetone(1)-benzene(2)-chloroform(3) mixture: (a) functions $K_i(x_3)$ on the side 2-3, (b) the function $x_{B3}(x^t_{rev})$ on the side 2-3. x^{t1}_{rev} and x^{t2}_{rev}, concentration of component 3 at first and second reversible trajectory tear-off points from the side 2-3 for given bottom point x_{B3}; A, tear-off point for end of possible bottom segment $\text{Reg}_B{}^1_{2,3}$ on the side 2-3 (x^{max}_{B3}).

It is expedient to discuss the influence of these peculiarities on the evolution of distillation trajectories at the concrete example of azeotropic mixture, such as acetone(1)-benzene(2)-chloroform(3). At side 2-3, there is *reversible distillation trajectory tear-off segment* $\text{Reg}^{t(2)}_{rev,s}$ of the bottom section from vertex 2 to α_{13}-point (Fig. 4.18a). Possible bottom segment $\text{Reg}_{rev,B}$, adjacent to vertex 2, corresponds to this segment. *Possible bottom segment* $\text{Reg}^{(2)}_B{}^j_i$ at adiabatic distillation of three-component mixtures (Petlyuk, Vinogradova, & Serafimov 1984) coincides with possible bottom segment at reversible distillation. To determine the end of this segment, it is necessary to scan the values of phase equilibrium coefficient of all components along side 2-3 (Fig. 5.15a) and to define maximum value of x_{B3}^{max} with the help of Eq. (4.20) (Fig. 5.15b). In the example under consideration, $x_{B3}^{max} = 0.266$. Reversible distillation trajectory tear-off point $x_{rev}^t = A = 0.44$ corresponds to this composition of the bottom product.

The existence of possible bottom segment $\text{Reg}_B\,_{2,3}^{\,1}$ at side 2-3 means that sharp split 1: 2,3 is feasible in two-section column at finite reflux, which is unfeasible according to the rule of connectedness (see Chapter 3) in the mode of infinite reflux. The feasibility of such separation was shown first by means of calculation in the work (Kondrat'ev et al., 1977).

For any point x_B located at possible bottom segment $\text{Reg}_B^{(2)} \equiv [0, x_{B3}^{max}]$, there are two trajectories of reversible distillation (Fig. 5.16) – with the node N_{rev} in

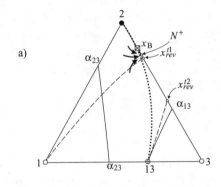

a)

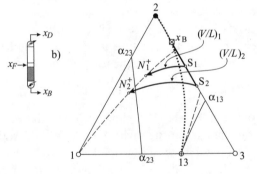

b)

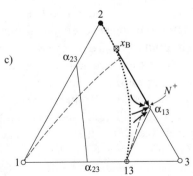

c)

Figure 5.16. The evolution of separatrix trajectory bundle and separatrix sharp split region $S \to N^+ \equiv \text{Reg}_{sep,s}^{sh,R}\,_{2,3}^{\,1}$ of the stripping section for the acetone(1)-benzene(2)-chloroform(3) mixture: (a) $V/L < (V/L)_{rev}^{t1}$, (b) $(V/L)_{rev}^{t1} < V/L < (V/L)_{rev}^{t2}$, (c) $V/L > (V/L)_{rev}^{t2}$. $(V/L)_{rev}^{t1}$ and $(V/L)_{rev}^{t2}$, ratios of flow rates V and L at first and second reversible trajectory tear-off points for bottom point x_B.

vertex $1(x_B \to x_{rev}^{t1} \to 1)$ and with the node N_{rev} in the point of azeotrope $13(x_B \to x_{rev}^{t2} \to 13)$.

In the tear-off point of the first of this trajectories x_{rev}^{t1}, the value of the parameter V/L (for the bottom section, it plays the same role as the parameter L/V for the top section) equals $(V/L)_{rev}^{t1} = 1/K_1(x_{rev}^{t1})$, and in tear-off point x_{rev}^{t2} of the second trajectory $(V/L)_{rev}^{t2} = 1/K_1(x_{rev}^{t2})$. At $V/L < 1/K_1(x_{rev}^{t1})$, there is no trajectory tear-off from side 2-3, and there is stable node N^+ at this side that at increase of V/L moves to the point x_{rev}^{t1} (Fig. 5.16a). At $V/L = (V/L)_{rev}^{t1}$, there is bifurcation. Distillation trajectory tear-off from side 2-3 becomes feasible. The stable node turns into saddle S. Distillation trajectory with the stable node N^+ at the first of reversible distillation trajectories (separatrix sharp split region $S \to N^+ \equiv \mathrm{Reg}_{ssep,s}^{sh,R} \overset{1}{\underset{2,3}{}}$)

appears (Fig. 5.16b). At further increase of the parameter V/L, the point S moves from the point x_{rev}^{t1} to the point x_{rev}^{t2} that reaches at $V/L = (V/L)_{rev}^{r2}$. In this moment, there is the second bifurcation – saddle S again turns into the stable node N^+, distillation trajectory tear-off from side 2-3 becomes unfeasible (Fig. 5.16c). Therefore, the distillation process is feasible in limited interval of the value of the parameter V/L (Petlyuk et al., 1984):

$$(V/L)_{max} > V/L > (V/L)_{min} \tag{5.17}$$

As far as the value V/L for the bottom section or the value L/V for the top section in the case under consideration is limited from above; the length of trajectory inside concentration triangle is also limited.

The analogous situation arises in the case if reversible distillation trajectory bundle does not have node points. For the mixture under consideration, such reversible distillation trajectories are available for the top product points, located at a small segment of side 1-2, adjacent to vertex 1 (Fig. 5.17b – $\mathrm{Reg}_{sep,r}^{sh,R} \overset{3}{\underset{1,2}{}}$).

Nonmonotony of change of the parameter L/V along the reversible distillation trajectory inside concentration triangle for the mixture under consideration become apparent if the top product point is located at side 1-3 (Fig. 5.18a). At some value of the parameter $L/V = (L/V)_{rev}^t$, there is a section distillation trajectory and reversible distillation trajectory tear-off from side 1-3 in point $x_{rev}^t = S_1$. Along with that, stable node N^+ comes into being at reversible distillation trajectory at some distance from side 1-3 (in point N_1^+). Therefore, at reversible distillation trajectory there is a segment at which there are no points N^+ at any values of the parameter L/V. In the example under consideration, $(L/V)_{rev}^t = K_2(x_{rev}^t) = 0,77$. Reversible distillation trajectory intersects two isolines $K_2 = 0,77$. The first point of intersection coincides with reversible distillation trajectory tear-off point x_{rev}^t and with adiabatic distillation trajectory tear-off x^t at $L/V = 0,77$ (i.e., with the saddle point S_1) and the second point of intersection coincides with the stable node N_1^+. This fact is connected with that the dependence $(L/V)_{rev} = K_2$ along reversible distillation trajectory has minimum (Fig. 5.19). In the two-section column, such

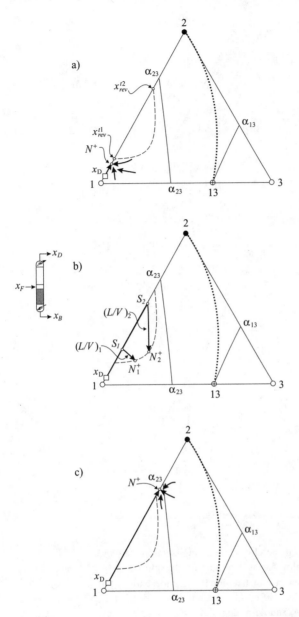

Figure 5.17. The evolution of separatrix trajectory bundle and separatrix sharp split region $S \rightarrow N^+ \equiv \text{Reg}_{sep,r}^{sh,R}$ of the rectifying section for the acetone(1)-benzene(2)-chloroform(3) mixture: (a) $L/V < (L/V)_{rev}^{t1}$, (b) $(L/V)_{rev}^{t1} < (L/V) < (L/V)_{rev}^{t2}$, (c) $L/V > (L/V)_{rev}^{t2}$.

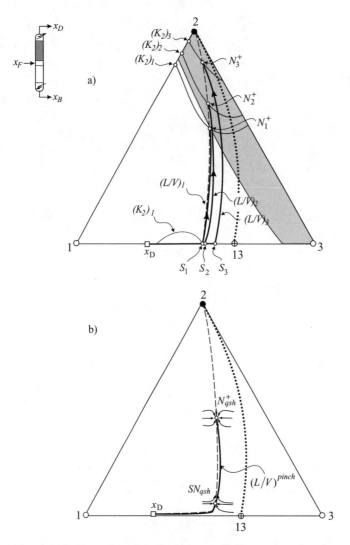

Figure 5.18. The tangential pinch in rectifying section for the acetone(1)-benzene(2)- chloroform(3) mixture for the split 1,3 : 2: (a) sharp separation (the tangential-pinch region $Reg_{tang}^{(3)}$ not shaded), (b) quasisharp separation. 1, 2, 3, different values of L/V and different iso-K_2 lines (thin lines); SN, saddle-node point.

nonmonotony leads to "*tangential pinch*." Figure 5.18 shows the region Reg_{tang} (it is not shaded), where points N^+ are absent at any points of the top product located at side 1-3 and at any values of parameter L/V (pinch region Reg_{tang}).

5.4.3. The Product Is a Three-Component Mixture ($k = 3$)

Three-component mixture is a separation product of initial three-component mixture at nonsharp separation.

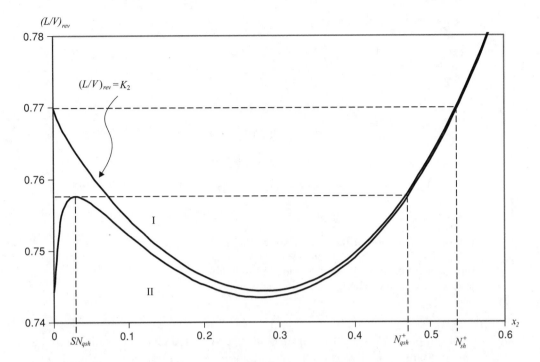

Figure 5.19. $(L/V)_{rev}$ as functions x_2 on the reversible-distillation trajectories in rectifying section for the split $1,3:2$ (for given x_D in the tangential-pinch region Reg_{tang}) of the acetone(1)-benzene(2)-chloroform(3) mixture: I, sharp separation $((L/V)^t_{rev} \equiv K^t_2 = K^N_2)$; II, quasisharp separation $((L/V)^{SN}_{rev} = (L/V)^N_{rev})$.

Previously, we discussed sharp splits in the column section, which are important theoretical abstraction useful at the solution of the main questions of optimal designing of distillation units.

If sharp separation is feasible, then product points at real separation can be located as much as one wants close to vertexes or sides of concentration triangle. We call such separation "quasisharp" one. If sharp separation is not feasible (i.e., at supposed sharp separation product point does not get into possible product segment $Reg_D^{(2)}$ or $Reg_B^{(2)}$), neither is *"quasisharp" separation*. In this case, only nonsharp separation is feasible, for that there is some minimum feasible content of admixture component or components. We call the separation at minimum feasible content of admixture component or components *"the best non-sharp"* one.

Let's discuss the structure and location of section trajectory bundles at quasisharp separation (Fig. 5.20). Let product point at quasisharp separation x_D be located in the vicinity of some product point at sharp separation that lies at the side of concentration triangle. At the set value of the parameter $(L/V)_1$, distillation trajectories at sharp and quasisharp separation are close to each other and, in particular, the stable nodes are close to each other. However, saddle point at quasisharp separation is located outside the concentration triangle in the vicinity of its side and inside the concentration triangle there is "quasisaddle" point qS, in the vicinity of which the change of concentration at neighboring plates is very

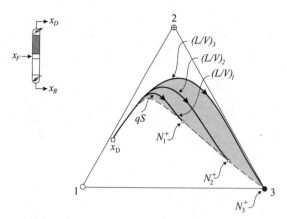

Figure 5.20. The trajectories of rectifying section for quasisharp separation of the ideal mixture. $(L/V)_3 = 1$, $(L/V)_3 > (L/V)_2 > (L/V)_1$, the region between the reversible-distillation trajectory and the distillation trajectory under infinite reflux is shaded. qS, quasistationary point.

small. In point qS, trajectory of quasisharp distillation changes its direction: between points x_D and qS it is almost parallel to the side of concentration triangle and between points qS and N^+ it goes inside $(x_D^{(3)} \rightarrow qS_r^{(3)} \rightarrow N_r^{+(3)})$.

Let's note, that *distillation trajectory of three-component mixture for the set product point is located between trajectory at infinite reflux (i.e., at $L/V = 1$, and reversible distillation trajectory)* (Kiva, 1976; Petlyuk & Serafimov, 1983; Wahnschafft et al., 1992; Castillo, Thong, & Towler, 1998).

The structure of the trajectory bundle has interesting peculiarities for quasisharp distillation at non-monotonous change of the parameter L/V along reversible distillation trajectory (i.e., at "tangential pinch"). Figure 5.19 shows for the mixture acetone(1)-benzene(2)-chloroform(3) dependence of the parameter $(L/V)_{rev}$ on concentration of component 2 along the trajectory of reversible quasisharp distillation (composition in the point x_D: 0,699; 0,001; 0,300). This dependence has two extremums – minimum and maximum. The point of maximum $(L/V = 0,758)$ corresponds to *stationary saddle-node point SN* inside concentration triangle (Fig. 5.18b). This stationary saddle-node point SN is a node point for the trajectories, located closer than this point to the side of triangle, and a saddle point for the rest of trajectories. The saddle-node point exists for the set product composition x_D at the unique value of the parameter L/V: $(L/V)^{pinch} = 0,758$. At the smaller value, there is no trajectory tear-off from the side and at the bigger value trajectory bundle has only node point N^+. The section's trajectory can be put in brief as follows: $x_D^{(3)} \rightarrow SN_r^{(3)} \rightarrow N_r^{+(3)}$.

The structure of trajectory bundle also has interesting peculiarities at the *best nonsharp separation*. In this case, trajectory bundle also has saddle-node and node points (Shafir et al., 1984). Figure 5.21a shows for the azeotropic mixture acetone(1)-benzene(2)-chloroform(3) the line of *best bottom product* (Poellmann & Blass, 1994), connecting the end of possible segment Reg_B^1 at side 2-3 at sharp

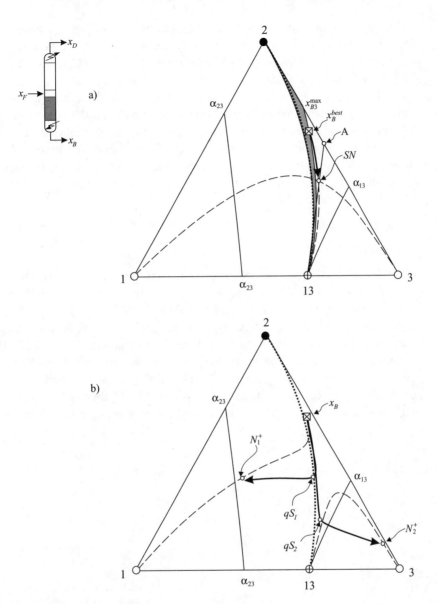

Figure 5.21. (a) The pitchfork region Reg_{pitch} for the acetone(1)-benzene(2)-chloro-form(3) mixture (shaded), the stripping section trajectories for given $x_B = x_B^{best}$; (b) the stripping section trajectories for given x_B on the pitchfork region ($x_B \in \mathrm{Reg}_{pitch}$). 1 and 2, different values of V/L; qS_1 and qS_2, quasistationary points; A, tear-off point for $x_B = x_{B,3}^{max}$ on side 2-3; SN, saddle-node point.

separation $x_{B,3}^{max}$ with the point of azeotrope 13 (line $x_{B,3}^{max} - 13$), and the line of the *saddle-node points SN* (line $A - 13$). Some point at the line of saddle-node points corresponds to each point at the line of the best product. Reversible distillation trajectory goes from product point to the point *SN* and then branches into three branches – to vertex 1, to vertex 3, and to azeotrope 13. The section's trajectory can be put in brief as follows: $x_B^{best(3)} \rightarrow SN_s^{(3)} \equiv x_{rev}^{branch}$.
$$\mathrm{Reg}_B^{best}$$

The *region of two-directed distillation (pitchfork region* Reg_{pitch}) is located between the line of the *best product compositions (pitchfork distillation boundary) and boundary of distillation regions at infinite reflux* Reg^∞ (Castillo et al., 1998; Davydyan, Malone, & Doherty, 1997; Wahnschafft et al., 1992).

For bottom product points x_B, located in this region, distillation trajectory at some values of the parameter V/L is directed to vertex 1 and at other values, bigger than them, it is directed to vertex 3 (Fig. 5.21b). It is connected with the fact that for these product points reversible distillation trajectory has two branches, one of the branches goes to vertex 1 and another branch goes to vertex 3. The point N^+ small values of the parameter V/L is located at one of the branches, and at bigger values it is located at other branches.

5.4.4. The Product Is Azeotrope

In some cases, at separation of azeotropic mixtures one of the product is binary or ternary azeotrope. In these cases, as in the case of the product being pure component, section working region $\text{Reg}_{w,r}^R \equiv \text{Reg}_{att}$ or $\text{Reg}_{w,s}^R \equiv \text{Reg}_{att}$ is two dimensional. The saddle and node points are located at different trajectories of reversible distillation inside concentration triangle. Figure 5.22a shows rectifying section region $\text{Reg}_{w,r}^R \equiv \text{Reg}_{att}^{23}$ for binary azeotrope 23 of the mixture acetone(1)-chloroform(2)-methanol(3). Figure 5.22b shows rectifying section region $\text{Reg}_{w,r}^R \equiv \text{Reg}_{att}^{123}$ for ternary azeotrope 123 of the mixture hexane(1)-methanol(2)-methylacetate(3). Azeotrope can be a separation product of the

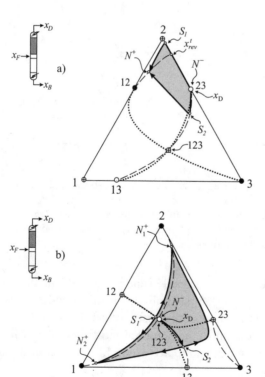

Figure 5.22. The rectifying section region $\text{Reg}_{w,r}^R \equiv \text{Reg}_{att}^{23}$ or Reg_{att}^{123} (shaded) for (a) the acetone(1)-chloroform(2)-methanol(3) mixture (azeotrope 23 is the product), (b) the hexane(1)-methanol(2)-methylacetate (3) mixture (azeotrope 123 is the product).

three-component mixture, if it is unstable ($x_D \equiv N^-$) or stable node ($x_B \equiv N^+$) of concentration space. Such splits are correspondingly direct or indirect splits as at the product – pure component. These splits can be used, for example, in the units, where entrainer, forming azeotrope with one of the components of the initial mixture, is used for separation of binary azeotropic mixture (see Chapter 8).

5.5. Structure and Evolution of Section Trajectory Bundles for Four- and Multicomponent Mixtures

5.5.1. Four-Component Mixture

General regularities of the evolutions of sections trajectory bundles, discussed in the previous section for three-component mixtures, are valid also for the mixtures with bugger number of components. Figure 5.23 shows evolution of top section trajectory bundle at separation of four-component ideal mixture, when the *product is pure component* (i.e., at direct split) ($K_1 > K_2 > K_3 >$

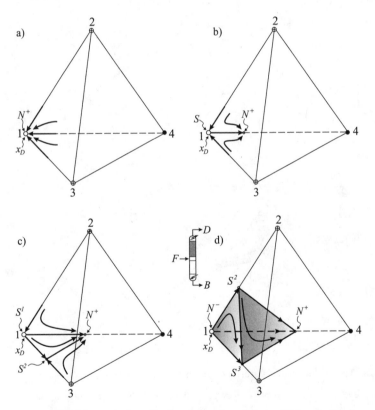

Figure 5.23. The evolution of trajectory region of the rectifying section for an ideal mixture with $K_1 > K_2 > K_3 > K_4$ (component 1 is the product): (a) $L/V < K_4^1$, (b) $K_4^1 < L/V < K_3^1$, (c) $K_3^1 < L/V < K_2^1$, (d) $L/V > K_2^1$ (the attraction region $\text{Reg}_{w,r}^{R} \equiv \text{Reg}_{att}^{1}$ shaded). K_2^1, K_3^1, and K_4^1, phase equilibrium coefficients of components 2, 3, and 4, respectively, in vertex 1.

K_4, $x_{D1} = 1$). At the increase of the parameter L/V, vertex 1 turns from stable node N^+ (Fig. 5.23a) into the saddle with two going in and one going out trajectories S (Fig. 5.23b), then it turns into the saddle with one going in and two going out trajectories S^1 (Fig. 5.23c) and, finally, it turns into the unstable node N^- (Fig. 5.23d). After that, *three-dimensional section region* $\text{Reg}_{w,r}^{R} \overset{2,3,4}{\underset{1}{\equiv}} \text{Reg}_{att}^1$ that increase at the increase of the parameter L/V appears.

Stationary points S^2, S^3, and N^+ ($S^1 \equiv N^-$) move along the edges of concentration tetrahedron. The section trajectory bundle may be presented in the following brief form (the bundle's direction is indicated by the double arrow, its stationary points around it):

$$x_D^{(1)} \equiv N_r^- \overset{S_r^{(2)}}{\underset{S_r^{(3)}}{\Rightarrow}} N_r^{+(2)}$$

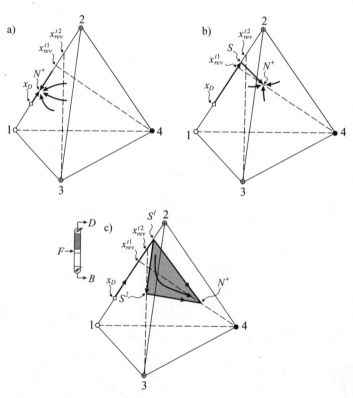

Figure 5.24. The evolution of separatrix trajectory bundle and separatrix sharp split region $S^1 \to S^2 \to N^+ \equiv \text{Reg}_{sep,r}^{sh,R}$ (shaded) of the rectifying section for an ideal mixture with $K_1 > K_2 > K_3 > K_4$ (mixture 1,2 is the product): (a) $(L/V) < (L/V)_{rev}^{t1}$, (b) $(L/V)_{rev}^{t1} < (L/V) < (L/V)_{rev}^{t2}$, (c) $(L/V) > (L/V)_{rev}^{t2}$. $(L/V)_{rev}^{t1}$ and $(L/V)_{rev}^{t2}$, values of L/V at reversible distillation trajectories tear-off points from the edge 1-2 to faces 1-2-4 and 1-2-3, respectively.

Figure 5.24 shows the evolution of *separatrix sharp split region* of top section $S^1 \rightarrow S^2 \rightarrow N^+ \equiv S^1 \Rightarrow N^+ \equiv \mathrm{Reg}_{sep,r}^{\overset{3,4}{sh,R}}{}_{1,2}$ at separation of four-component ideal mixture, when the *product is binary mixture* 1,2 (i.e., at the intermediate split) ($K_1 > K_2 > K_3 > K_4, x_{D1} + x_{D2} = 1$).

At small values of the parameter L/V, the stable node N^+ that at the increase of the parameter L/V moves away from the product point x_D in the direction to vertex 2 appears at edge 1-2 (Fig. 5.24a). After this node reaches reversible distillation trajectory tear-off point x_{rev}^{t1} into face 1-2-4, it turns into the saddle with one trajectory going out (Fig. 5.24b). After reaching reversible distillation trajectory tear-off point x_{rev}^{t2} into face 1-2-3, it turns into the saddle with two trajectories going out (Fig. 5.24c). After that *two-dimensional separatrix sharp split region* $\mathrm{Reg}_{sep,r}^{\overset{3,4}{sh,R}}{}_{1,2}$ that increases at the increase of the parameter L/V appears. The stationary points S^1, S^2, and N^+ move correspondingly along reversible distillation trajectories at edge 1-2, at face 1-2-3, and at face 1-2-4. Operating [Eq. (5.9)] and structural [Eq. (5.15)] conditions of trajectory tear-off from the boundary elements of concentration simplex should be valid for the points S^1 and S^2. The section's trajectory can be put in brief as follows: $x_D^{(2)} \underset{\mathrm{Reg}_D}{\rightarrow} S_r^{1(2)} \underset{\mathrm{Reg}_r^t}{\overset{S_r^{2(3)}}{\Rightarrow}} N_r^{+(3)}$.

Figure 5.25 shows the evolution of top section trajectory bundle at separation of four-component ideal mixture, when the *product is ternary mixture* 1,2,3 (i.e., at indirect split) ($K_1 > K_2 > K_3 > K_4, x_{D1} + x_{D2} + x_{D3} = 1$). At small values of the parameter L/V, the stable node N^+ that at the increase of the parameter L/V moves away from the product point along reversible distillation trajectory appears at face 1-2-3 (Fig. 5.25a). After this node reaches reversible distillation trajectory tear-off point x_{rev}^t from face 1-2-3 inside concentration tetrahedron, it turns into the saddle S and *one-dimensional separatrix sharp split region* $\mathrm{Reg}_{sep,r}^{\overset{4}{sh,R}}{}_{1,2,3} \equiv S^1 \rightarrow N^+$ (one trajectory) appears inside concentration tetrahedron (Fig. 5.25b). At further increase of the parameter L/V, stationary points S and N^+ move correspondingly along reversible distillation trajectories at face 1-2-3 and inside concentration tetrahedron. The section's trajectory can be put in brief as follows: $x_D^{(3)} \underset{\mathrm{Reg}_D}{\rightarrow} S_r^{1(3)} \underset{\mathrm{Reg}_r^t}{\rightarrow} N_r^{+(4)}$.

Deviations from the described evolution for nonideal zeotropic and azeotropic mixtures are analogous to those that were discussed before for three-component mixtures. As an example of such deviation, let's examine separation of four-component mixture, the top product of which is component 1 and inside concentration tetrahedron there is α_{23}-*surface*, that divides it into component-order regions $\mathrm{Reg}_{ord}^{1,2,3,4}$ and $\mathrm{Reg}_{ord}^{1,3,2,4}$. Figure 5.26 shows the evolution of top section region $\mathrm{Reg}_{w,r}^{\overset{2,3,4}{R}}{}_1 \equiv \mathrm{Reg}_{att}^1$ for this case.

We examine other deviations from ideal behavior at the example of the mixture acetone(1)-benzene(2)-chloroform(3)-toluol(4). Figures 4.15 and 4.16 show

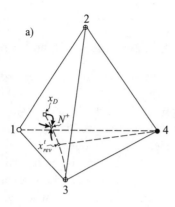

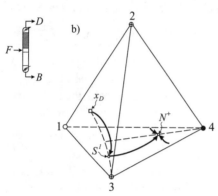

Figure 5.25. The evolution of separatrix trajectory bundle and separatrix sharp split region $S^1 \rightarrow N^+ \equiv \mathrm{Reg}_{sep,r}^{sh,R} \underset{1,2,3}{\overset{4}{}}$ of the rectifying section for an ideal mixture with $K_1 > K_2 > K_3 > K_4$ (mixture 1,2,3 is the product): (a) $L/V < (L/V)_{rev}^t$, (b) $L/V > (L/V)_{rev}^t$. $(L/V)_{rev}^t$, values of L/V at reversible-distillation trajectoriy tear-off point from the face 1-2-3.

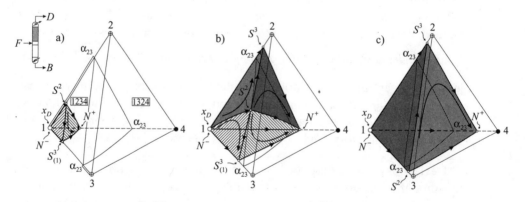

Figure 5.26. The evolution of trajectory regions, $\mathrm{Reg}_{w,r}^{R} \underset{1}{\overset{2,3,4}{}} \equiv \mathrm{Reg}_{att}^1$ of the rectifying section (hatched and shaded) for an mixture with α_{23} surface (component 1 is the product): (a) $L/V > K_2^1 > K_3^1 > K_4^1$, (b) $L/V > K_2$ at point α_{23} on side 1-2, (c) $L/V > K_2$ at point α_{23} on side 1-3. Double line, tear-off segments $\mathrm{Reg}_r^t \underset{1,2}{\overset{3,4}{}}$ and $\mathrm{Reg}_r^t \underset{1,3}{\overset{2,4}{}}$.

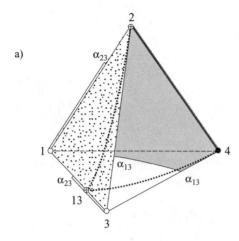

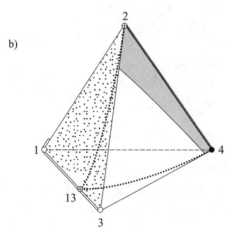

Figure 5.27. Tear-off regions $\mathrm{Reg}_r^{t\,^{3,4}_{1,2}}$, $\mathrm{Reg}_r^{t\,^{2,4}_{1,3}}$, $\mathrm{Reg}_s^{t\,^{1,3}_{2,4}}$, $\mathrm{Reg}_r^{t\,^{4}_{1,2,3}}$, and $\mathrm{Reg}_s^{t\,^{1}_{2,3,4}}$ (a) and possible product regions $\mathrm{Reg}_D^{^{3,4}_{1,2}}$, $\mathrm{Reg}_D^{^{2,4}_{1,3}}$, $\mathrm{Reg}_B^{^{1,3}_{2,4}}$, $\mathrm{Reg}_D^{^{4}_{1,2,3}}$ and $\mathrm{Reg}_B^{^{1}_{2,3,4}}$ (b) for the acetone(1)-benzene(2)-chloroform(3)-toluene(4) mixture on the edges and faces (dotty for top section and shaded for bottom section) of concentration tetrahedron.

correspondingly *component-order segments* Reg_{ord}^{ijk} at the edges of concentration tetrahedron and *component-order* regions Reg_{ord}^{ijk} inside it. *Trajectory tear-off segments and regions* $\mathrm{Reg}_r^{t\,^{3,4}_{1,2}}$, $\mathrm{Reg}_r^{t\,^{2,4}_{1,3}}$, $\mathrm{Reg}_s^{t\,^{1,3}_{2,4}}$, $\mathrm{Reg}_r^{t\,^{4}_{1,2,3}}$, and $\mathrm{Reg}_s^{t\,^{1}_{2,3,4}}$ at the edges and faces of concentration tetrahedron, shown in Fig. 5.27a, correspond to these component-order segments and regions. Let's find now *possible product segments and regions* $\mathrm{Reg}_D^{^{3,4}_{1,2}}$, $\mathrm{Reg}_D^{^{2,4}_{1,3}}$, $\mathrm{Reg}_B^{^{1,3}_{2,4}}$, $\mathrm{Reg}_D^{^{4}_{1,2,3}}$, and $\mathrm{Reg}_B^{^{1}_{2,3,4}}$ (Fig. 5.27b) at the edges and faces of concentration tetrahedron. The segment at the edge is possible product segment $\mathrm{Reg}_D^{(2)\,j}$ or $\mathrm{Reg}_B^{(2)\,j}{}_i$, if it is reversible distillation product segment $\mathrm{Reg}_{rev,D}^{(2)}$ or $\mathrm{Reg}_{rev,B}^{(2)}{}_i$ in two faces, adjacent with the edge under consideration. It is seen from Fig. 5.27 that there are possible overhead segments $\mathrm{Reg}_D^{(2)\,^{3,4}_{1,3}}$ and $\mathrm{Reg}_D^{(2)\,^{2,4}_{1,2}}{}_{1,3}$ only at edges 1-2 and 1-3 and that of bottom segment $\mathrm{Reg}_B^{(2)}{}_{2,4}$ are at edge 2-4. Possible product regions $\mathrm{Reg}_D^{(3)\,j}{}_i$

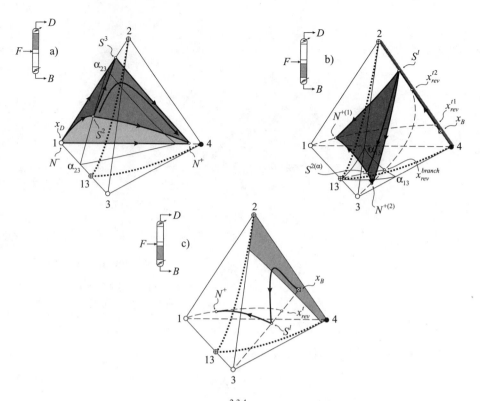

Figure 5.28. (a) Working trajectory region $\text{Reg}_{w,r}^{R}{}^{2,3,4}_{1} \equiv \text{Reg}_{att}^{1}$ (shaded, component 1 is product) for the acetone(1)-benzene(2)-chloroform(3)-toluene(4) mixture for rectifying section, (b) separatrix sharp split regions $S^1 \to S^{2\alpha} \to N^{+(1)} \in \text{Reg}_{sep,s}^{sh,R}{}^{1,3}_{2,4}$ and $S^1 \to S^{2(\alpha)} \to N^{+(2)} \in \text{Reg}_{sep,s}^{sh,R}{}^{1,3}_{2,4}$ for stripping section (shaded, mixture 2,4 is product), (c) separatrix sharp split region $S^1 \to N^+ \equiv \text{Reg}_{sep,s}^{sh,R}{}^{1}_{2,3,4}$ for stripping section (mixture 2,3,4 is product) and $\text{Reg}_{B}^{(3)}$ (shaded).

and $\text{Reg}_{B}^{(3)}{}^{j}_{i}$ in the face of concentration tetrahedron is limited by some segments at the sides of this face. These segments should be possible product segments $\text{Reg}_{rev,D}^{(2)}$ or $\text{Reg}_{rev,B}^{(2)}$ of reversible distillation in the faces adjacent with the face under consideration. The boundaries of possible product regions in the faces of concentration tetrahedron for adiabatic $\text{Reg}_{D}^{(3)}{}^{j}_{i}$ and $\text{Reg}_{B}^{(3)}{}^{j}_{i}$ and reversible $\text{Reg}_{rev,D}^{(3)}$ and $\text{Reg}_{rev,B}^{(3)}$ distillation coincide with each other.

Figure 5.28 shows examples of the structure of the top section region $\text{Reg}_{w,r}^{R}{}^{2,3,4}_{1} \equiv \text{Reg}_{att}^{1}$ for possible one-component product (second region as in Fig. 5.26b not shown in Fig. 5.28a) and bottom separatrix sharp split regions $\text{Reg}_{sep,s}^{sh,R}{}^{1,3}_{2,4}$ and $\text{Reg}_{sep,s}^{sh,R}{}^{1}_{2,3,4}$ for possible two and three-component products.

If the top product point coincides with vertex 1 $(x_{D1} = 1)$ (i.e., $N^- \equiv S^1 \equiv x_{D1}$) (Fig. 5.28a), then, as the value of L/V parameter is increased, the point S^2 first moves along the edge 1-2 up to the α_{23}-point, then along the α_{23}-*line* in the face 1-2-3, then along the edge 1-3 up to azeotrope 13 (at $L/V = 1$).

If the top product point lies on the edge 1-3, it is the case of *tangential pinch*, which we are going to consider in the next section (not shown in Fig. 5.28).

If the bottom product point lies on the edge 2-4 (Fig. 5.28b), the point S^2, as the value of V/L parameter is increased, first goes along the reversible distillation trajectory within the face 2-3-4 until it meets the α_{13}-*line* (x_{rev}^{branch}), then along the reversible distillation trajectory within the α_{13}-*surface* up to azeotrope 13 (at $V/L = 1$). Simultaneously, in the face 2-3-4, in the trajectory of reversible distillation, after point x_{rev}^{branch} a stable node $N^{+(2)}$ arises, and the point $S^{2(\alpha)}$ engenders a separatrix $S^1 - S^{2(\alpha)}$, that divides the whole separatrix bundle $\text{Reg}_{sep,s}^{sh,R}$ into two separate trajectory bundles.

If the bottom product point lies in the *possible composition region* $\text{Reg}_{B}^{(3)} \underset{i}{\overset{j}{}}$ in the face 2-3-4 (Fig. 5.28c), the point $N^+ \equiv S^2$ moves, as the value of V/L goes up, along the reversible distillation trajectory inside the concentration tetrahedron up to vertex 1 $(V/L < (V/L)_{max})$.

5.5.2. Mixtures with Any Number of Components

Let's turn now to the mixtures with any number of components, and let's discuss general conditions of existence of sections trajectory bundles and their structure.

It follows from Eqs. (5.15) and (5.16) that distillation trajectory tear-off at finite reflux from k-component product boundary element inside concentration simplex is feasible in that case, if in tear-off point x^t conditions of tear-off into all the $(k + 1)$-component boundary elements, adjacent with the product boundary element are valid.

To check conditions that possible product point at some k-component boundary element$(C_{bound}^{(k)})$ should meet, it is necessary: (1) for the product point $x_D^{(k)}$ or $x_B^{(k)}$ under examination to construct reversible distillation trajectory inside the product boundary element; and (2) to define all the first and second (if they are) reversible trajectory tear-off points x_{rev}^{t1} and x_{rev}^{t2} from the product boundary element into all the adjacent $(k + 1)$-component boundary elements. If there is only one reversible distillation trajectory tear-off point x_{rev}^{t1} into each adjacent boundary element, the point under examination is possible product point and part of reversible distillation trajectory from the most remote from it tear-off point max x_{rev}^{t1} to the end of reversible distillation trajectory $N_{rev}^{(k)}$ is trajectory tear-off segment $\text{Reg}_{t}^{(k)} \underset{i}{\overset{j}{}} \equiv [\max x_{rev}^{t1}, N_{rev}^{(k)}]$.

If there are two points of reversible distillation trajectory tear-off into, at least, one of the adjacent boundary elements and if there is segment of reversible distillation trajectory, limited by the most distant from the product point under examination the first tear-off point max x_{rev}^{t1} and by the closest second tear-off point min x_{rev}^{t2}, then the point under examination is possible product point and mentioned segment is trajectory tear-off segment $\underset{i}{\overset{j}{\operatorname{Reg}}}_t^{(k)} \equiv [\max_i x_{rev}^{t1}, \min_j x_{rev}^{t2}]$.

At marked in such a way trajectory tear-off segments $\underset{i}{\operatorname{Reg}}_t^{(k)}$, Eqs. (5.13) ÷ (5.16) are valid.

Let's note that besides the mentioned cases distillation trajectory tear-off at finite reflux from k-component boundary element to $(k+2)$- component boundary element, if there is α-hypersurface which indexes dont include components of k-component boundary element under examination.

All the possible product points in the boundary elements form possible product region $\underset{i}{\overset{j}{\operatorname{Reg}}}_D^{(k)}$ or $\underset{i}{\overset{j}{\operatorname{Reg}}}_B^{(k)}$: at $k = 1$, it is vertex of simplex; at $k = 2$ it is segment, in the face $(k = 3)$ it is polygon, in the hyperface $(k > 3)$ it is polihedron or hyperpolyhedron.

The definition of components concentrations in the boundary points of possible product compositions regions $\underset{i}{\overset{j}{\operatorname{Reg}}}_D^{(k)}$ or $\underset{i}{\overset{j}{\operatorname{Reg}}}_B^{(k)}$ (in the ends of the segments, in the vertexes of the polygons, of the polyhedrons or hyperpolyhedrons) is main step of the algorithm of azeotropic mixtures separation flowsheets synthesis that is described in Chapter 8. To define these concentration, Eqs. (4.19) and (4.20) connecting concentrations in product points and in reversible distillation trajectory tear-off points are used.

If the product point of sharp distillation is located in possible product composition region ($x_D \in \underset{i}{\overset{j}{\operatorname{Reg}}}_D^{(k)}$ or $x_B \in \underset{i}{\overset{j}{\operatorname{Reg}}}_B^{(k)}$) and if the value of the parameter L/V lies inside the interval of the values of the parameter L/V, for which distillation trajectory tear-off from the boundary product element of concentration simplex is feasible($(L/V)_{min} < L/V < (L/V)_{max}$), then rectifying or stripping bundle appears inside this simplex $\operatorname{Reg}_{w,r}^R$ or $\operatorname{Reg}_{w,s}^R$.

The stationary points of this bundle are located both in the boundary elements of simplex and inside it, at reversible distillation trajectories. The number of such stationary points of the bundle is equal to the difference between the number of the components of the mixture being separated n and the number of the components of section product k plus one. Stationary points of the bundle of top or bottom section are one unstable node N^- (it exists inside the simplex only in the product point, if product is a pure component or an azeotrope); one stable node N^+ (it is located at the boundary element, containing one component more than the product if $K < n - 1$); the rest of the stationary points of the bundle are saddle points S. The first (in the course of the trajectory) saddle point (S^1) is located at the product boundary element (if product is pure component or azeotrope, then the saddle point S^1 coincides with the unstable node N^- and with product point). The second saddle point (S^2) is located at the boundary element, containing product components and one additional component, closest to product components in phase equilibrium coefficient in point S^1. Each of the other saddle points (if they exist) is located at the boundary element, formed by product components and by one of the additional components, except the heaviest (for top section) or the lightest one (for bottom section) among the absent in the product components. The stable node N^+ is located at the boundary element, formed by product components and by additional component, that is the heaviest (for top section) or the lightest one (for bottom section). If the product contains number of components smaller by one than separated mixture, then the stable node N^+ is located inside simplex, there is tear-off point S^1, point S^2 coincides with point N^+, and the rest of stationary points are absent. The described regularities are explained by the fact that reversible distillation trajectories at which all the stationary points of the bundle are located can be found only at the boundary elements mentioned above. For nonideal mixtures (especially for azeotropic), saddle points S^1 or S^2 can be located not only at the boundary elements, but also at α-lines, α-surfaces, or α-hypersurfaces inside simplex. Only at those α-lines, α-surfaces, or α-hypersurfaces, where phase equilibrium coefficients of the components, absent in the product, are equal. Only in this case reversible distillation trajectory, at which the point S^1 or S^2 can be located, goes through the mentioned lines, surfaces, or hypersurfaces.

So far, discussing distillation trajectories and their bundles, we proceeded from the fact, that separation stages are equilibrium ("theoretical") plates. In real separation process at plates of distillation columns equilibrium is not achieved and the degree of nonequilibrium is different for different components. That leads to decrease of difference between compositions at neighboring plates and to change of curvature of distillation trajectories (Castillo & Towler, 1998), but does not influence the location of stationary points of distillation trajectory bundles because in the vicinity of stationary points equilibrium and nonequilibrium trajectories behave equally. Therefore, implemented above analysis of the structure and of evolution of section trajectory bundles is also valid for nonequilibrium trajectory bundles.

At sharp split separatrix sharp split region $S^1 \to S^2 \to \cdots \to N^+ \equiv S^1 \Rightarrow N^+ \equiv$ $\mathrm{Reg}_{sep}^{sh,R}$ (below simply $S^1 - N^+$), that is the boundary element of working section region $\mathrm{Reg}_{w,r}^R$, appears in concentration simplex. Its trajectories, including the working one, go through the product point x_D or x_B and tear-off point S^1. The dimensionality of this separatrix bundle is equal to the difference $(n - k)$ between dimensionality of concentration simplex $(n - 1)$ and dimensionality of the product boundary element $(k - 1)$. In bundle $S^1 - N^+$, point S^1 is its unstable node. As we see below, at discussion of joining of section trajectory bundles of two-section columns (see next section), not only separatrix bundle $S^1 - N^+$ will be of great importance for us, but also its boundary element, the most remote from product point – separatrix bundle $S^2 \to \cdots \to N^+$ (below simply $S^2 - N^+$), having dimensionality $(n - k - 1)$ smaller by one than dimensionality of bundle $S^1 - N^+$. Point S^2 is the unstable node of this bundle (separatrix min-reflux region $\mathrm{Reg}_{sep}^{min,R}$).

At nonsharp separation tear-off point S^1 is absent because it is located outside the concentration simplex. However, for practical purposes, it is expedient to examine the separation, close to sharp one, which assumes the content of admixture components in products to be small (quasisharp separation). With this approach, one can examine the same stationary points $S^1, S^2 \ldots N^+$, as for the corresponding absolutely sharp separation, bearing in mind, that trajectories of quasisharp process go not through the stationary points themselves, but close to them (through quasistationary points). That allows to use the theory of sharp separation trajectories bundles for the solution of practical tasks, for which distillation process cannot be absolutely sharp (i.e., besides product components, each product contains also admixture components).

Section trajectory bundle in its general form may be put in brief as follows:

$$x_D^{(k)} \rightarrow S_r^{1(k)} \overset{S_r^{(k+1)}}{\Rightarrow} N_r^{+(k+1)}, \text{ or } x_B^{(k)} \rightarrow S_s^{1(k)} \overset{S_s^{(k+1)}}{\Rightarrow} N_s^{+(k+1)}.$$
$$\text{Reg}_D \quad \text{Reg}_r' \qquad\qquad\qquad \text{Reg}_B \quad \text{Reg}_s^1$$

5.6.   Conditions of Section Trajectories Joining and Methods of Minimum Reflux Calculating

5.6.1.   Two Models of Feed Tray

So the distillation process in two-section column may be feasible, it is necessary that sections trajectories are joined with each other (i.e., that there is material balance between sections flows at the plates above and below feed cross-section).

The mixture of two flows of liquid goes into the plate, located below feed cross-section: the liquid part of feeding and of liquid, following down from top section bottom plate (Fig. 5.29a). Therefore, between liquid leaving top section and liquid going into bottom section, there is a leap of concentrations in accordance with the equation of material balance in feed cross-section:

$$L_r x_{f-1} + L_F x_F = L_s x_f, \tag{5.18}$$

where x_{f-1} and x_f should belong to trajectory bundles of top and bottom sections correspondingly ($x_{f-1} \in \text{Reg}_{w,r}^R$ and $x_f \in \text{Reg}_{w,s}^R$). Hereinafter, we use the following symbol for a leap of concentrations in the feed cross-section: $x_{f-1} \overset{\Downarrow}{\Rightarrow} x_f$.

The simplified model of feed tray, based on the assumption that feed plate is common for both sections and that the process of mixing and the process of equilibrium achievement go on simultaneously (Fig. 5.29b), is used in a number of works (Levy et al., 1985; Julka & Doherty, 1990). According to this model, the composition x_f can be determined from the equations of both sections (i.e., point x_f should lie at the intersection of two sections trajectories) ($x_f \in \text{Reg}_{w,r}^R \bullet \text{Reg}_{w,s}^R$).

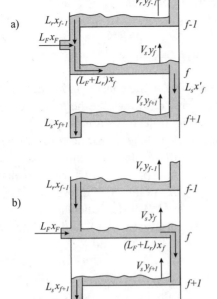

Figure 5.29. Models of feed tray: (a) first mixing then attain equilibrium, and (b) mixing and attain equilibrium simultaneously.

5.6.2. Conditions of Section Trajectories Joining

As we saw in the previous sections, at the increase of the parameter $(L/V)_r$ in top section and of the parameter $(V/L)_s$ in bottom section trajectory bundles of sections $\text{Reg}_{w,r}^R$ and $\text{Reg}_{w,s}^R$ increase, filling up bigger and bigger parts of concentration simple. Along with that the increase of the parameter $(L/V)_r$ leads to the certain increase of the parameter $(V/L)_s$ in accordance with the equations of material and thermal balance of the column at given x_D and x_B.

At some value of parameter $(L/V)_r^{\min}$, trajectory bundles of sections Reg_r and Reg_s adjoin each other by their boundary elements – separatrix min-reflux regions $\text{Reg}_{sep,r}^{\min,R} \equiv (S_r^2 \Rightarrow N_r^+,$ shortly $S_r^2 - N_r^+)$ and $\text{Reg}_{sep,s}^{\min,R} \equiv (S_s^2 \Rightarrow N_s^+,$ shortly $S_s^2 - N_s^+)$, mostly remote from product points x_D and x_B, if one uses for determination of $(L/V)_r^{\min}$ the model in Fig. 5.29b, or if validity of condition (Eq. [5.18]) is achieved between some points of these boundary elements, if one uses the model in Fig. 5.29a. At this value of the parameter $(L/V)_r^{\min}$, the distillation process becomes feasible in infinite column at set product compositions. Such distillation mode is called the mode of minimum reflux. It follows from the analysis of bundle dimensionality $S_r^2 - N_r^+$ and $S_s^2 - N_s^+$ that, at separation without distributed components, points x_{f-1} and x_f can be located in these bundles only at one value of the parameter $(L/V)_r$.

Really, the split without distributed components $1, 2 \ldots k : k+1, \ldots n$ the dimensionality of the bundle $S_r^2 - N_r^+$ is equal to $(n-k-1)$ and dimensionality of bundle $S_s^2 - N_s^+$ is equal $(k-1)$ (see section 5.5). Therefore, total dimensionality of those bundles is equal to $(n-2)$ at dimensionality of concentration simplex $(n-1)$.

Therefore, points x_{f-1} and x_f cannot lie in bundles $S_r^2 - N_r^+$ and $S_s^2 - N_s^+$ at arbitrary values of the parameter $(L/V)_r$, but only at one definite value of this parameter.

The question about location of points x_{f-1} and x_f for splits with distributed component is discussed below.

The task of calculation of minimum reflux mode consists in the determination of parameter $(L/V)_r^{min}$ and of compositions x_{f-1} and x_f at the joining of sections trajectories. Conditions of joining of sections trajectories are different for different splits: for direct and indirect ones, for intermediate ones, and for splits with distributed component. Therefore, algorithms of calculation of minimum reflux mode are different for these splits but include common preliminary stages: (1) calculation of coordinates of sections bundles stationary points $S_r^1, S_r^2, S_s^1, S_s^2, \ldots N_r^+, N_s^+$ at gradually increasing value of the parameter $(L/V)_r$ (i.e., calculation of reversible distillation trajectories of sections for set product points), and (2) linearization of *separatrix trajectory bundles* $\mathrm{Reg}_{sep,r}^{sh,R} \equiv (S_r^1 \Rightarrow N_r^+)$ and $\mathrm{Reg}_{sep,s}^{sh,R} \equiv (S_s^1 \Rightarrow N_s^+)$ *(rectifying and stripping sharp split regions)* and $\mathrm{Reg}_{sep,r}^{min,R} \equiv (S_r^2 \Rightarrow N_r^+)$ and $\mathrm{Reg}_{sep,s}^{min,R} \equiv (S_s^2 \Rightarrow N_s^+)$ *(rectifying and stripping min-reflux regions)* (i.e., calculation of linear equation coefficients, describing the straight lines, planes, or hyperplanes, going through the stationary points of these bundles at different values of the parameter $[L/V]_r$).

The method of reversible distillation trajectories calculation is described above in Section 4.4. To determine coefficients of linear equations, describing straight lines, planes, and hyperplanes, going trough stationary points, by coordinates of these points well-known formulas of analytic geometry are used.

Let's now examine posterior stages for various splits.

5.6.3. Direct and Indirect Splits (One of the Products Is Pure Component or Azeotrope)

Taking into consideration the symmetry of these splits, we confine the discussion to the direct split. In the mode of minimum reflux, point x_f should coincide with the stable node N_s^+, and point x_{f-1} should belong to rectifying minimum reflux bundle $S_r^2 - N_r^+$ (Fig. 5.30). Along with that, Eq. (5.18) should be valid. The search for the value $(L/V)_r^{min}$ is carried out in the following way: at different values $(L/V)_r$, the coordinates of point $x_f \equiv N_s^+$ are determined by means of the method "tray by tray" for bottom section and then the coordinates of point x_{f-1} are determined by Eq. (5.18). At $(L/V)_r < (L/V)_r^{min}$, points x_{f-1} and x_D are located on different sides from the plane or hyperplane $S_r^2 - N_r^+$ and, at $(L/V)_r > (L/V)_r^{min}$, these points are located on one side. That finds approximate values $(L/V)_r^{min}$ (not taking into consideration curvature of bundle $S_r^2 - N_r^+$) and approximate coordinates of points x_{f-1} and x_f. To determine exact values $(L/V)_r^{min}$ and coordinates x_{f-1} and x_f, one varies the values of $(L/V)_r$ in the vicinity of found approximate value $(L/V)_r^{min}$. Then one realizes trial calculations of top section trajectories by means of the method "tray by tray" from feed cross-section to the product. If at

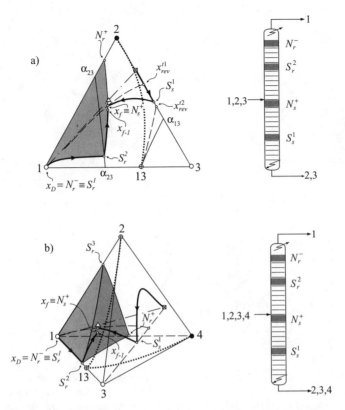

Figure 5.30. The joining of section trajectories under minimum reflux for the direct split of (a) the acetone(1)-benzene(2)-chloroform(3) mixture, and (b) the acetone(1)-benzene(2)-chloroform(3)-toluene(4) mixture. The attraction region Reg_{att}^1 is shaded ($x_f = N_s^+, x_{f-1} \in \mathrm{Reg}_{sep,r}^{\min,R}\overset{2,3}{\underset{1}{}}$ or $x_{f-1} \in \mathrm{Reg}_{sep,r}^{\min,R}\overset{2,3,4}{\underset{1}{}}$).

some trial value of $(L/V)_r$ point x_{f-1} turns out to be inside the rectifying region $\mathrm{Reg}_{w,r}^R \equiv \mathrm{Reg}_{att} \equiv (N_r^- \Rightarrow N_r^+)$, then calculation trajectory achieves top product point x_D (i.e., set vicinity of vertex of concentration simplex). Otherwise, it does not achieve. This allows to find exact value of $(L/V)_r^{\min}$ and exact coordinates of points x_{f-1} and x_f, at which point x_{f-1} is at bent line, surface, or hypersurface $x_{f-1} \in \mathrm{Reg}_{sep,r}^{\min,R}\underset{i}{\overset{j}{}} \equiv S_r^2 - N_r^+$, by means of series of trial calculations (Fig. 5.31). The column trajectory, the calculation direction taken into consideration, may be presented in the following brief form:

$$\begin{array}{ccccccc} x_B^{(n-1)} & \to & S_s^{1(n-1)} & \to & x_f^{(n)} & \Downarrow \Rightarrow & x_{f-1}^{(n)} & \to & S_r^{2(2)} & \to & x_D^{(1)} \\ \mathrm{Reg}_B & & \mathrm{Reg}_s^t & & N_s^+ & & \mathrm{Reg}_{sep,r}^{\min R} & & \mathrm{Reg}_r^t & & N_r^- \end{array}$$

For azeotropic mixtures there is a set of attraction regions Reg_{att} and a set of separatrix min-reflux regions $\mathrm{Reg}_{sep,r}^{\min R}$ (see Fig. 5.13b and 5.26b). The working region $\mathrm{Reg}_{sep,r}^{\min R}$ is one of these. It is determined for calculated point x_{f-1} by means

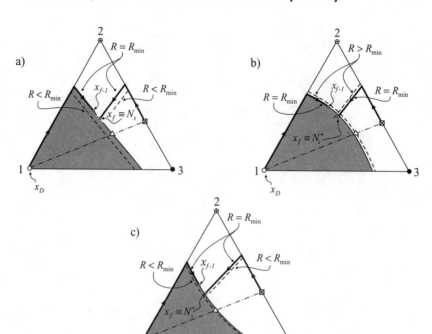

Figure 5.31. The joining of section trajectories under minimum reflux for the direct split of (a) the ideal mixture, (b) the nonideal mixture with the convex boundary of the attraction region Reg^1_{att}, and (c) the nonideal mixture with the concave boundary of the attraction region Reg^1_{att}.

of the method "product simplex" (see Chapter 3) taking into consideration all the stationary points. Then the working point $S_r^{2(2)}$ identified.

5.6.4. Intermediate Splits

For the intermediate splits, it is possible to take into consideration in the best way the regularities of location of section trajectory bundles, using two- or three-stage algorithm of search for $(L/V)_r^{\min}$ with gradual precise of the value of this parameter.

At the first stage, the simplified model of feed tray (Fig. 5.29b) and assumption about linearity of minimum reflux bundles are accepted. At this stage, the value $(L/V)_r^{\min}$ is determined taking into consideration the coordinates of stationary points from the condition of intersection of linearized minimum reflux bundles $S_r^2 - N_r^+$ and $S_s^2 - N_s^+$ (i.e., the smallest value of parameter $[L/V]_r$ is found at which there is intersection of linearized bundles $S_r^2 - N_r^+$ and $S_s^1 - S_s^2 - N_s^+$ or $S_s^2 - N_s^+$ and $S_r^1 - S_r^2 - N_r^+$). The coordinates of intersection point should meet the linear equation system, describing each of two intersecting bundles. Along with that, if the bundles intersect each other, then the point of intersection should be located inside corresponding linear manifolds, but not at their continuation. This criterion is checked at each trial value of $(L/V)_r$, gradually increasing this value of with the step $\Delta(L/V)_r$. If at some value of $(L/V)_r$ the bundles do not intersect each other and at the next value they do intersect each other, then the value of $(L/V)_r^{\min}$ lies between these trial values.

At the second stage, one turns to more exact model of feed plate (Fig. 5.29a), using the result of the first stage as good initial approximation.

As far as it follows from Eq. (5.18) that points x_{f-1}, x_f and x_F should be located at one straight line, point x_{f-1} should lie in rectifying bundle $S_r^2 - N_r^+$ and point x_f should lie in stripping bundle $S_s^2 - N_s^+$. Point x_{f-1} is found as intersection point of linear manifolds $S_r^2 - N_r^+$ and $x_F - S_s^2 - N_s^+$ and point x_f is found as intersection point of linear manifolds $S_s^2 - N_s^+$ and $x_F - S_r^2 - N_r^+$. After that, one calculates discrepancy of material balance in feed cross-section:

$$\varphi(L_r/V_r) = L_r x_{f-1} + L_F x_F - L_s x_f \tag{5.19}$$

This discrepancy is monotonous function of the value of the parameter $(L/V)_r$ in small vicinity of the value $(L/V)_r$, found at the first stage of the algorithm. Therefore, determination of more precise value of $(L/V)_r^{\min}$ at the second stage of the algorithm does not cause any calculation difficulties (Fig. 5.33).

At necessity, at the third stage of the algorithm, one takes into consideration nonlinearity of rectifying and stripping trajectory bundles $S_r^2 - N_r^+$ and $S_s^2 - N_s^+$ (Petlyuk & Danilov, 2002), which is not considerable even for azeotropic mixtures. As an example, Fig. 5.32 shows nonlinearity of the bundle $S_r^1 - S_r^2 - N_r^+ \equiv \mathrm{Reg}_{sep,r}^{sh,R}{}^{2,4}_{1,3}$ for split $1,3:2,4$ of the azeotropic mixture acetone(1)-benzene(2)-chloroform(3)-toluene(4).

At nonsharp separation at minimum reflux, the only one impurity component in each product is the key nonproduct component: in the case of split $1, 2 \ldots k : k+1, \ldots n$, it will be component $k+1$ in the top product and component k in the bottom product. At set product purities η_D and η_B,

$$x_{D,k+1} = 1 - \eta_D$$
$$x_{B,k} = 1 - \eta_B \tag{5.20}$$

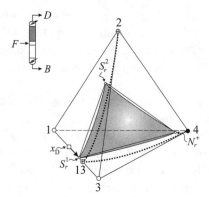

Figure 5.32. The curvature of the separatrix sharp split region for rectifying section $S_r^1 - S_r^2 - N_r^+ \equiv \mathrm{Reg}_{sep,r}^{sh,R}{}^{2,4}_{1,3}$ (shaded) for the split $1,3:2,4$ of the acetone(1)-benzene(2)-chloroform(3)-toluene(4) mixture.

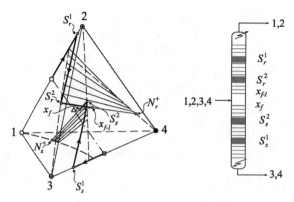

Figure 5.33. The joining of section trajectories under minimum reflux (thick lines with arrows) for the intermediate split of the ideal mixture. A progressive change in position of rectifying $\text{Reg}_{sep,r}^{\min,R} \overset{3,4}{\equiv} S_r^2 \to N_r^+$ and stripping $\text{Reg}_{sep,s}^{\min,R} \overset{1,2}{\underset{3,4}{\equiv}} S_s^2 \overset{1,2}{\to} N_s^+$ lines (thin lines) as L/V increases.

To make the calculation of $(L/V)_r^{\min}$ accurate, one must take into consideration the concentration of these components when determining the coordinates of points $S_r^2, S_r^3, \dots N_r^+, S_s^2, S_s^3, \dots N_s^+$. At high product purities, one can determine the coordinates of all stationary points for sharp separation. The column trajectory at nonsharp separation, the calculation direction taken into consideration, may be presented in the following brief form:

$$x_D^{(k+1)} \to qS_r^{1(k+1)} \to S_r^{2(k+1)} \to x_{f-1}^{(n)} \Downarrow \Rightarrow x_f^{(n)} \leftarrow S_s^{2(n-k+1)}$$
$$\text{Reg}_D \quad\quad \text{Reg}_r^t \quad\quad \text{Reg}_r^t \quad\quad \text{Reg}_{sep,r}^{\min,R} \quad \text{Reg}_{sep,s}^{\min,R} \quad \text{Reg}_s^t$$

$$\leftarrow qS_s^{1(n-k+1)} \leftarrow x_B^{(n-k+1)}$$
$$\text{Reg}_s^t \quad\quad \text{Reg}_B$$

Figure 5.33 shows movement of the lines $\text{Reg}_{sep,r}^{\min,R} \overset{3,4}{\equiv} S_r^2 - N_r^+$ and $\text{Reg}_{sep,s}^{\min,R} \overset{1,2}{\equiv}$ $S_s^2 - N_s^+$ at the increase of the parameter $(L/V)_r$ for separation of the equimolar mixture pentane(1)-hexane(2)- heptane(3)-octane(4) at intermediate split 1,2 : 3,4. Found values of $(L/V)_r^{\min}$ are after the first stage 0,468, after the second stage 0,471 (for comparison at "criterion of the smallest angle" [Koehler et al., 1991] it is 0,463).

The algorithm of calculation of minimum reflux mode at tangential pinch has some peculiarities. At tangential pinch in top section $(L/V)_r^{\min} = (L/V)_r^{pinch} = \max(L/V)_{rev}$ along reversible distillation trajectory (see Fig. 5.19), if along with that there is intersection of separatrix bundles $S_r^2 - N_r^+$ and $x_F - S_s^2 - N_s^+$ or $S_s^2 - N_s^+$ and $x_F - S_r^2 - N_r^+$. If there is no such intersection at $(L/V)_r^{pinch}$, then general algorithm, described above, is used. In this case, the phenomenon of tangential pinch does not become apparent, in spite of the availability of values L/V maximum along reversible distillation trajectory (Koehler et al., 1991).

At tangential pinch, points x_{f-1} and x_f are located not in minimum reflux bundles but inside rectifying and stripping bundles, as at a reflux bigger than the minimum one (see Chapter 7). In the section where tangential pinch takes place, there is a zone of constant concentrations where the composition at plates corresponds to the composition in the point of pinch, and the second section is finite.

The example of *tangential pinch* for four-component mixture is quasisharp separation of azeotropic mixture acetone (1)-benzene (2)-chloroform (3)-toluol (4) of composition x_f (0,350; 0,250; 0,150; 0,250) at intermediate split 1,3(2) : 2,4(3) (admixture components heavy and light key are in brackets correspondingly) at the following composition the products x_D (0,699; 0,001; 0,300, 0) and x_B (0; 0,500; 10^{-8}; 0,500). The same top product composition, as in the previous example (Fig. 5.18b) of separation of three-component mixture in the top section, is accepted for convenience of analysis. In this case, the boundary elements of top section trajectory bundle, located in face 1-2-3, completely coincides with top section trajectory bundle at separation of previously mentioned three-component mixture.

Figure 5.34 shows rectifying trajectory bundle $SN_r - S_r^2 - N_r^+ \equiv \mathrm{Reg}_{ssep,r}^{sh,R}{}_{1,3}^{2,4}$, obtained by means of calculation. In face 1-2-3, pinch point SN_r is stable node for the trajectories, located closer than it to edge 1-3, and is saddle for the rest of trajectories.

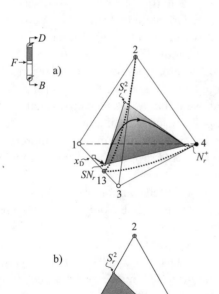

Figure 5.34. (a) The tangential pinch in rectifying section for the acetone(1)-benzene(2)-chloroform(3)-toluene(4) mixture for the split 1,3 : 2,4, and (b) natural projection. Separatrix sharp split region for rectifying section $SN_r - S_r^2 - N_r^+ \equiv \mathrm{Reg}_{ssep,r}^{sh,R}{}_{1,3}^{2,4}$ (shaded).

In rectifying bundle $SN_r - S_r^2 - N_r^+$, pinch point SN_r is an unstable node. At $(L/V)_r < (L/V)_r^{\min}$, there is no top section trajectory tear-off from face 1-2-3 inside concentration tetrahedron and, at $(L/V)^r = (L/V)_r^{\min}$, there is trajectory tear-off and two stationary points appear in face 1-2-3: the pinch point SN_r and point S_r^2.

5.6.5. Splits with Distributed Component

Besides splits without distributed components, we also discuss splits with one distributed component $1, 2, \ldots k - 1, k : k, k + 1, \ldots n$. The significance of these splits is conditioned, first, by the fact that they can be realized for zeotropic mixtures at any product compositions, while at two or more distributed components only product compositions, belonging to some unknown regions of boundary elements of concentration simplex, are feasible. Let's note that for ideal mixtures product composition regions at distribution of several components between products can be determined with the help of the Underwood equation system (see, e.g., Fig. 5.4). This method can be used approximately for nonideal mixtures. From the practical point of view, splits with one distributed component in a number of cases maintain economy of energy consumption and capital costs (e.g., so-called "Petlyuk columns," and separation of some azeotropic mixtures [Petlyuk & Danilov, 2000]).

The analysis of dimensionality of sections trajectory separatrix bundles shows that for splits with one distributed component trajectory of only one section in the mode of minimum reflux goes through corresponding stationary point S_r^2 or S_s^2 (there is one exception to this rule, it is discussed below). The dimensionality of bundle $S_s^2 - N_s^+$ is equal to $k - 2$, that of bundle $S_r^2 - N_r^+$ is equal to $n - k - 1$. The total dimensionality is equal to $n - 3$. Therefore, points x_{f-1} and x_f cannot belong simultaneously to minimum reflux bundles at any value of $(L/V)_r$. If only one of the composition points at the plate above or below the feed cross-section belongs to bundle $S^2 - N^+$ and the second point belongs to bundle $S^1 - S^2 - N^+$, then the total dimensionality of these bundles will become equal $n - 2$; therefore, such location becomes feasible at unique value of $(L/V)_r$ (i.e., in the mode of minimum reflux).

At quasisharp separation with one distributed component in the mode of minimum reflux zone of constant concentrations is available only in one of the sections (in that, trajectory of which goes through point S^2).

The following cases of location of composition points at plates above and below feed cross-section x_{f-1} and x_f: (1) point x_{f-1} lies in rectifying minimum reflux bundle $S_r^2 - N_r^+$, and point x_f lies inside the working trajectory bundle of the bottom section (at nonsharp separation) or in separatrix bundle $S_s^1 - S_s^2 - N_s^+$ (at sharp separation) – Fig. 5.35a ($x_{f-1} \in \mathrm{Reg}_{sep,r}^{\min,R}$, $x_f \in \mathrm{Reg}_{sep,s}^{sh,R}$) with superscripts 3,4 and 1 and subscripts 1,2 and 2,3,4;

(2) point x_f lies in stripping minimum reflux bundle $S_s^2 - N_s^+$, and point x_{f-1} lies inside the working trajectory bundle of the top section (at nonsharp

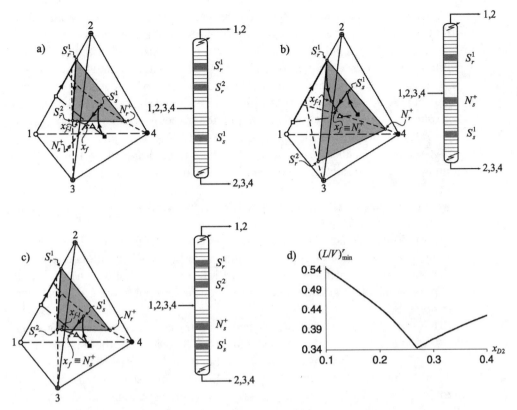

Figure 5.35. The joining of section trajectories under minimum reflux for the split $1,2 : 2,3,4$ of the ideal mixture with $K_1 > K_2 > K_3 > K_4$: (a) case a when $x_{f-1} \in S_r^2 \to N_r^+ \equiv \mathrm{Reg}_{sep,r}^{\min,R} \, {}_{3,4}, x_f \in S_s^1 \to N_s^+ \equiv \mathrm{Reg}_{sep,r}^{sh,R} \, {}_{2,3,4} {}^{1}$, (b) case a when $x_f = N_s^+ \equiv \mathrm{Reg}_{sep,s}^{\min,R} \, {}_{2,3,4} {}^{1}, x_{f-1} \in S_r^1 \to S_r^2 \to N_r^+ \equiv \mathrm{Reg}_{sep,r}^{sh,R} \, {}_{1,2} {}^{3,4}$, (c) case a when $x_{f-1} \in S_r^2 \to N_r^+ \equiv \mathrm{Reg}_{sep,r}^{\min,R} \, {}_{1,2} {}^{3,4}$ and $x_f = N_s^+ \equiv \mathrm{Reg}_{sep,s}^{\min,R} \, {}_{2,3,4} {}^{1}$, and (d) dependence on $(L/V)_r^{\min}$ on $x_{D,2}$. Separatrix sharp split region for rectifying section $\mathrm{Reg}_{sep,r}^{sh,R}$ shaded.

separation) or in separatrix bundle $S_r^1 - S_r^2 - N_r^+$ (at sharp separation) – Fig. 5.35b $(x_f = N_s^+ \equiv \mathrm{Reg}_{sep,s}^{\min,R} \, {}_{2,3,4} {}^{1}, x_{f-1} \in \mathrm{Reg}_{sep,r}^{sh,R} \, {}_{1,2} {}^{3,4})$.

At some ratio of amounts of the distributed component in the separation products, there is a transitional split between above-mentioned ones: both points x_{f-1} and x_f belong correspondingly to minimum reflux bundles $S_r^2 - N_r^+$ and $S_s^2 - N_s^+ (x_{f-1} \in \mathrm{Reg}_{sep,r}^{\min,R} \, {}_{1,2} {}^{3,4}$ and $x_f = N_s^+ \equiv \mathrm{Reg}_{sep,s}^{\min,R} \, {}_{2,3,4} {}^{1}$; Fig. 5.35c). In contrast to the general case, for this split the trajectories of both sections go through the corresponding points S^2. When designing columns with one distributed component, one of the tasks is to find out this distribution coefficient because the smallest value of the parameter $(L/V)_r^{\min}$ corresponds to it (Fig. 5.35d).

Different types of joining of section trajectories at different component distribution coefficients reflect the fact that split with one distributed component $1, 2, \ldots k-1, k : k, k+1, \ldots n$ occupies intermediate position between two splits without distributed components: $1, 2, \ldots k-1 : k, k+1, \ldots n$ and $1, 2, \ldots k-1,$ $k : k+1, \ldots n$. Number and location of stationary points for rectifying minimum reflux bundle $S_r^2 - N_r^+$ for the split with one distributed component is the same as for split $1, 2, \ldots k-1, k : k+1, \ldots n$ and, for stripping minimum reflux bundle $S_s^2 - N_s^+$, it is the same as for split $1, 2, \ldots k-1 : k, k+1, \ldots n$.

At relatively small content of the distributed component k in top product, joining of section trajectories goes on at type, characteristic for splits $1, 2, \ldots k-1 : k, k+1, \ldots n$ (i.e., point x_f lies in bundle $S_s^2 - N_s^+$) and at big content joining goes on at type, characteristic for split $1, 2, \ldots k-1, k : k+1, \ldots n$, (and point x_{f-1} lies in bundle $S_r^2 - N_r^+$).

At some intermediate ("boundary") content of the component k in top product joining of section trajectories goes on simultaneously at two mentioned types.

The algorithm of calculation of minimum reflux mode for splits with distributed component includes the same stages as for intermediate splits without distributed components.

The value of $(L/V)_r^{\min}$, at which there is intersection of linearized bundles $S_r^2 - N_r^+$ and $S_s^1 - S_s^2 - N_s^+$ or $S_s^2 - N_s^+$ and $S_r^1 - S_r^2 - N_r^+$ (i.e., the smallest value of $[L/V]_r$, at which there is intersection of bundles $S_r^1 - S_r^2 - N_r^+$ and $S_s^1 - S_s^2 - N_s^+$, is determined at the first stage). The point of intersection can be located both inside bundles $S_r^2 - N_r^+$ and $S_s^1 - S_s^2 - N_s^+$, and inside bundles $S_s^2 - N_s^+$ and $S_r^1 - S_r^2 - N_r^+$, which determines the type of joining of sections trajectories in the mode of minimum reflux (see Fig. 5.35a,b).

The coordinates of points x_{f-1} and x_f are defined at the second stage in accordance with determined at the first stage type of joining of sections trajectories. If, for example, point x_{f-1} lies in bundle $S_r^2 - N_r^+$ and point x_f lies in bundle $S_s^1 - S_2^2 - N_s^+$, then point x_{f-1} can be found as intersection point of linear manifolds $S_r^2 - N_r^+$ and $x_F - S_s^1 - S_s^2 - N_s^+$ and point x_f can be found as intersection point of linear manifolds $S_s^1 - S_s^2 - N_s^+$ and $x_F - S_r^2 - N_r^+$. In other respects, the second stage of search for $(L/V)_r^{\min}$ for splits with distributed components remains the same as for splits without distributed components.

Nonlinearity of separatrix trajectory bundles is taken into consideration only at the third stage, if it is necessary to determine precisely the value of $(L/V)_r^{\min}$. Usually to solve practical tasks, it is sufficient to confine oneself to the first two stages of the algorithm.

Figure 5.35 is carried out according to the results of calculation of $(L/V)_r^{\min}$ for equimolar mixture pentane(1)-hexane(2)-heptane(3)-octane(4) were made at separation of it with distributed component at split $1,2 : 2,3,4$ at different distribution coefficients of component 2 between products. This figure shows the location of rectifying plane $S_r^1 - S_r^2 - N_r^+$ and of bottom section trajectory in minimum reflux mode at several characteristic values of distribution coefficient of component 2: (1) at joining "at the type of direct split" $(1 : 2,3,4)$ (Fig. 5.35b; $x_{D2} = 0.1, x_f \equiv N_s^+$, zone of constant concentrations is located in feed cross-section in bottom section); (2) at joining "at the type of intermediate split without distributed components" $1,2:3,4$ (Fig. 5.35a; $x_D = 0.4$, x_{f-1} lies at line $S_r^2 - N_r^+$, zone of constant concentrations is located in the middle part of the top section); (3) at joining with optimal distribution of component 2 between products (Fig. 5.35c; $x_{D2} = 0,268$, $x_f \equiv N_s^+$, x_{f-1} lies at line $S_r^2 - N_r^+$, zones of constant concentrations are available in both sections: in the middle part of the top section and in feed cross-section in the bottom section, $[L/V]_r^{\min}$ is the smallest comparing with any splits at items 1 and 2).

Figure 5.35d shows for this example the change of the value of $(L/V)_r^{\min}$, depending on the parameter x_{D2} (the first two stages of general algorithm were used for calculations). The column trajectories, the calculation direction taken into consideration, may be presented in the following brief form:

$$\begin{array}{ccccccc} x_D^{(2)} & \to S_r^{1(2)} & \to S_r^{2(3)} & \to x_{f-1}^{(4)} & \Downarrow \Rightarrow x_f^{(4)} & \leftarrow S_s^{1(3)} & \leftarrow x_B^{(3)} \\ \text{Reg}_D & \text{Reg}_r^t & \text{Reg}_r^t & \text{Reg}_{sep,r}^{\min,R} & \text{Reg}_{sep,s}^{sh,R} & \text{Reg}_s^t & \text{Reg}_B \end{array} \quad \text{(Fig. 5.35a)}$$

$$\begin{array}{cccccc} x_D^{(2)} & \to S_r^{1(2)} & \to x_{f-1}^{(4)} & \Downarrow \Rightarrow x_f^{(4)} & \leftarrow S_s^{1(3)} & \leftarrow x_B^{(3)} \\ \text{Reg}_D & \text{Reg}_r^t & \text{Reg}_{sep,r}^{sh,R} & N_s^+ & \text{Reg}_s^t & \text{Reg}_B \end{array} \quad \text{(Fig. 5.35b)}$$

$$\begin{array}{ccccccc} x_D^{(2)} & \to S_r^{1(2)} & \to S_r^{2(3)} & \to x_{f-1}^{(4)} & \Downarrow \Rightarrow x_f^{(4)} & \leftarrow S_s^{1(3)} & \leftarrow x_B^{(3)} \\ \text{Reg}_D & \text{Reg}_r^t & \text{Reg}_r^t & \text{Reg}_{sep,r}^{\min,R} & N_s^+ & \text{Reg}_s^t & \text{Reg}_B \end{array} \quad \text{(Fig. 5.35c)}$$

5.6.6. Equations of Thermal Balance

Stated above algorithms include two assumptions: (1) liquid and vapor flows in column sections are accepted as constant, and (2) it is accepted that plates are equilibrium.

Small complication of the algorithms excludes the first assumption by means of entering into algorithms of equations of thermal balance. These equations for each section should be constructed at contour, embracing part of the column from cross-section in a zone of constant concentration to the end of the column. It is necessary to examine all the zones of constant concentrations – real and fictitious (i.e., corresponding to all the stationary points: S^1, S^2, $S^3 \ldots N^+$). For the top section, the equation of thermal balance looks like

$$V_r^{st} H_r^{st} + L_r^{top} h_r^{top} = L_r^{st} h_r^{st} + V_r^{top} H_r^{top} \tag{5.21}$$

The similar equation can be constructed for the bottom section. In Eq. (5.21), H_r^{top} and h_r^{top} are enthalpies of liquid and vapor at the top of the column, depending on their compositions, and H_r^{st} and h_r^{st} are enthalpies of vapor and liquid in the cross-section of the zone of constant concentrations, depending on composition in the stationary point. Knowing these compositions and setting $(L/V)_r^{top}$, it is possible to determine $(L/V)_r^{st}$ from Eq. (5.21) and from the equation of material balance and then to specify composition in the stationary point, located at reversible distillation trajectory for $(L/V)_r^{st}$.

Therefore, to determine $(L/V)_r^{st}$ and compositions in stationary points, it is necessary to carry out several iterations. In other respects, the algorithms of calculation of minimum reflux mode, taking into consideration the change of flows of liquid and vapor in the column sections, do not differ from those described previously.

As far as the second assumption is concerned, as was mentioned in Section 5.5, it does not influence the compositions in the stationary points. Therefore, it does not influence the first two stages of the described algorithms of calculation of minimum reflux mode. This assumption could have some influence only at the third stage of the algorithms, when curvature of separatrix trajectory bundles should be taken into consideration. Therefore, the assumption about equilibrium plates at calculation of minimum reflux mode is even more justified than at calculation of finite columns.

Therefore, the stated algorithm of calculation of minimum reflux mode, based on the geometry of the trajectory bundles in concentration space, are potentially as one likes precise and most general, because they embrace any splits on mixtures with any components number and any degree of nonideality.

5.6.7. Visualization of Section Trajectories

Visualization is of great importance for understanding calculation results, and for understanding peculiarities and difficulties in separation of any mixture (it is especially important for azeotropic mixtures). Geometric approach to the solution of calculation task is closely connected with visualization of its results. We already demonstrated it at certain examples of separation of three- and four-component mixtures. Separation of mixtures with a component number above four does not allow for such complete visualization. However, less complete and more rough visualization can be used for such mixtures, considering the components lighter than light key component as one component, having in each point of the trajectory the concentration, equal to the sum of concentrations of the components lighter than the light key component. It should be done similarly with all components heavier than the heavy key one. The use of such an approach allows to present separation of mixture with any components number at any intermediate split without distributed components as separation of the four-component mixture containing two key components, one pseudocomponent lighter than the light key component and one pseudocomponent heavier than the heavy key component.

Location of sections trajectories of such conditional mixture in concentration tetrahedron should follow general described above regularities for usual four-component mixtures. Therefore, such visualization allows us to understand and to foresee the designing peculiarities and difficulties of separation of mixtures with large numbers of component.

5.7.   **Necessary and Sufficient Conditions of Separability of Mixtures**

5.7.1.   Adiabatic Columns

We discussed before in Section 5.3 the necessary conditions of separability of mixtures. The main necessary condition is belonging of product points to possible product regions. At this condition and at the sufficient for trajectory tear-off values of parameters L/V in the top section and V/L in the bottom section, the trajectory bundles of the sections arise inside the concentration simplex.

We saw in the previous section that these conditions are not sufficient for separation in two-section column.

The process of separation becomes possible only if trajectory bundles of the sections become sufficiently big for the fact that the conditions of sections joining could be valid (i.e., that the material balance would be valid in feed cross-section of the column between the certain trajectory points of the sections).

For splits without distributed components and with one distributed component, the conditions of joining can always be valid at sufficiently large values of parameters L/V and V/L in the top and bottom sections, respectively, if the sizes of trajectory bundles unlimitedly grow at the increase of these parameters.

But we saw in Section 5.4 that the values of parameters L/V and V/L and the sizes of trajectory bundles of adiabatic columns sections are limited, if for product point there are two reversible distillation trajectory tear-off points. Therefore, necessary conditions of separability in adiabatic columns can be insufficient if for one or for both product points there are two reversible distillation trajectory tear-off points from boundary elements to which points S^1 belong. In these cases, to check separability it is necessary to verify whether corresponding separatrix sections trajectory bundles join at the maximum possible value of the parameter L/V or V/L.

5.7.2. Nonadiabatic Columns

Usage of nonadiabatic columns (i.e., column with intermediate at height inputs or outputs of heat), broadens conditions of separability of mixtures, having two reversible distillation trajectory tear-off points. If, for example, heat is brought and drawn off in feed cross-section (Fig. 5.36), as it was offered in the work (Poellman & Blass, 1994), then it is possible in one of the sections of the column, where there is limitation at the value of the parameter L/V or V/L, to keep the corresponding allowed value of this parameter, and joining of trajectories of the sections can be maintained at the expense of increase of vapor and liquid flows in the second section (Petlyuk & Danilov, 1998). In a more general case, when there are limitations of the values of the parameters L/V and V/L in both sections, it is possible to use intermediate inputs and outputs of heat in the middle cross-sections of both sections.

However, in some cases, even usage of nonadiabatic columns does not maintain separability. These are the cases, when reversible distillation trajectories for both product points do not have node points. In these cases, section trajectory bundles not only of adiabatic, but also of nonadiabatic columns, are limited because reversible distillation trajectories at which section trajectory bundles stationary point lie are located in limited parts of concentration space, adjacent to product boundary elements (see Fig. 5.17b for x_D). To check whether it is possible to separate the mixture of this kind into a set product, it is necessary to examine the bundles $S_r^2 - N_r^+$ and $S_s^2 - N_s^+$ for points S_r^2 and S_s^2, the most remote along

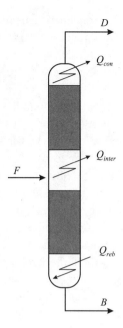

Figure 5.36. The column with the heat output (Q_{inter}) in the feed cross-section.

reversible distillation trajectories from product points. If these bundles intersect each other, the separation is feasible. Otherwise, it is not feasible.

5.8.   Conclusion

At sharp distillation with finite reflux, the product points should belong to possible product regions Reg_D^j and Reg_B^j at those boundary elements of concentration simplex at which trajectory tear-off regions $\text{Reg}_r^t{}_i^j$ and $\text{Reg}_s^t{}_i^j$ are available. In the tear-off points equilibrium coefficients of absent components j should be smaller (bigger) than those of present components i in the top (bottom) section ($K_{ri} > K_{rj}$, $K_{si} < K_{sj}$). Section trajectories tear off from the boundary elements at the value of the parameter L/V greater than the values of phase equilibrium coefficients of the absent components ($L/V > K_j$) in tear-off point x^t.

In the mode of minimum reflux, $R_{\min}$ at sharp distillation without distributed components trajectory of the top (bottom) section goes from the product point x_D (x_B) to the trajectory tear-off point $S_r^1(S_s^1)$ into the boundary element, containing one additional component referring to product components, that is the closest one by phase equilibrium coefficient, then it goes from point $S_r^1(S_s^1)$ to the point of trajectory tear-off $S_r^2(S_s^2)$ inside concentration simplex, then it goes from point $S_r^2(S_s^2)$ to point x_{f-1} (x_f) in the feed cross-section of the column. Along with that, material balance should be valid in the feed cross-section.

Separatrix sharp region of the section trajectories $\text{Reg}_{sep}^{sh,R} \equiv S^1 - S^2 - N^+$ passing through product point has the dimensionality equal to the difference between the number of components in feeding and the number of components in the product: $(n - k)$. Separatrix min-reflux region of the section $\text{Reg}_{sep}^{min,R} \equiv S^2 - N^+$ that is the most remote from product point has the dimensionality smaller by one than the dimensionality of separatrix sharp region of the section. In the mode of minimum reflux at separation without distributed components, the composition points in the feed cross-section belong to these boundary elements $S_r^2 - N_r^+$ and $S_s^2 - N_s^+ (x_{f-1} \in \text{Reg}_{sep,r}^{min,R}, x_f \in \text{Reg}_{sep,s}^{min,R})$ and, at separation with one distributed component, the composition point in feed cross-section in one of the sections belongs to trajectory bundle $S^1 - S^2 - N^+$ and, in the other section, it belongs to the boundary element of this bundle $S^2 - N^+ (x_{f-1} \in \text{Reg}_{sep,r}^{sh,R}, x_f \in \text{Reg}_{sep,s}^{min,R}$ or $x_{f-1} \in \text{Reg}_{sep,r}^{min,R}, x_f \in \text{Reg}_{ssep,s}^{sh,R})$.

Stationary points of trajectory bundles $S^1 - S^2 - N^+$ are located at reversible distillation trajectories in the boundary elements of concentration simplex or in the α-lines, α-surfaces, and α-hypersurfaces. Their coordinates can be calculated for each value of parameter L/V. Trajectory bundles of the sections $S^1 - S^2 - N^+$ and their boundary elements $S^2 - N^+$ can be accepted to be linear for practical purposes. This calculates minimum reflux mode for any mixtures and any splits with sufficient precision. Found values $(L/V)^{min}$ can also be used for quasisharp separation at sufficient purity of the products.

Phase equilibrium coefficients field of each concrete nonideal or azeotropic mixture determines boundaries of various regions at the boundaries of concentration simplex and inside it (of component order regions $\text{Reg}_{ord}^{i,j,k}$, of sharp split regions $\text{Reg}_{sh}^{i:j}$, of trajectory tear-off regions Reg^t, of possible product regions Reg_D and Reg_B, of tangential pinch regions Reg_{tang}, and of pitchfork regions Reg_{pitch}, etc.). These regions are polygons, polyhedrons, or hyperpolyhedrons with curvilinear boundaries, vertexes of which are located at edges of concentration simplex. Coordinates of these vertexes can be determined by helping to calculate values of phase equilibrium coefficients of the components at edges of concentration simplex.

This solves the task of determination of possible splits for any mixture and synthesizes its separation flowsheet.

At quasisharp separation, possible product composition regions Reg_D and Reg_B grow at the decrease of purity of the products. Boundaries found for sharp separation deliberately ensure possible splits for quasisharp separation, but, if it is necessary, the widened boundaries for the set purity can be found.

5.9. Questions

1. Can the top product of the column contain components 4, 5, and 6 if in face 4-5-6 there is component order region Reg_{ord}^{56432}? Can the bottom product contain components 3 and 4 if at the edge there is segment Reg_{ord}^{15243}?

2. Is it possible to separate mixture at split 2,5,4: 5,1,3 if in face 2-5-4 there is component order region Reg_{ord}^{45231} and in face 5-1-3 there is region Reg_{ord}^{42135}?

3. Will tangential pinch arise if the top product contains components 4, 5, and 6, tear-off point belongs to region Reg_{ord}^{56432}, and in the vicinity of tear-off point component 3 has a phase equilibrium coefficient that is smaller than in tear-off point itself?

4. Name known to you regions in concentration simplex influencing distillation process. Give definitions of these regions, and describe their properties and significance.

5. What is the difference between working region of the section Reg_w^R, rectifying or stripping separatrix sharp split region $\text{Reg}_{sep}^{sh,R}$, and rectifying or stripping separatrix min-reflux region $\text{Reg}_{sep}^{min,R}$? What is the dimensionality of these bundles? What is their significance?

5.10. Exercises with Software

1. For mixture acetone(1)-benzene(2)-chloroform(3)-toluene(4) of composition (0,2; 0,4; 0,1; 0,3), calculate the mode of minimum reflux for the split 1,3 : 2,4 and find location of trajectories of the sections in the concentration simplex.

2. Determine other possible sharp splits for this mixtures, calculate the minimum reflux mode for each.

3. Do the same as in Exercise 1 for mixture pentane(1)-hexane(2)-heptane(3)-octane(4) of composition (0,35; 0,25; 0,15; 0,25) at split 1,2 : 3,4.

4. Do the same for the latter mixture at split 1,2 : 2,3,4 and at $x_{D2} = 0,15$.

5. Do the same for this mixture for the split 1,2,3 : 3,4 at equal and optimal distribution of component 3 among the products.

6. Do the same for this mixture for the splits 1 : 2,3,4 and 1,2,3 : 4.

References

Acrivos, A., & Amundson, N. R. (1955). On the Steady State Fractionation of Multicomponent and Complex Mixture in an Ideal Cascade. *Chem. Eng. Sci.*, 4, 29–38, 68–74, 141–8, 159–66, 206–8, 249–54.

Castillo, F. J. L., Thong, Y. C., & Towler, G. P. (1998). Homogeneous Azeotropic Distillation. Design Procedure for Single-Feed Columns at Nontotal Reflux. *Ind. Eng. Chem. Res.*, 37, 987–97.

Castillo, F. G. L., & Towler, G. P. (1998). Influence of Multicomponent Mass Transfer on Homogeneous Azeotropic Distillation. *Chem. Eng. Sci.*, 53, 963–76.

Chien, H. H. Y. (1978). A Rigorous Method for Calculating Minimum Reflux Rates in Distillation. *AIChE J.*, 24, 606–13.

Davydyan, A. G., Malone, M. F., & Doherty, M. F. (1997). Boundary Modes in a Single-Feed Distillation Column for Separation of Azeotropic Mixtures. *Theor. Found. Chem. Eng.*, 31, 327–38.

Erbar, R. C., & Maddox, R. N. (1962). Minimum Reflux Rate for Multicomponent Distillation Systems by Rigorous Plate Calculations. *Can. J. Chem. Eng.*, 2, 25–30.

Fidkowski, Z. T., Doherty, M. F., & Malone, M. F. (1993). Feasibility of Separations for Distillation of Nonideal Ternary Mixtures. *AIChE J.*, 39, 1303–21.

Fidkowski, Z. T., Malone, M. F., & Doherty, M. F. (1991). Nonideal Multicomponent Distillation: Use of Bifurcation Theory for Design. *AIChE J.*, 37, 1761–79.

Franklin, N. L. (1986). Counterflow Cascades: Part I. *Chem. Eng. Res. Des.*, 64, 56–66.

Franklin, N. L. (1988a). Counterflow Cascades: Part II. *Chem. Eng. Res. Des.*, 66, 47–64.

Franklin, N. L. (1988b). The Theory of Multicomponent Countercurrent Cascades. *Chem. Eng. Res. Des.*, 66, 65–74.

Franklin, N. L., & Forsyth, J. S. (1953). The Interpretation of Minimum Reflux Conditions in Multicomponent Distillation. *Trans. Inst. Chem. Eng.*, 31, 363–88.

Hausen, H. (1934). Einfluss des Argons auf die Rektifikation der Luft. *Forsc. Geb. Ingenieurwes*, 6, 290–97 (Germ.).

Hausen, H. (1935). Rektifikation von Dreistoffgemischen - Insbesondere von Sauerstoff-Stickstoff-Luft. *Forsch. Geb. Ingenieurwes*, 6, 9–22 (Germ.).

Hausen, H. (1952). Rektifikation Idealer Dreistoffgemische. *Z. Angew. Phys.*, 4, 41–51 (Germ.).

Holland, C. D. (1963). *Multicomponent Distillation*. New York: Prentice Hall.

Julka, V., & Doherty, M. F. (1990). Geometric Behavior and Minimum Flows for Nonideal Multicomponent Distillation. *Chem. Eng. Sci.*, 45, 1801–22.

Kiva, V. N. (1976). Qualitative Analysis of Distillation by Means of Weak Mathematical Model. In *Physical-Chemical Investigation of Mass-Transfer Processes*. Leningrad: VNIISK (Rus.).

Koehler, J., Aguirre, P., & Blass, E. (1991). Minimum Reflux Calculations for Nonideal Mixtures Using the Reversible Distillation Model. *Chem. Eng. Sci.*, 46, 3007–21.

Kondrat'ev, A. A., Frolova, L. N., Serafimov, L. A., & Hasanov, Z. K. (1977). Peculiarities of Distillation of Azeotropic Mixtures with Intersection of Boundaries of Distillation Regions. *Theor. Found. Chem. Eng.*, 11, 907–12.

Lee, E. S. (1974). Estimation of Minimum Reflux in Distillation and Multipoint Boundary Value Problems. *Chem. Eng. Sci.*, 29, 871–5.

Levy, S. G., & Doherty, M. F. (1986). A Simple Exact Method for Calculating Tangent Pinch Points in Multicomponent Nonideal Mixtures by Bifurcation Theory. *Chem. Eng. Sci.*, 41, 3155–60.

Levy, S. G., Van Dongen, D. B., & Doherty, M. F. (1985). Design and Synthesis of Homogenous Azeotropic Distillation. 2. Minimum reflux Calculations for Nonideal and Azeotropic Columns. *Ind. Eng. Chem. Fundam.*, 24, 463–74.

McCabe, W. L., & Thiele, E. W. (1925). Graphical Design of Fractionating Columns. *Ind. Eng. Chem.*, 17, 606–11.

McDonough, J. A., & Holland, C. D. (1962). Figure Separations This New Way–Part 9 – How to Figure Minimum Reflux. Hydrocarbon Process. *Petrol. Refin.*, 41, 153–60.

Petlyuk, F. B. (1978). Rectification of Zeotropic, Azeotropic and Continuous Mixtures in Simple and Complex Infinite Columns at Finite Reflux. *Theor. Found. Chem. Eng.*, 12, 671–8.

Petlyuk, F. B. (1998). Simple Predicting Methods for Feasible Sharp Separations of Azeotropic Mixtures. *Theor. Found. Chem. Eng.*, 32, 245–53.

Petlyuk, F. B., Avet'yan, V. S., & Platonov, V. M. (1968). Research of Multicomponent Distillation at Minimum Reflux. *Theor. Found. Chem. Eng.*, 2, 155–68.

Petlyuk, F. B., & Danilov, R. Yu. (1998). Calculations of Distillation Trajectories at Minimum Reflux for Ternary Azeotropic Mixtures. *Theor. Found. Chem. Eng.*, 32, 548–59.

Petlyuk, F. B., & Danilov, R. Yu. (1999). Feasible Separation Variants and Minimum Reflux Calculations. *Theor. Found. Chem. Eng.*, 33, 571–83.

Petlyuk, F. B., & Danilov, R. Yu. (2000). Synthesis of Separation Flowsheets for Multicomponent Azeotropic Mixtures on the Basis of the Distillation Theory. Synthesis: Finding Optimal Separation Flowsheets. *Theor. Found. Chem. Eng.*, 34, 444–56.

Petlyuk, F. B., & Danilov, R. Yu. (2001a). Few-Step Iterative Methods for Distillation Process Design Using the Trajectory Bundle Theory: Algorithm Structure. *Theor. Found. Chem. Eng.*, 35, 224–36.

Petlyuk, F. B., & Danilov, R. Yu. (2001b). Theory of Distillation Trajectory Bundles and Its Application to the Optimal Design of Separation Units: Distillation Trajectory Bundles at Finite Reflux. *Trans IChemE*, 79, Part A, 733–46.

Petlyuk, F. B., & Sezafimov, L. A. (1983). Multicomponent Distillation. Theory and Calculation. Moscow: Khimiya (Rus).

Petlyuk, F. B., & Vinogradova, E. I. (1980). Theoretical Analysis of Minimum Reflux Regime for Ternary Azeotropic Mixtures. *Theor. Found. Chem. Eng.*, 14, 413–18.

Petlyuk, F. B., Vinogradova, E. I., & Serafimov, L. A. (1984). Possible Compositions of Products of Ternary Azeotropic Mixture Distillation at Minimum Reflux. *Theor. Found. Chem. Eng.*, 18, 87–94.

Poellmann, P., & Blass, E. (1994). Best Products of Homogeneous Azeotropic Distillations. *Gas Separation and Purification*, 8, 194–228.

Poellmann, P., Glanz, S., & Blass, E. (1994). Calculating Minimum Reflux of Nonideal Multicomponent Distillation Using Eigenvalue Theory. *Comput. Chem. Eng.*, 18, 549–53.

Schreinemakers, F. A. H. (1901). Dampfdrucke ternarer Gemische. *Z. Phys. Chem.*, 36, 413–49 (Germ.).

Serafimov, L. A., Timofeev, V. S., & Balashov, M. I. (1973a). Rectification of Multicomponent Mixtures. 3. Local Characteristics of the Trajectories Continuous Rectification Process at Finite Reflux Ratios. *Acta Chimica Academiae Scientiarum Hungarical*, 75, 235–54.

Serafimov, L. A., Timofeev, V. S., & Balashov, M. I. (1973b). Rectification of Multicomponent Mixtures. 4. Non-Local Characteristics of Continuous Rectification, Trajectories for Ternary Mixtures at Finite Reflux Ratios. *Acta Chimica Academiae Scientiarum Hungarical*, 75, 255–70.

Shafir, A. P., Petlyuk, F. B., & Serafimov, L. A. (1984). Change of Composition of Azeotropic Mixtures Distillation Products in Infinite Columns at Increase of Reflux Rate. In *The Calculation Researches of Separation for Refining and Chemical Industry* (pp. 55–75). Moscow: Zniiteneftechim (Rus.).

Shiras, R. N., Hanson, D. N., & Gibson, G. H. (1950). Calculation of Minimum Reflux in Distillation Columns. *Ind. Eng. Chem.*, 42, 871–6.

Stichlmair, J., Offers, H., & Potthoff, R. W. (1993). Minimum Reflux and Minimum Reboil in Ternary Distillation. *Ind. Eng. Chem. Res.*, 32, 2438–45.

Stichlmair, J. G., Offers, H., & Potthoff, R. W. (1993). Minimum Reflux and Reboil in Ternary Distillation. *Ind. Eng. Chem. Res.*, 32, 2438–45.

Tavana, M., & Hanson, D. N. (1979). The Exact Calculation of Minimum Flows in Distillation Columns. *Ind. Eng. Chem. Process Des. Dev.*, 18, 154–6.

Underwood, A. J. V. (1945). Fractional Distillation of Ternary Mixtures. Part I. *J. Inst. Petrol.*, 31, 111–18.

Underwood, A. J. V. (1946a). Fractional Distillation of Ternary Mixtures. Part II. *J. Inst. Petrol.*, 32, 598–613.

Underwood, A. J. V. (1946b). Fractional Distillation of Multicomponent Mixtures (Calculation of Minimum Reflux Ratio). *J. Inst. Petrol.*, 32, 614–26.

Underwood, A. J. V. (1948). Fractional Distillation of Multicomponent Mixtures. *Chem. Eng. Prog.*, 44, 603–14.

Vogelpohl, A. (1964). Rektifikation von Dreistoffgemischen (Teil 1: Rektifikation als Stoffaustauschvorgang und Rektifikationslinien Idealer Gemische). *Chem.-Ing.-Tech.*, 36, 1033–45 (Germ.).

Vogelpohl, A. (1970). Rektifikation Idealer Vielstoffgemische. *Chem.-Ing.-Tech.*, 42, 1377–82 (Germ.).

Wahnschafft, O. M., Koehler, J. W., Blass, E., & Westerberg, A. W. (1992). The Product Composition Regions of Single-Feed Azeotropic Distillation Columns. *Ind. Eng. Chem. Res.*, 31, 2345–62.

White, R. R. (1953). Stripping, Rectifying and Distillation of Ternary, Quaternary and Multicomponent Mixtures. *Petrol. Process.*, 8, 357–62, 539–43, 704–9, 892–6, 1026–31, 1174–9, 1366–69, 1533–6, 1705–7.

6

Distillation Trajectories in Infinite Complex Columns and Complexes

6.1. Introduction

This chapter extends the geometric description of the distillation process to infinite complex columns and complexes, and then on this basis to develop methods of their calculation.

Here we understand by complex columns a countercurrent cascade without branching of flows, without recycles and bypasses, which, in contrast to simple columns, contains more than two sections. The complex column is a column with several inputs and/or outputs of flows. The column of extractive distillation with two inputs of flows – feed input and entrainer input – is an example of a complex column.

We understand by distillation complex a countercurrent cascade with branching of flows, with recycles or bypasses of flows. Columns with side stripping or side rectifier and columns with completely connected thermal flows (the so-called "Petlyuk columns") are examples of distillation complexes with branching of flows. A column of extractive distillation, together with a column of entrainer regeneration, make an example of a complex with recycle of flows. Columns of this complex work independently of each other; therefore, we do not examine it in this chapter, and the questions of its usage in separation of azeotropic mixtures and questions of determination of entrainer optimal flow rate are discussed in the following chapters.

The fundamental difference between complex columns and complexes and simple columns lies in the availability of intermediate sections (besides the top and the bottom ones). The intermediate sections exchange vapor and liquid flows with other sections or with the decanter.

Therefore, for the intermediate sections, the equations of material balance should be transformed in such a way that the flow rate and composition of a pseudoproduct – that is, the difference between the outgoing and the incoming flows – should be substituted there for the flow rate and composition of the product. Figure 6.1 shows an example of a closed contour for obtaining material balance equations for the intermediate section.

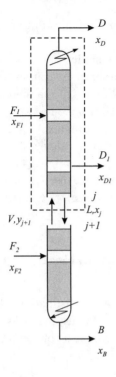

Figure 6.1. A complex column with two feeds and one side product. Control volume (dotted line) for obtaining material balance equations for the intermediate section.

For the section between points of feed input F_2 and product output D_1, the equation of material balance is as follows:

$$Vy_{j+1} = Lx_j + Dx_D + D_1x_{D1} - F_1x_{F1} \tag{6.1}$$

Let's designate the flow rate of the pseudoproduct:

$$D' = D + D_1 - F_1 \tag{6.2}$$

and the composition of the pseudoproduct:

$$x'_D = (Dx_D + D_1x_{D1} - F_1x_{F1})/(D + D_1 - F_1) \tag{6.3}$$

In contrast to the product point, the pseudoproduct point can be located not only inside or at the boundary of the concentration simplex, but also outside it. The latter case refers to columns of extractive distillation with two feeds, which leads to new regularities of location of trajectory bundles and their stationary points, that differ from regularities of location of top and bottom section trajectories. Therefore, we pay a lot of attention in this chapter to trajectory bundles of intermediate sections in extractive distillation columns.

As far as distillation complexes with flows branching are concerned, their main peculiarity is that they can be described as sets of several two-section columns interacting with one another.

Pseudoproduct points of intermediate sections of these complexes are always located inside or at the boundary of the concentration simplex. Therefore, the location of stationary points of these section trajectory bundles does not differ from their location for two-section columns.

To describe geometrically the distillation process in complex columns and complexes, we use, as for simple columns, the conception of sharp separation, and then we turn from a sharp separation to a quasisharp one. We need to examine the conditions of joining intermediate sections to the top and bottom ones.

In contrast to simple columns, complex columns and complexes have greater degrees of freedom of designing, which complicates their calculation and designing.

For the main types of complex columns and complexes, we discuss their sphere of application, history of investigation, geometric description of trajectory bundles, and methods of calculation.

6.2.   Columns with Intermediate Inputs and Outputs of Heat: "Pinch Method"

Columns under consideration are columns of nonadiabatic distillation (that can also be used in simple two-section columns, in complex columns, and in distillation complexes). The application of simple nonadiabatic columns for separation of azeotropic mixtures was examined in Chapter 5, Section 5.7, when separation in adiabatic columns is unfeasible.

Here we examine another application of nonadiabatic columns – to decrease energy consumption in separation. Nonadiabatic columns are widely used for this purpose in petroleum refining (heat output by "pumparounds").

In the mode of minimum reflux adiabatic sections trajectories intersect reversible distillation trajectories in points S^2. Therefore, the separation process between product point and point S^2 can be carried out in principle, maintaining phase equilibrium between meeting flows of vapor and liquid in the cross-section at the height of the column by means of differential input or output of heat. We call such a separation process, with the same product compositions as at adiabatic distillation, a partially reversible one. A completely reversible process is feasible only for the preferable split that is rarely used in practice. Nonadiabatic distillation used in industry is a process intermediate between adiabatic and partially reversible distillation. Summary input and output of heat at nonadiabatic and adiabatic distillation are the same, and the energetic gain at nonadiabatic distillation is obtained at the transfer of a part of input or output heat to more moderate temperature level, which uses cheaper heat carriers and/or coolants.

We examine the column with one intermediate input of heat in the bottom section and one intermediate output of heat in the top section. Figure 6.2a shows the change of internal liquid flows along the height of such a nonadiabatic column, depending on the value inverse to absolute temperature $(1/T)$. Figure 6.2b shows the distillation trajectory of nonadiabatic column:

$$\underset{\text{Reg}_D}{x_D^{(1)}} \to \underset{Q_{con}^{int}}{N_r^{int(2)}} \to \underset{\text{Reg}_r^t}{S_r^{(2)}} \to \underset{\text{Reg}_{sh}}{x_{f-1}^{(3)}} \Downarrow \Rightarrow \underset{\text{Reg}_{sh}}{x_f^{(3)}} \leftarrow \underset{Q_{reb}^{int}}{N_s^{int(3)}} \leftarrow \underset{\text{Reg}_s^t}{S_s^{(2)}} \leftarrow \underset{\text{Reg}_B}{x_B^{(2)}}.$$

a)

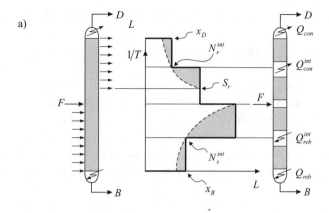

b)

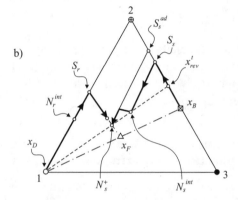

Figure 6.2. (a) A liquid profile for a column with direct split 1: 2,3 with intermediate heat input and heat output (solid line), and for partially reversible column (dotted line). (b) Section trajectories for a column with intermediate heat input and heat output.

Liquid flows saltatory increases or decreases in the points of intermediate output or input of heat at the temperature T_{con}^{int} and T_{reb}^{int}. The minimum possible value of liquid flows at parts from column ends to the points of intermediate input and output of heat is equal to the value of liquid flow at partially reversible process in those cross-sections, where $T_{rev} = T_{con}^{int}$ and $T_{rev} = T_{reb}^{int}$. Calculation of reversible distillation trajectory at parts from column ends to points S_r and S_s determines the function $L_{rev} = f(1/T)$ for these parts and then determines such optimal values opt T_{con}^{int} and opt T_{reb}^{int}, at which summary cost of inputs and outputs energy is minimum.

Such an approach was introduced in the work (Terranova & Westerberg, 1989; Dhole & Linnhoff, 1993) and was named "pinch method."

If it is accepted that the price for input heat is proportional to the value $(1/T_0 - 1/T_{reb}^{int})$ and the price for output heat is proportional to the values $(1/T_{con}^{int} - 1/T_0)$, where T_0 is the ambient temperature, and amount of input or output heat is proportional to liquid flow, then the cost of energy consumption in the main and intermediate reboilers will turn out to be proportional to the hatched area in the bottom section in Fig. 6.2. In the main and intermediate condensers, it is proportional to the hatched area in the top section. Under the assumptions mentioned, the values opt T_{con}^{int} and opt T_{reb}^{int} will correspond to the minimum of these areas.

Minimum values of the parameters $(L/V)_r$ and $(V/L)_s$ in the feed cross-section of the column and compositions at trays above and below this cross-section x_{f-1} and x_f at adiabatic and nonadiabatic distillation remain the same. The stationary points S_r and S_s also coincide, but at parts of reversible distillation trajectories between column ends and stationary points S_r and S_s the additional stationary points N_r^{int} and N_s^{int}, corresponding to the points of intermediate inputs and output of heat (Fig. 6.2a), appear.

Therefore, the conceptual calculation of infinite column with intermediate input and/or output of heat consists in two stages: (1) calculation of minimum reflux mode for adiabatic column, and (2) determination of opt T_{con}^{int}, opt Q_{con}^{int}, opt T_{reb}^{int}, and opt Q_{reb}^{int} ("pinch method").

Figure 6.2 shows the results of such calculation at the example of direct separation of ideal three-component mixtures. However, this approach can also be easily used in the most general case for any kinds of mixtures, including azeotropic ones, at any component numbers and for any splits.

6.3.   Distillation Trajectories and Minimum Reflux Mode in Two-Feed Columns with Nonsharp Separation in Intermediate Sections

Columns with several inputs of feed are used in a number of different cases: (1) when flows with the same set of components but different compositions come to the unit; (2) when the raw materials are gradually warmed and are put into the column in several flows different in temperature, or when they are gradually evaporated or condensed, and after separation into liquid and vapor phases they are put into the column in several flows, different in temperature, composition, and phase state (units of petroleum refining, units of productions of ethylene and propylene); (3) when an absorbent is used for separation (units having absorbers or fractioning absorbers); and (4) when an entrainer that is put into the column of extractive distillation in a separate flow is used for separation.

Gradual heating and evaporation is used in the case of separation of mixtures with a wide interval of boiling, when heat is put in at a lower, and cold is put in at a higher temperature, compared with their input in the reboiler and condenser. This allows for a decrease of total energy consumption in separation.

Absorption is used in the case of extraction of liquid components from the gas phase, when the usage of distillation is unprofitable because of the necessity for too low temperatures in condensers.

Extractive distillation is used to increase the relative volatility of components being separated of nonideal mixtures and to separate azeotropic mixtures that cannot be separated by means of simple distillation.

Columns with several inputs of feed have one or several intermediate sections, located between these inputs of feed. To calculate minimum reflux mode of such columns at separation of mixtures with constant relative volatilities and molar flows, the Underwood method (Barnes, Hanson, & King, 1972; Nikolaides & Malone, 1987) was used.

For nonideal three-component mixtures, the methods of calculation of minimum reflux mode was developed in the works (Glanz & Stichlmair, 1997; Levy & Doherty, 1986). The simplified method that was offered before for the columns with one feed (Stichlmair, Offers, & Potthof, 1993) was developed in the work (Glanz & Stichlmair, 1997).

It follows from general thermodynamic considerations that at one and the same product compositions the column with several feed flows of different composition should require less energy for separation than the column with one feed flow formed by mixing all the feed flows. It follows from the fact that summary entropy of all feed flows should be smaller than that of the mixed flow because the mixing of flows of different composition increases the entropy and the separation of flows decreases it. Therefore, the minimum reflux number for the column with several feed inputs should be smaller than that for the column with one mixed feed flow (i.e., it is unprofitable to mix flows before their separation).

In Chapter 5, to develop a general algorithm of calculation of minimum reflux mode for columns with one feed, we had to understand the location of reversible distillation trajectories and the structure of top and bottom section trajectory bundles.

As in that case, to develop a general algorithm of calculation of minimum reflux mode for columns with several feed inputs, we need to understand the location of reversible distillation trajectories of intermediate sections and the structure of trajectory bundles for these sections.

6.3.1. Location of Reversible Distillation Trajectories of Intermediate Sections

Locations of reversible distillation trajectories depends on position of pseudoproduct point (i.e., on compositions and on flow rates of feeds and of separation products, as is seen from Eq. [6.3]). Difference from the top and bottom sections appears, when the pseudoproduct point of the intermediate section is located outside the concentration simplex (i.e., if concentrations of some components x'_{Di} obtained from Eq. [6.3], are smaller than zero or bigger than one), which in particular takes place, if concentration of admixture components in separation products are small components (i.e., at sharp separation in the whole column). The location of reversible distillation trajectories of the intermediate sections at $x'_{Di} < 0$ or $x'_{Di} > 1$ differs in principle from location of ones for top and bottom sections, as is seen from Fig. 6.3 for ideal three-component mixture ($K_1 > K_2 > K_3$) and from Fig. 6.4 for ideal four-component mixture ($K_1 > K_2 > K_3 > K_4$).

As far as pseudoproduct point x'_D and liquid-vapor tie-line in all points of reversible distillation trajectory should lie at one straight line, pseudoproduct point x'_D at Fig. 6.3, can lie behind side 2-3 or side 1-2 and at Fig. 6.4, they can lie behind face 1-2-3 or face 2-3-4.

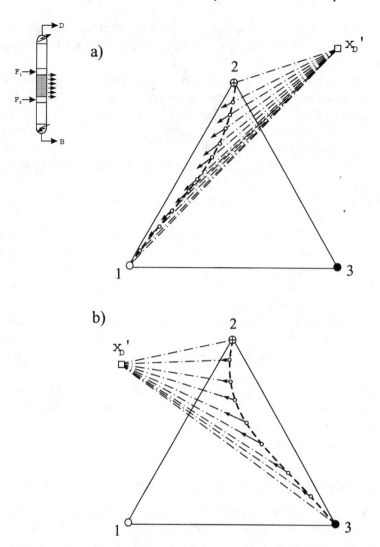

Figure 6.3. Reversible distillation trajectories of ideal ternary mixtures $(K_1 > K_2 > K_3)$ for intermediate section of two-feed column: (a) $x'_{D,1} < 0$; (b) $x'_{D,3} < 0$. Solid lines with arrows, tie-lines liquid–vapor; $x'_{D,1}$ and $x'_{D,3}$, concentrations of components 1 and 3, respectively, in pseudoproduct point.

Reversible distillation trajectories at Fig. 6.3 should connect vertexes 1 and 2 or 2 and 3, and at Fig. 6.4, they should connect vertexes 2 and 4 or 1 and 3.

In this section, we examine only the nonsharp distillation in an intermediate section, when all flows of the feed contain all the components (in the sections to follow, we examine sharp extractive distillation in intermediate section, when entrainer and main feed have different sets of components).

At nonsharp distillation in the intermediate section, as in top and bottom sections, there is only one reversible distillation trajectory, but in the intermediate section it has two node points N_{rev} in vertexes of concentration simplex, and in the top and bottom sections it has one node point in one vertex.

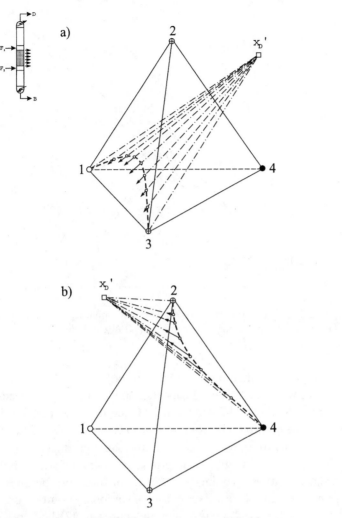

Figure 6.4. Reversible distillation trajectories of ideal four-component mixtures ($K_1 > K_2 > K_3 > K_4$) for intermediate section of two-feed column: (a) $x'_{D,1} < 0$; (b) $x'_{D,4} < 0$. Solid lines with arrows, tie lines liquid–vapor.

6.3.2. The Structure of Trajectory Bundles of Intermediate Sections

We examine the structure of trajectory bundles of intermediate sections (i.e., location and character of the stationary points of these bundles).

In Chapter 5, we saw that the distillation process in a column section is feasible only if there are reversible distillation trajectories inside concentration simplex and/or at several of its boundary elements, because only in this case a section trajectory bundle with stationary points lying at these trajectories of reversible distillation arises in concentration simplex. This condition of feasibility of the process in the section has general nature and refers not only to the top and the bottom, but also to intermediate sections. Therefore, pseudoproduct points x'_D can be located only the way it is shown in Figs. 6.3 and 6.4, (it is result of direction of

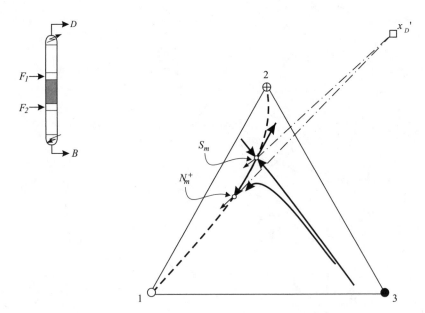

Figure 6.5. Intermediate section trajectories of ideal ternary mixtures ($K_1 > K_2 > K_3$) for two-feed column. x'_D, pseudoproduct point, S_m and N_m^+, stationary points of intermediate section regions $\text{Reg}_{\text{int}}^R$.

tie-lines liquid-vapor) and the points of top x_D or bottom product x_B, along with that, in accordance with Eq. (6.3) can be located only in the vicinity of sides 1-2 or 2-3 in Fig. 6.3 or facets 1-2-3 or 2-3-4 in Fig. 6.4. Hence, it follows that feasible splits for columns with one or two feeds are the same (i.e., if the flows of several feeds are mixed before separation, we can only get the same products as in a column with several feeds, but the energy consumption for separation will be bigger).

Figure 6.5 shows stationary points S_m and N_m^+ of trajectory bundles of intermediate section $\text{Reg}_{\text{int}}^R$ and separatrixes of saddle stationary point S_m obtained by means of calculation for ideal mixture pentane(1)-hexane(2)-heptane(3) at the composition of pseudoproduct $x'_{D,1} = -1.0$; $x'_{D,2} = 1.5$; $x'_{D,3} = 0.5$ and at the value of $L/V = 1.2$.

In contrast to nonsharp separation in the top and bottom sections, the intermediate section has at reversible distillation trajectory not just one node stationary point, but there are saddle point S_m and node point N_m. Separatrixes of the saddle points S_m divide concentration triangle into four regions $\text{Reg}_{\text{int}}^R$ filled trajectory bundles of intermediate section, one of which is the working one $\text{Reg}_{w,\text{int}}^R$.

6.3.3. Control Feed at Minimum Reflux Mode

The trajectory of the intermediate section in a three-section column connects the trajectories of the top and bottom sections, and should join them in the cross-sections of the top and the bottom feeds (i.e., at trays above and below these cross-sections, the material balance should be valid). In the mode of minimum reflux, only two of three sections adjacent to one of the feeds x_{F1} or x_{F2}, called "control" one, should be infinite. One of the sections (the top or bottom one) remains finite in this mode. At the increase of reflux number (i.e., of the parameter $[L/V]_r$) the stationary points of three-section trajectory bundles move along reversible distillation trajectories for set points of products x_D and x_B and of pseudoproduct x'_D. The general regularities of this movement for top and bottom sections were examined in detail in Chapter 5. The stationary points of intermediate section move along the unique trajectory of reversible distillation. At some value of $(L/V)_r = (L/V)^1_r$, there is joining of intermediate section trajectory with the trajectory of the first of the rest of the sections. At some greater value of $(L/V)_r = (L/V)^2_r$, there is joining of the intermediate section trajectory with the remaining section at infinite number of trays in these sections ($N_m = \infty$ and $N_r = \infty$ or $N_s = \infty$). The found values of $(L/V)^2_r$ are minimum for the separation in a three-section column. The feed located between the intermediate and the last section, with which there was a joining, is the *control* one. Along with that, the first section, with which there was a joining, is finite and its joining at $(L/V)^2_r$ with the intermediate section goes on as at a reflux bigger than minimum (regularities of joining at a reflux bigger than minimum are examined in Chapter 7).

Because in the mode of minimum reflux the intermediate section should be infinite, its trajectory should pass though one of its stationary points S_m or N^+_m. Therefore, the following cases are feasible in minimum reflux mode: (1) point N^+_m coincides with the composition at the tray above or below the cross-section of control feed; (2) composition point at the trays of the intermediate section in the cross-section of control feed lies on the separatrix line, surface, or hypersurface of point S_m (i.e., in separatrix min-reflux region of intermediate section $\text{Reg}^{min,R}_{sep,int}$ filled of trajectory bundle $S_m - N^+_m$). In both cases, composition point at the tray of the top or bottom section, adjacent to the control feed, should lie in the separatrix min-reflux region of this section $\text{Reg}^{min,R}_{sep}$ ($S^2 - N^+$).

6.3.4. General Algorithm of Calculation of Minimum Reflux Mode

This develops the general algorithm of calculation of minimum reflux mode for the columns with two feed inputs at distillation of nonideal zeotropic and azeotropic mixtures with any number of components. The same way as for the columns with one feed, the coordinates of stationary points of three-section trajectory bundles are defined at the beginning at different values of the parameter $(L/V)_r$. Besides that, for the intermediate section *proper values of the system of distillation differential equations* are determined for both stationary points from the values of phase equilibrium coefficients. From these proper values, one finds which of the stationary points is the saddle one S_m, and states the direction of *proper vectors* for the saddle point. The directions of the proper vectors obtain linear equations describing linearized boundary elements of the working trajectory bundle of the intermediate section. We note that, for sharp separation in the top and bottom sections, there is no necessity to determine the proper vectors of stationary points in order to obtain linear equations describing boundary elements of their trajectory bundles, because to obtain these linear equations it is sufficient to have

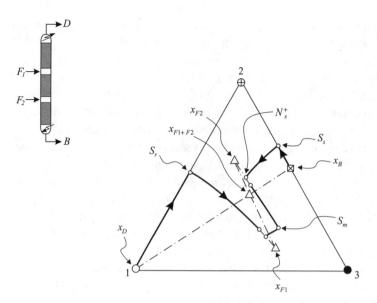

Figure 6.6. Section trajectories of ideal ternary mixtures ($K_1 > K_2 > K_3$) for two-feed column with direct split $1 : 2,3$.

coordinates of the stationary points that are located inside and at the boundary elements of the concentration simplex. Such necessity exists for the intermediate section because, at nonsharp separation, part of the stationary points located outside the concentration simplex and their coordinates cannot be determined.

The rest of the algorithm is similar to that for columns with one feed. Conditions of joining of trajectories in cross-sections of both feeds are checked at various values of the parameter $(L/V)_r$ and the value $(L/V)_r^2$, at which there is a joining in the cross-section of control feed, is found. Both feasible cases of joining described above (see Subsection 6.3.3) are checked. The first case corresponds to direct or indirect split in the column with one feed, and the second case corresponds to intermediate split.

It was shown in the work (Glanz & Stichlmair, 1997) that in some cases expenses on separation are smaller if feed with a higher bubble temperature is brought into the higher feed cross-section of the column. Figure 6.6 features such a case. In this figure part of the trajectory of the intermediate section is directed into the side, opposite the part of the top section trajectory (control feed is bottom, the trajectory passes through point S_m). At this part of the trajectory of the intermediate section, there is an increase of temperature at the trays of the column in the upward direction, which is indicative of the process inverse to distillation (see Chapter 2). The trajectory in Fig. 6.6 may be briefly described as follows:

$$x_D^{(1)} \rightarrow S_r^{(2)} \rightarrow x_{f-1,1}^{(3)} \Downarrow \Rightarrow \quad x_{f,1}^{(3)} \leftrightarrow S_m^{(3)} \leftrightarrow x_{f-1,2}^{(3)} \Downarrow \Rightarrow x_{f,2}^{(3)} \leftarrow S_s^{(2)} \leftarrow x_B^{(2)}$$

$$\text{Reg}_D \quad \text{Reg}_r^t \quad \text{Reg}_{att}^t \quad \text{Reg}_{sep,int}^{min,R} \quad \text{Reg}_{sep,int}^{min,R} \quad \text{Reg}_{sep,int}^{min,R} \quad N_3^+ \quad \text{Reg}_s^t \quad \text{Reg}_B$$

Therefore, the conceptual design calculation of columns with several feeds includes the determination of the best succession of bringing in these feeds along the height of the column, which requires the calculation of the minimum reflux mode at different successions. For the column with two feeds, one has to begin the calculations with the regular succession (that means bringing in the feed with lower bubble temperature into the higher cross-section of the column). If it turns out that the energy consumption at separation is smaller than in the column with one mixed feed, then one can leave the inverse succession unexamined. Otherwise, one has to carry out the calculations for the inverse succession.

6.4.   Trajectories of Intermediate Sections of Extractive Distillation Columns

Columns of extractive distillation represent an important particular case of columns with two feeds. The peculiarity·consists of the fact that the entrainer is their top feed and the mixture being separated is their bottom feed. The entrainer can consist of one or more components, included or not included in the mixture under separation. In the first case, the separation process is called autoextractive distillation; in the second case, it is called just extractive distillation.

Columns of extractive and autoextractive distillation are widely used in industry (Benedict & Rubin, 1945; Drew, 1979; Happe, Cornell, & Eastman, 1946; Hoffman, 1964; Kogan, 1971). The separation process of binary mixtures with azeotrope, having minimum bubble temperature, with the help of heavy entrainer brought into the column higher than the main feed was discovered empirically and started to be used in the 1940s in connection with military needs – in particular, for extraction of butadiene and toluene.

The feasibility of azeotropic mixture separation and bringing in light entrainer in vapor phase into the cross-section lower than the main feed was shown along with regular extractive distillation (Kiva et al., 1983).

Theoretical analysis of the separation of azeotropic mixtures with the help of extractive distillation was carried out in the works (Levy & Doherty, 1986; Knight & Doherty, 1989; Knapp & Doherty, 1990; Knapp & Doherty, 1992; Wahnschafft & Westerberg, 1993; Knapp & Doherty, 1994; Wahnschafft, Kohler, & Westerberg, 1994; Bauer & Stichlmair, 1995; Rooks, Malone, & Doherty, 1996; Stichlmair & Fair, 1998; Doherty & Malone, 2001). Characteristic peculiarities of the process of extractive distillation of binary azeotropic mixtures were investigated in these works. More general conception of the processes of extractive and autoextractive distillation on the basis of the theory of intermediate section trajectory tear-off from boundary elements of concentration simplex was introduced in the works (Petlyuk, 1984; Petlyuk & Danilov, 1999). Trajectory bundles of intermediate section for multicomponent mixtures were examined in the latter work.

6.4.1.   Sharp Extractive Distillation of Three-Component Mixtures

In contrast to columns with two feeds examined in the previous section for which composition and amount of each feed flow were set, only the main feed (i.e., the mixture under separation F) is set in columns of extractive distillation. The amount and composition of the entrainer E can be chosen so the separation of main feed is best conducted.

This allows us to actively influence the location of the pseudoproduct point x'_D of the intermediate section in order to maintain sharp separation (i.e., separation at which the intermediate section trajectory ends at some boundary element of the concentration simplex). This is feasible in the case when inside concentration simplex there is one trajectory of reversible distillation for pseudoproduct point x'_D that ends at mentioned boundary element, and there is the second trajectory inside this boundary element. To maintain these conditions, pseudoproduct point x'_D of the intermediate section should be located at the continuation of the mentioned boundary element, because only in this case can liquid–vapor tie–lines in points of reversible distillation trajectory located in this boundary element lie at the lines passing through the pseudoproduct point x'_D. We discuss these conditions in Chapter 4. It was shown that in reversible distillation trajectory tear-off point $x^t_{rev,e}$ from the boundary element the component absent in it should be intermediate at the value of the phase equilibrium coefficient between the components of the top product and of the entrainer ($K^t_{rev,D} > K^t_{rev,j} > K^t_{rev,E}$). This condition is the structural condition of reversible distillation trajectory tear-off for the intermediate section. Mode condition of tear-off as for other kinds of sections consists of the fact that in tear-off point the value of the parameter $(L/V)_m$ should be equal to the value of phase equilibrium coefficient of the component absent at the boundary element in tear-off point of reversible distillation trajectory ($(L/V)_m = K_j(x^t_{rev})$).

At sharp adiabatic distillation in the intermediate section, several components may be absent at the pseudoproduct boundary element at which section trajectory ends, and the structural conditions of reversible distillation trajectory tear-off into all the adjacent boundary elements having dimensionality bigger by one should be valid in the trajectory tear-off point x^t_e from the boundary element ($K^t_{rev,D} > K^t_{rev,j} > K^t_{rev,E}$) for all the components of top product i^D, entrainer i^E, and the components, absent at the pseudoproduct boundary element, j. This condition is similar to the corresponding condition for the top and bottom sections. It should be valid not only in trajectory tear-off point, but in all its points (i.e., intermediate section trajectory can be located only in the region, where $K_{Di} > K_{DEj} > K_{Ei}$ [sharp split region of the intermediate section $\text{Reg}^{i:j(E)}_{sh,e}$]).

The condition $(L/V)_m > K_j(x^t_e)$ (mode condition of section trajectory tear-off) should be valid in trajectory tear-off point x^t_e instead of the condition for reversible distillation $(L/V)_m = K_j(x^t_{rev})$. Therefore, the minimum value of the parameter $(L/V)_m$ at which intermediate section trajectory bundle can arise is $(L/V)^{min}_m > \max_j K_j(x^t_e)$.

Intermediate section trajectory tear-off point x^t_e should lie on the reversible distillation trajectory in the boundary element $\text{Reg}^{i:j(E)}_{sh,e}$ ($x^t_e \in \text{Reg}^{i:j(E)}_{sh,e}$) farther from pseudoproduct point x'_D than all tear-off points $x^t_{rev,e}$ from it of reversible distillation trajectories into adjacent boundary elements.

These conditions are also similar to the corresponding conditions for the top and bottom sections.

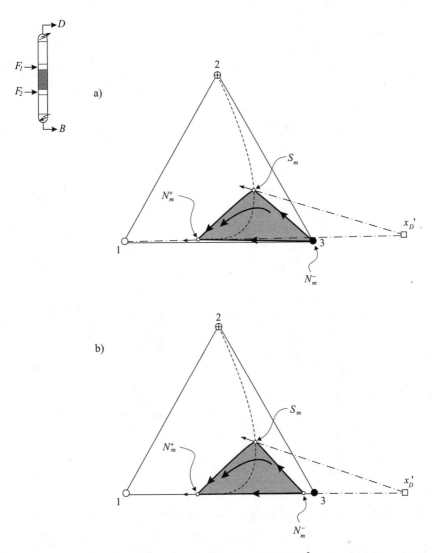

Figure 6.7. A region (bundle) of intermediate section $\text{Reg}^R_{w,int}\,^2$ (shaded) of ideal ternary mixtures $(K_1 > K_2 > K_3)$ for two-feed column^{1,3}: (a) quasisharp split $(x'_{D,2} > 0)$; (b) sharp split $(x'_{D,2} = 0)$.

Figure 6.7 shows for comparison the trajectories of quasisharp (a) and sharp (b) reversible distillation in the intermediate section for ideal mixture $(K_1 > K_2 > K_3)$. This figure shows that, at movement of pseudoproduct point from the vicinity of continuation of side 1-3 to this continuation itself, there is transformation of reversible distillation trajectory: it disintegrates into two parts – into one that lies inside the triangle and into the part that lies at side 1-3. We note that similar transformation also takes place at passage from quasisharp distillation to sharp one for the top and bottom sections. The stationary point S_m at this transformation stays at the internal part of reversible distillation trajectory. Point N^+_m passes to side 1-3. Point N^-_m passes from vertex 3 to its vicinity at side 1-3.

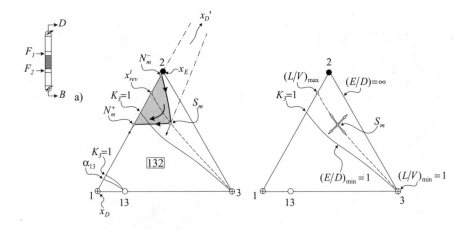

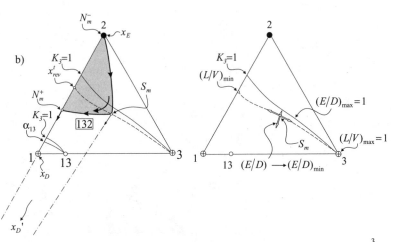

Figure 6.8. Evolution of the region (bundle) of intermediate (extractive) section $\text{Reg}^R_{w,e}{}^{3}$ (shaded) located in component order region $\text{Reg}^{1,3,2}_{ord}$ ($\text{Reg}^R_{w,e}{}^{3}{}^{1,3,2} \in \text{Reg}^{1,2}_{ord}$) with a variation of E/D and L/V for the acetone(1)-water(2)-methanol(3)^{1,2} azeotropic mixture: (a) $K_3 > 1$ at point S_m; (b) $K_3 < 1$ at point S_m. Short solid lines with arrows, tie lines liquid–vapor; double arrows, a movement of point S_m with a variation of E/D and L/V; component 1, top product; component 2, entrainer.

Therefore, working trajectory region of the intermediate section $\text{Reg}^R_{w,\text{int}}{}^{2}{}_{1,3}$ contains the separatrixes $N^-_m - S_m$ and $S_m - N^+_m$.

We now use general regularities of the location of the intermediate section trajectories to analyze the evolution of trajectory bundle $\text{Reg}^R_{w,e}{}^{3}{}^{1,2}$ for azeotropic mixtures on the example of a mixture acetone(1)-water(2)-methanol(3) (the separated mixture, 1,3 with azeotrope 13; entrainer, 2) (Fig. 6.8).

We note that this mixture cannot be separated into the components in a column with one feed. It is interesting for sharp extractive distillation to examine the evolution of working trajectory bundle of intermediate section at change of two parameter $(L/V)_m$ and E/D (for the top and bottom sections we examined in Chapter 5, the evolution of trajectory bundles at change of unique parameter $[L/V]_r$). In accordance with the structural condition, the segment α_{13}-2, where $K_1 > K_3 > K_2$ is the intermediate section trajectory tear-off segment Reg_e^t at side 1-2. As for the top and bottom sections, working trajectory bundle of the intermediate section should be located in a sharp distillation region where the order of the components is the same as at the trajectory tear-off region i.e. segment $\underset{1,2}{\overset{3}{\text{Reg}}}_e^t$ (in our case, it is region $\overset{1,3,2}{\text{Reg}_{ord}}$). The line $K_3 = 1$ divides region *132* into two subregions, where $K_3 > 1$ and $K_3 < 1$. Location of pseudoproduct point x_D' higher or lower than the concentration triangle and corresponding to that location of reversible distillation trajectory in the subregion $K_3 > 1$ or $K_3 < 1$ depend on the value of the parameter E/D: the first case takes place at the big values of E/D $(1 < E/D)$ the second case takes place at the small values E/D $(ED < 1)$. It follows from Eq. 6.3, where $D_1 = 0$ and $F_1 = E$. We examine the evolution of trajectory bundle for the first case $(E/D > 1)$. At $E/D \to 1$, point D' goes away into infinity and reversible distillation trajectory coincides with line $K_3 = 1$. At $E/D = \infty$, point x_D' coincides with vertex 2 and reversible distillation trajectory coincides with side 2-3 (see Eq. 6.3). The parameter $(L/V)_m$ for the case under consideration is bigger than one because $(L/V)_m = K_3$ in the points of reversible distillation trajectory (3 - absent component in the pseudoproduct). At $(L/V)_m = 1$, point S_m coincides with vertex 3 and, at $(L/V)_m = (L/V)_m^{max}$, point S_m coincides with reversible distillation trajectory tear-off point from side 1-2 (point $x_{rev,e}^t$). Therefore, the square, filled up with intermediate section trajectory bundle (region $\text{Reg}_{W,e}^R$), is maximum at $(L/V)_m = 1$ and is equal to zero at $(L/V)_m = (L/V)_m^{max}$.

We now examine the evolution of the trajectory bundle for the second case $(E/D < 1)$. At $E/D \to 1$, point x_D' goes into infinity and reversible distillation trajectory coincides with the line $K_3 = 1$. At decrease of the parameter E/D, point x_D' comes nearer to vertex 1, inside concentration triangle besides with reversible distillation trajectory, passing through vertex 3 and ending at side 1-2, there is fictitious trajectory, passing through points 3 and 13. At decrease of the parameter E/D, these two trajectories are brought together and, at $E/D = (E/D)_{min}$, they intersect each other in point of branching of reversible distillation trajectories x_{rev}^{branch}. At the smaller values of the parameter (E/D), sharp distillation in the intermediate section becomes unfeasible.

The parameter $(L/V)_m$ for the case under consideration is smaller than one. At $(L/V)_m = 1$, point S_m coincides with vertex 3 and, at $(L/V)_m = (L/V)_m^{min}$, it coincides with reversible distillation trajectory tear-off point $x_{rev,e}^t$ from side 1-2.

Therefore, the square, filled up with the trajectory bundle of the intermediate section, is maximum at $(L/V)_m = 1$ and is equal to zero at $(L/V)_m = (L/V)_m^{min}$.

We note that, for the top and bottom sections, the square filled up with the trajectory bundle is also maximum at $(L/V)_r = 1$ and $(L/V)_s = 1$ (the mode of infinite reflux).

6.4.2. Sharp Extractive Distillation of Four- and Multicomponent Mixtures

We now examine the structure of intermediate section trajectory bundle for four-component mixtures at the example of ideal mixture ($K_1 > K_2 > K_3 > K_4$).

In accordance with the structural conditions of trajectory tear-off, three following variants of extractive distillation are feasible for such a mixture: (1) the top product is component 1, the entrainer is component 4, and the pseudoproduct point x'_D lies in the continuation of edge 1-4 (Fig. 6.9a); (2) the top product is mixture 1,2, the entrainer is component 4, and the pseudoproduct point x'_D lies in continuation of face 1-2-4 (Fig. 6.9b); and (3) the top product is component 1, the entrainer is mixture 3,4, and the pseudoproduct point x'_D lies in the continuation of face 1-3-4 (Fig. 6.9c).

The working trajectory region $\text{Reg}^R_{w,e}$ at Fig. 6.9a is three dimensional. It is limited by the saddle points S^1_m and S^2_m at reversible distillation trajectories in faces 1-3-4 and 1-2-4 and by node points N^-_m and N^+_m at edge 1-4.

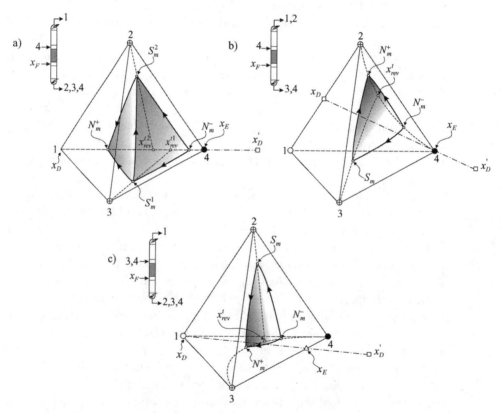

Figure 6.9. Regions extractive section $\text{Reg}^R_{w,e}$ (shaded) for the extractive distillation of an ideal four-component mixtures ($K_1 > K_2 > K_3 > K_4$): (a) $\text{Reg}^{R\,2,3}_{w,e\,3}$ (component 1, top product; component 4, entrainer); (b) $\text{Reg}^{R\,1,4}_{w,e\,1,2,4}$ (mixture 1,2, top product; component 4, entrainer); (c) $\text{Reg}^{R\,2}_{w,e\,1,3,4}$ (component 1, top product; mixture 3,4, entrainer).

The working trajectory regions $\overset{3}{\text{Reg}}_{w,e}^{R}$ at Fig. 6.9b and $\overset{2}{\text{Reg}}_{w,e}^{R}$ at Fig. 6.9c are two dimensional. They are limited by the saddle point S_m at the $\overset{1,2,4}{}$ reversible distillation trajectory inside concentration tetrahedron and by the node points N_m^- and N_m^+ at reversible distillation trajectories in face 1-2-4 (at Fig. 6.9b) or in face 1-3-4 (at Fig. 6.9c). We note that the location of reversible distillation trajectories and of the stationary points in these faces is similar to their location at nonsharp distillation in the intermediate section of corresponding three-component mixtures (Fig. 6.5) because the location of the pseudoproduct point referring these faces is similar.

The analysis carried out above (see Figs. 6.8 and 6.9) allows for a general conclusion about dimensionality, structure, and evolution of trajectory bundles (regions) of intermediate section at sharp extractive distillation for any multicomponent mixtures.

The dimensionality of intermediate section trajectory bundle is equal to $n - m + 1$, where n is total number of components, and m is summary number of components of top product and entrainer. Pseudoproduct point x'_D is located at the continuation of the boundary element, formed by all the components of the top product and entrainer. Reversible distillation trajectories and the stationary points are located at the mentioned pseudoproduct boundary element and at all boundary elements whose dimensionality is bigger by one (at $m = n - 1$, they are located inside concentration simplex).

In this section, we discuss only regular extractive distillation, when heavy entrainer is brought in higher than cross-section of input of the separated mixture. Similar analysis can also be done for indirect extractive distillation, when light entrainer is brought in lower than cross-section of input of the separated mixture.

6.5. Conditions of Separability in Extractive Distillation Columns and Minimum Reflux Mode

6.5.1. Conditions of Separability in Extractive Distillation Columns

Let's examine the conditions of sharp separation in each of the three sections of the extractive distillation column: (1) intermediate section trajectory tear-off region $\overset{j}{\underset{i=D+E}{\text{Reg}}_{e}^{t}}$ (region, in points of which $K_{i,D}^{t} > K_{j,DE}^{t} > K_{i,E}^{t}$) should exist at the boundary element of the concentration simplex of the mixture; (2) the top product region $\overset{j=E}{\underset{i=D}{\text{Reg}}_{D}}$ should exist at the boundary element of the concentration simplex of the mixture of top product and entrainer $(D + E)$; (3) the bottom product region $\overset{j=D}{\underset{i=B}{\text{Reg}}_{B}}$ should exist at the boundary element of concentration simplex of the mixture $(F + E)$; and (4) sharp split region of extractive section $\text{Reg}_{sh,e}^{i:j}$ should intersect sharp split regions of top section $\text{Reg}_{sh,r}^{D:E}$ and of bottom section $\text{Reg}_{sh,s}^{i:j}$.

Conditions $1 \div 3$ are necessary for the fact that at big enough values of parameters $(L/V)_r, (L/V)_m$, and $(L/V)_s$, in each of the three sections, trajectory bundles of these sections arise. Condition 4 maintains the potential feasibility of joining of these trajectory bundles. Therefore, the conditions $(1 \div 4)$ are necessary ones of mixture separability in a three-section column. These conditions not valid; sharp separation with obtaining set products is unfeasible.

6.5.2. Three-Component Mixtures

For the mixture acetone(1)-water(2)-methanol(3) (the entrainer is water, see Fig. 6.8) these four conditions are valid: (1) at side 1-2 there is intermediate section trajectory tear-off region Reg_e^t, where $K_1 > K_3 > K_2$ (the segment $[\alpha_{13}, 2]$); (2) vertex 1 is the unstable node N_r^- at side 1-2 (region $\mathrm{Reg}_{sh,r}^{1:2}$) (i.e., it can be the top product point $x_D = N_r^- \equiv \mathrm{Reg}_D$ at separation of the mixture 1,2 $[K_1 > K_2]$); (3) the whole side 2-3 is bottom product segment Reg_B for the mixture 1,2,3 ($K_1 > K_3 > K_2$); and (4) intermediate section sharp split region $\mathrm{Reg}_{sh,e}^{1,2:3}$ (region $K_1 > K_3 > K_2$) coincides with the bottom section sharp split region $\mathrm{Reg}_{sh,s}^{1:3,2}$ and intersects the top section sharp split region $\mathrm{Reg}_{sh,r}^{1:2}$ at side 1-2 (the whole side 1-2).

We now examine the conditions of joining of sections trajectories at a set flow rate of entrainer (i.e., at set value of the parameter E/D) for a three-component mixture in the mode of minimum reflux. Each of two feeds can be the control one, and the intermediate section trajectory in the mode of minimum reflux in both cases should pass through the saddle point S_m because this trajectory passes through the node point N_m^+ not only in the mode of minimum reflux, but also at reflux bigger than minimum (point N_m^+ arises at the boundary element of the concentration simplex because the extractive distillation under consideration is sharp).

Therefore, point x_{f-1} (composition at the tray of the intermediate section that is higher than the feed cross-section) in the mode of minimum reflux should lie on the separatrix $N_m^- - S_m$ (Fig. 6.10). If it turns out that point x_{f-1} is outside the working trajectory bundle of the intermediate section $\mathrm{Reg}_{w,e}^R$ $[(L/V)_m < (L/V)_m^{\min}]$ then, as this figure shows, the trajectory of the intermediate sections deviates into the side opposite to point N_m^+. At a reflux bigger than minimum $[(L/V)_m > (L/V)_m^{\min}]$, point x_{f-1} gets inside the working trajectory bundle of the intermediate section and the trajectory of this section does not pass through point S_m. Figure 6.10 shows the case when the bottom feed is the control one. In this case, point x_f (composition at the tray of the bottom section that is lower than the feed cross-section) should coincide with point N_s^+. Along with that, the condition of material balance should be valid in the feed cross-section:

$$L_m x_{f-1} + L_F x_F = L_s x_f, \quad \text{where} \tag{6.4}$$

$$L_m + L_F = L_s \tag{6.5}$$

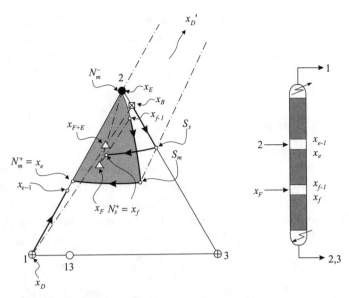

Figure 6.10. Joining of the stripping, intermediate, and rectifying section trajectories of extractive distillation of the acetone(1)-water(2)-methanol(3) azeotropic mixture at minimum reflux (bottom feed is the control one); x_{F+E}, total composition of inital feed F and entrainer E, region extractive section $\text{Reg}^{R}_{w,e}$ (shaded).

Therefore, the joining of trajectories of the bottom and the intermediate sections in the mode of minimum reflux for the case when the bottom feed is the control one is similar to that of section trajectories of two-section column at direct split (see Section 5.6). In this mode, zones of constant concentration arise in the bottom and in the intermediate sections. The column trajectory may be put in brief as follows:

$$
\underset{\text{Reg}_B}{x_B^{(2)}} \to \underset{\text{Reg}_s^t}{S_s^{(2)}} \to \underset{N_s^+}{x_f^{(3)}} \Leftarrow\Downarrow \underset{\text{Reg}_{sep,e}^{min,R}}{x_{f-1}^{(3)}} \to \underset{\text{Reg}_{sep,e}^{min,R}}{S_m^{(3)}} \to \underset{\text{Reg}_e^t}{x_e^{(2)}} \Leftarrow\Downarrow \underset{\text{Reg}_{att}^1}{x_{e-1}^{(2)}} \to \underset{\text{Reg}_D}{x_D^{(1)}}.
$$

If the top feed is the control one, two variants are feasible in the mode of minimum reflux. For the first of them, point x_{e-1} (composition at the tray higher than input of entrainer) should coincide with point N_r^+ and point x_e (composition at the tray of the intermediate section lower than input of entrainer) as at any mode should coincide with point N_m^+. For this variant, a zone of constant concentration arises in the top section in the cross-section of the input of the entrainer, and a pseudozone of constant concentrations, caused by the sharpness of separation but not by the value of parameter $(L/V)_m$, arises in the intermediate section.

For the second variant of minimum reflux mode with top control feed — $(L/V)_m = K_j$; that is, not a pseudozone arises in the intermediate section, but a true zone of constant concentrations, caused by the value of the parameter $(L/V)_m$. In the top section, for this variant, no zone of constant concentrations arises in the cross-section of input of the entrainer. Along with that, the conditions of material balance should be valid in the cross-section of input of the entrainer:

$$L_r x_{e-1} + E x_E = L_m x_e, \quad \text{where} \tag{6.6}$$

$$L_r + E = L_m \tag{6.7}$$

6.5.3. The Four- and Multicomponent Mixtures

We now examine the four-component mixture for $m_m = 2$ (Fig. 6.9a). The conditions of joining of section trajectories at a set flow rate of the entrainer in the mode of minimum reflux in the cases of top or bottom control feed do not differ from the conditions for three-component mixtures discussed above. In the case of bottom control, feed point x_{f-1} should lie in the separatrix min-reflux region $\text{Reg}_{sep,e}^{min,R}$ $(N_m^- - S_m^1 - S_m^2)$ and $x_f \equiv N_s^+$. The column trajectory may be put as follows:

$$
\begin{array}{ccccccccccccccc}
x_B^{(3)} & \to & S_s^{(3)} & \to & x_f^{(4)} & \Leftarrow\Downarrow & x_{f-1}^{(4)} & \to & S_m^{2(4)} & \to & x_e^{(2)} & \Leftarrow\Downarrow & x_{e-1}^{(2)} & \to & x_D^{(1)} \\
1 & & 1 & & 2,3 & & 2,3 & & 2,3 & & 2,3 & & & & 4 \\
\text{Reg}_B & & \text{Reg}_{s}^{t} & & N_s^+ & & \text{Reg}_{sep,e}^{min,R} & & \text{Reg}_{sep,e}^{min,R} & & \text{Reg}_{e}^{t} & & \text{Reg}_{att}^{1} & & \text{Reg}_D \\
2,3,4 & & 2,3,4 & & & & 1,4 & & 1,4 & & 1,4 & & & & 1
\end{array}.
$$

With the top feed for the control one, $x_{e-1} \equiv N_r^+$ and $x_e \equiv N_m^+$. The equations of material balance (6.4) $\div$ (6.7) should be valid in both cases.

For four-component mixtures at $m_m = 3$ and at two components in the bottom product (Fig. 6.9b), the conditions of joining in the case of bottom control feed are defined by the dimensionality of trajectory bundles $N_m^- - S_m$ $(d = 1)$ and $S_s^2 - N_s^+$ $(d = 1)$ and are similar to those of joining of sections trajectories of two-section column in the mode of minimum reflux at intermediate split (see Section 5.6). Point x_{f-1} should lie on the separatrix min-reflux region $\text{Reg}_{sep,e}^{min,R}$ $(N_m^- - S_m)$ and point x_f should lie on the separatrix min-reflux region $\text{Reg}_{sep,s}^{min,R}$ $(S_s^2 - N_s^+)$.

At top control feed $x_{e-1} \equiv N_r^+$ and $x_e \equiv N_m^+$; that is, point N_r^+ should lie on the continuation of the straight line $4 - N_m^+$ (it follows from Eq. [6.6]).

It is shown below that such a joining variant of sections trajectories at arbitrary compositions of the top product and of the pseudoproduct is unfeasible. Joining is feasible if point x_{e-1} belongs to working trajectory bundle of top section $\text{Reg}_{w,r}^{R}$ $(S_r - N_r^+)$. In other words, the top feed cannot be the control one.

Finally, at $m_m = 3$ and at three components in the bottom product (Fig. 6.9c), with the bottom feed for the control one, the analysis of dimensionality of trajectory bundles of the bottom and the intermediate sections shows that at any value of the parameter $(L/V)_m$ point x_{f-1} cannot belong to the separatrix min-reflux region $\text{Reg}_{sep,e}^{min,R}(N_m^- - S_m)$ and at the same time cannot $x_f \equiv N_s^+$. Only the following variants are feasible: (1) point x_{f-1} belongs to the bundle $N_m^- - S_m - N_m^+$ and $x_f \equiv N_s^+$, or (2) point x_{f-1} belongs to the separatrix min-reflux region $\text{Reg}_{sep,e}^{min,R}$ $(N_m^- - S_m)$ and point x_f belongs to the separatrix sharp split region $\text{Reg}_{sep,s}^{sh,R}$ $(S_s - N_s^+)$. Therefore, in the case under consideration, the conditions of joining of trajectories of bottom and intermediate section are similar to the condition of joining of section trajectories of two-section column in the mode of minimum reflux at split with one distributed component (see Section 5.6).

In the case of top control feed $x_e \equiv N_m^+$, and point x_{e-1} should lie on the separatrix $S_r - N_r^+$ in face 1-3-4 ($\text{Reg}_{sep,r}^{sh,R}$).

We now examine the general case of separation of a multicomponent mixture by means of sharp extractive distillation in a column with two feeds at a set flow rate of entrainer in the mode of minimum reflux. The conditions of sections joining are similar to the conditions of sections joining of the two-section column and depend on the number of components in the product or in the pseudoproduct of each section (m_r, m_m, and m_s) (i.e., on the dimensionality of the working and separatrix bundles of the sections).

Trajectory bundles of bottom and intermediate sections in the mode of minimum reflux should join with each other in the concentration space of dimensionality $(n-1)$. Therefore, joining is feasible at some value of the parameter $(L/V)_m^{min}$ if the summary dimensionality of these bundles is equal to $(n-2)$.

The dimensionality of the working region of the intermediate section $\text{Reg}_{w,e}^{R}$ ($N_m^- - S_m - N_m^+$), as it is shown in Fig. 6.9, is equal to $(n - m_m + 1)$ and that of the separatrix min-reflux region $\text{Reg}_{sep,e}^{min,R}$ ($N_m^- - S_m$) is smaller by one, that is, $(n - m_m)$.

The dimensionality of the sharp split region of the bottom section $\text{Reg}_{sep,s}^{sh,R}$ ($S_s^1 - S_s^2 - N_s^+$), as it is shown in Section 5.6, is equal to $(n - m_s)$, and that of the separatrix min-reflux region $\text{Reg}_{sep,s}^{min,R}$ ($S_s^2 - N_s^+$) is smaller by one, that is, $(n - m_s - 1)$ (if $m_s = n - 1$, then this bundle degenerates into point N_s^+).

Therefore, the conditions of joining of two separatrix min-reflux bundles is as follows:

$$2n - m_s - m_m - 1 = n - 2, \tag{6.8}$$

and the condition of joining of separatrix min-reflux bundle of one of the sections and of separatrix sharp split bundle of the second section is as follows:

$$2n - m_s - m_m = n - 2 \tag{6.9}$$

Therefore, if the control feed is the bottom one and Eq. (6.8) is valid, then in the general case the joining takes place as at the intermediate split in two-section columns and, in the particular case of $m_s = n-1$, it takes place as at the direct split.

If Eq. (6.9) is valid, then the joining takes place as at the split with one distributed component in two-section columns.

Finally, if

$$2n - m_s - m_m < n - 2, \tag{6.10}$$

then the joining is similar to that in two-section columns with several distributed components (i.e., in this case the compositions of the bottom product and of the pseudoproduct should meet some limitations).

If the control feed is the top one, then the trajectory bundle of the top section should join the stationary point N_m^+ (i.e., the bundle of zero dimensionality) in the concentration space of dimensionality $m_m - 1$. Therefore, joining is feasible at some value of the parameter $(L/V)_m^{min}$ if the dimensionality of the trajectory bundle of the top section $\text{Reg}_{sep,r}^{sh,R}$ is equal to $m_m - 2$.

The dimensionality of the separatrix sharp split bundle of the top section $\text{Reg}_{sep,r}^{sh,R}$ $(S_r^1 - S_r^2 - N_r^+)$ is equal to $m_m - m_r$ and that of the separatrix min-reflux bundle $\text{Reg}_{sep,r}^{min,R}$ $(S_r^2 - N_r^+)$ is equal, correspondingly, to $m_m - m_r - 1$ (if $m_r = 1$ then $S_r^1 \equiv N_r^-$, if $m_r = m_m - 1$ then the separatrix bundle degenerates into point N_r^+).

Therefore, the condition of joining of the top section separatrix min-reflux bundle $\text{Reg}_{sep,r}^{min,R}$ with the trajectory of the intermediate section is $m_m - m_r - 1 = m_m - 2$, that is,

$$m_r = 1 \tag{6.11}$$

and the condition of joining of the top section separatrix sharp split bundle $\text{Reg}_{sep,r}^{sh,R}$ with the trajectory of the intermediate section is $m_m - m_r = m_m - 2$, that is,

$$m_r = 2 \tag{6.12}$$

We have a considerable limitation of sharp extractive distillation process in the column with two feeds: the process is feasible if the top product components number is equal to one or two. This limitation arises because, in the boundary element formed by the components of the top product and the entrainer, there is only one point, namely, point N_m^+, that belongs to the trajectory bundle of the intermediate section. If Eq. (6.11) is valid, then the joining of the trajectories of the intermediate and top sections takes place as at direct split in two-section columns in the mode of minimum reflux. If Eq. (6.12) is valid then joining goes on as at split with one distributed component.

Finally, if $m_r > 2$, then joining is similar to joining in two-section columns with several distributed components (i.e., composition of pseudoproduct should meet some limitations).

Therefore, extractive distillation at three and more components in the top product is feasible in principle, but requires a search for an allowed composition of the pseudoproduct. If it is necessary to design a sharp extractive distillation column at $m_r = 3$, then it is not allowed to set the rate of the entrainer arbitrarily, but it is necessary to determine it from the conditions of joining of trajectories of the top and intermediate sections. The parameter E/D ensures an additional degree of freedom, which allows to increase by one the number of product components in the top product m_r.

The analysis carried out before for three- and four-component mixtures completely corresponds to the above-formulated general conditions of trajectories joining for multicomponent mixtures.

The general algorithm of calculation of the minimum reflux mode for columns of extractive distillation with two feeds requires the check-up of the conditions of trajectories joining for the cases of bottom and top control feed and requires the determination of the values of $(L/V)_{m2}^{min}$ and $(L/V)_{m1}^{min}$. The bigger of these two values corresponds to the mode of minimum reflux, and the corresponding feed is the control one.

To determine the values of $(L/V)_{m1}^{\min}$ and $(L/V)_{m2}^{\min}$, the general algorithm of calculation of the minimum reflux mode for two-section columns at various splits is used.

In particular, for the most widespread split with one-component entrainer and one-component top product ($m_m = 2, m_r = 1$), the joining of intermediate section trajectories with the trajectories of the top and the bottom sections goes on the way it is at direct split in two-section columns. This uses the simplest modification of the algorithm of calculation of the minimum reflux mode.

6.6. Determination of Minimum Flow Rate of Entrainer

The second important parameter, besides the parameter $(L/V)_m^{\min}$, at designing of sharp extractive distillation columns with two feeds is the parameter $(E/D)_{\min}$. The theory of trajectory tear-off easily determines this parameter at any splits in an extractive distillation column.

We express the value of the parameter (E/D) through limit conditions when the point of tear-off of reversible distillation trajectory coincides with the saddle point of trajectory bundle of the intermediate section (e.g., $x_{rev,e}^t = S_m$ at Fig. 6.8, $x_{rev,e}^t = S_m^2$ at Fig. 6.9a or $x_{rev,e}^t = S_m$ at Fig. 6.9b,c). The conditions in the tear-off point of this trajectory establish connections between coordinates of the tear-off point $x_{rev,e}^t$ and of the pseudoproduct point x_D' the way it was done earlier for the connection between coordinates of the tear-off point and of the product point in two-section columns (see Eq. 4.20). The corresponding equation for the intermediate section looks the same way as Eq. 4.20:

$$x_{D,i}' = x_i^t(K_i^t - K_j^t)/(1 - K_j^t) \tag{6.13}$$

Here, j is the component not entering into the number of components of the top product and the entrainer.

We now express the parameter E/D through $x_{D,i}'$ with the help of material balance equation of the intermediate section (Eq. [6.3]). After transformations, we get the following:

$$(1 - E/D)x_{D,i}' = x_{D,i} - (E/D)x_{E,i} \tag{6.14}$$

If, for example, the top product is component 1 and the entrainer does not contain component 1, that is, $x_{D,1} = 1$ and $x_{E,1} = 0$, then we get the following from Eq. (6.14):

$$E/D = 1 - 1/x_{D,i}' \tag{6.15}$$

After substitution into Eq. (6.13), we get:

$$E/D = 1 - (1 - K_j^t)/x_i^t(K_i^t - K_j^t) \tag{6.16}$$

For the mixture acetone(1)-water(2)-methanol(3) at side 1-2, at which there is an intermediate section trajectory tear-off segment Reg_e^t, the dependence of phase

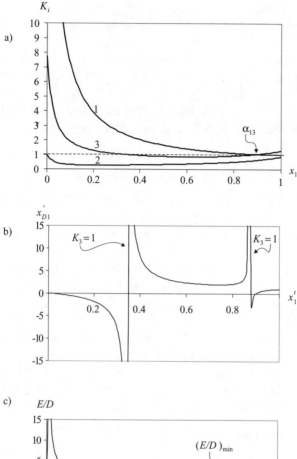

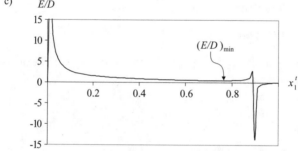

Figure 6.11. About calculation minimum entrainer flow rate $(E/D)_{min}$. K_i, $x'_{D,1}$ and E/D as functions $x^t_{e,1}$ (a,b,c, respectively) for extractive distillation of the acetone(1)-water(2)-methanol(3) azeotropic mixture. $x^t_{e,1} = x^t_{rev,e}$ concentration of component 1 in tear-off point of intermediate section reversible distillation trajectory on side 1-2; K_i, phase equilibrium coefficient of component i in point $x^t_{e,1}$, $x'_{D,1}$ and E/D, concentration of component 1 in pseudoproduct point and ratio of entrainer and overhead flow rates, respectively, if tear-off point of intermediate section trajectory $x^t_{e,1}$ on side 1–2 coincide with point $S_{m,1}$.

equilibrium coefficients of components $1 \div 3$ on the concentration of component 1 is shown in Fig. 6.11a, dependence of the concentration of component 1 in pseudo-product point x'_D on its concentration in reversible distillation trajectory tear-off point $x^t_{rev,e}$, according to Eq. (6.13), is shown in Fig. 6.11b and the dependence of the relative rate of the entrainer E/D on this concentration $x^t_{rev,e}$, according to Eq. (6.16), is shown in Fig. 6.11c.

It is seen from Fig. 6.11a that $K_3 = 1$ in points $x_1 = 0.35$ and $x_1 = 0.85$ and that $K_3 = K_1$ in point $x_1 = 0.9$. In accordance with this (see Eq. 6.13) in point $x^t_{rev,1} = 0.35$ $x'_{D,i} = \pm\infty$, in point $x^t_{rev,1} = 0.85$ $x'_{D,i} = +\infty$, and in point $x^t_{rev,1} = 0.9$ $x'_{D,i} = 0$ (Fig. 6.11b). Along with that, in the segment $[0; 0.35]$ $x'_{D,i} < 0$ (i.e., point x'_D lies on the continuation of side 1-2 behind vertex 2) in the segment $[0.35; 0.85]$ $x'_{D,i} > 1$ (i.e., point x'_D lies on the continuation of side 1-2 behind vertex 1) and in the segment $[0.9; 1.0]$ $0 < x'_{D,i} < 1$ (i.e., point x'_D lies on side 1-2) (Fig. 6.8).

According to Fig. 6.11c, the parameter $E/D \to \infty$ at $x^t_{rev,1} \to 0$ and $(E/D)_{min} = 0.5$ at $x^t_{rev,1} = 0.75$.

As it becomes evident from this example, to determine $(E/D)_{min}$ at $m_m = 2$ it is sufficient to scan the values of phase equilibrium coefficients of the components in the intermediate section trajectory tear-off segment Reg^t_e and to use Eq. (6.16).

In a more general case at $m_m > 2$ instead of the trajectory tear-off segment, there is a tear-off region, in points of which it is necessary to determine the phase equilibrium coefficients and then to determine the values of the parameter $(E/D)_{min}$. For example, for the split at Fig. 6.9c the parameter $(E/D)_{min}$ depends on composition of entrainer (i.e., on location of point x_E at side 3-4). Therefore, to determine $(E/D)_{min}$ it necessary to scan the phase equilibrium coefficients of the components in all the points of face 1-3-4. As a result, the value $(E/D)_{min}$ and the optimal composition of the entrainer x_E (i.e., optimal correlation in it of concentrations of components 3 and 4) will be found.

For the split in Fig. 6.9b, the composition of the top product x_D, as a rule, is fixed because it is conditioned by the correlation of components 1 and 2 in the mixture under separation. Therefore, to determine $(E/D)_{min}$ it is sufficient to scan the phase equilibrium coefficients of the components in tear-off points x^t_{rev} for the pseudoproduct points x'_D on continuation of line $x_D - x_E$.

6.7. Distillation Complexes with Thermal Coupling Flows

6.7.1. Kinds of Distillation Complexes with Thermal Coupling Flows

Three kinds of distillation complexes with thermal coupling flows (with branching of liquid and/or vapor flows) – columns with side stripping, columns with side rectifiers and complexes with full thermal coupling flows, called Petlyuk column – are used in industry at present.

For three-component mixtures, these kinds of complexes can be easily obtained from three splits of three-component mixture in the first in motion column of the

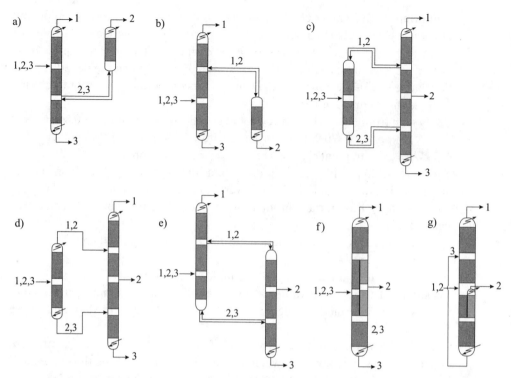

Figure 6.12. Some complex columns for ternary mixtures: (a) with side rectifying; (b) with side stripping; (c) Petlyuk column; (d) with prefractionator; (e) more operable Petlyuk column; (f) with divided wall; and (g) with divided wall for extractive distillation.

sequence: (1) of direct split (1 : 2,3); (2) of indirect split (1,2 : 3); and (3) of preferable split (1,2 : 2,3).

To transform each separation sequence into distillation complex, it is sufficient to exclude one reboiler, one condenser, or both and to replace them by liquid or vapor flows from the other column (Fig. 6.12). This figure for the sequence with preferable separation of flows in the first column shows some additional modifications (Fig. 6.12e,f).

Columns with side stripping (Fig. 6.12b) are used at refinery beginning with the first decades of the twentieth century (see Watkins, 1979).

Columns with side rectifiers (Fig. 6.12a) are used at air separation with obtaining of oxygen, nitrogen, and argon. The extractive column with side rectifier in one body (Fig. 6.12g) was offered recently (Emmrich, Gehrke, & Ranke, 2001). Such column takes the place of extractive distillation column and of entrainer regeneration column. It was shown that the application of such a column for recovery of benzene from the mixture with nonaromatic hydrocarbons compared with two columns effects energy saving of 15% ÷ 20% and capital costs saving of approximately 20%. The fraction of hydrocarbons C_6 with high content of benzene (from catalytic reformate or from hydrogenated product of benzine pyrolisis) is the feed of the extractive column and N-formylmorpholine is the entrainer.

6.7.2. Petlyuk Columns

Columns with completely thermal coupling flows were patented in different modifications by a number of authors: sequence with prefractionator (Brugma, 1942; Fig. 6.12d); column with dividing wall (Wright, 1949; Fig. 6.12f); and column with dividing wall for separation of four-component mixtures (Cahn et al., 1962).

Later, these columns were independently rediscovered (Petlyuk, Platonov, & Slavinskii, 1965; Platonov, Petlyuk, & Zhvanetskiy, 1970) on the basis of theoretical analysis of thermodynamically reversible distillation because this distillation complex by its configuration coincides with the sequence of thermodynamically reversible distillation of three-component mixture (see Chapter 4), but in contrast to this sequence it contains regular adiabatic columns. The peculiarities of Petlyuk columns for multicomponent mixtures are (1) total number of sections is $n(n-1)$ instead of $2(n-1)$ in regular separation sequences; (2) it is sufficient to have one reboiler and one condenser; (3) the lightest and the heaviest components are the key components in each two-section constituent of the complex; and (4) n components of a set purity are products.

The modifications of Petlyuk column in Fig. 6.12c,e,f are thermodynamically equivalent. The modification in Fig. 6.12e has the advantage that vapor flow from the second three-section column is selected in the top and some middle cross-section, and is directed to the first three-section column (i.e., there are no vapor flows of different directions passing from one column into the other). This allows to keep the pressure in the second column at the level slightly above that of the first one. Therefore, valves allowing to regulate splitting of vapor flows can be installed at one or both vapor flows. This modification was introduced in the works (Kaibel, 1987; Smith & Linnhoff, 1988; Agrawal & Fidkowski, 1998).

Petlyuk columns decrease energy expenditures for separation of three-component mixtures, on the average, by 30% due to their thermodynamical advantages: (1) in the preliminary column, the composition of flows in feed cross-section is close to feed composition (i.e., thermodynamic losses at mixing of flows are nearly absent); (2) these losses at the mixing of flows at the ends of the columns are nearly absent; (3) absence of reboiler or condenser at output of component 2 decreases energy expenditures due to the fact that liquid and vapor flows are used twice in the sections located above and below output of component 2; and (4) thermodynamic losses for the reason of repeated mixing of flows in the second column at regular separation sequence are absent (the concentration of component 2 at the end of the first column at direct split along distillation trajectory decreases, which requires additional expenditures of energy in the second column for obtaining pure component 2).

An interesting new application of Petlyuk columns is mentioned in the work (Agrawal, Woodward, & Modi, 1997) – in order to remove microadmixtures and obtain products of very high purity at air separation.

The modification in Fig. 6.12f (column with a dividing wall) allows, besides that, to decrease capital expenses, on the average, by 30%.

It is just lately that Petlyuk columns came into industrial use. BASF AG was the first firm since 1985 to use them in industry (25 industrial columns with a dividing wall) (Becker et al., 2001). Firm MW Kellog Limited, in collaboration with BP, applied a column with a dividing wall in 1998, at the petroleum refinery in Coriton (Great Britain) for the separation of benzines. This column was modernized from that with side withdrawal of aircraft gasoline, which increased its manufacture by more than 50%. In 1998, the firms Sumitomo Heavy Industries Co. and Kyowa Yuka installed a column with a dividing wall for obtaining various chemical products of high purity (right up to 99.999%) (Parkinson et al., 1999) and has been involved in designing at least six columns (Parkinson, 2000).

The firm Krupp Uhde (Ennenbach, Kolbe, & Ranke, 2000) started to use columns with a dividing wall in order to reduce from benzine fraction C_6 with big content of benzene (50%) for obtaining of benzine with small content of benzene and for posterior obtaining of pure benzene. Fraction C_5 is the top product, and fraction C_{7+} is the bottom product. Another sphere of application is recovery of benzene-toluene fraction from reformate or from benzine pyrolisis product. The economy of energy expenditures compared with regular sequence of two column constitutes 35% and economy of capital costs constitutes 20%. New installation was put into operation in the year 2000, and the first reconstruction of a simple column into a column with a dividing wall was realized in 1999 (replacement of the middle part of the column required only 10 days). Linde AG constructed the largest column with a dividing wall for Sasol, estimated to be 107 m tall and 5 m in diameter (Parkinson, 2000). UOP designed columns with a dividing wall for new linear alkyl benzene complex. One its application is the prefractionation of kerosene (top product being C_{10} and lighter; side(main) product, $C_{11} \div C_{13}$; bottom product, C_{14} and heavier). Column with a dividing wall yields an energy savings of 30% and a capital savings of 28%. UOP also applied a column with a dividing wall for an untypical separation: it removes C_{7+} aromatics from a desired C_{7+} olefin/paraffin mixture. In this case, the column with a dividing wall has three product flows, two feedings, and two external reflux streams. Column with a dividing wall yields an energy savings of 50% and a capital savings of 45% (Schultz et al., 2002).

Therefore, wide introduction of columns with a dividing wall into fine chemical industry, into oil-refining and petrochemistry, is occurring.

Petlyuk columns were also introduced (Petlyuk et al., 1965) for separation of mixtures with more than three products (Fig. 6.13a). Similar sequences of consecutive separation were examined in the work (Sargent & Gaminibandara, 1976). Other configurations of these columns with satellite shells of the columns were introduced in the work (Agrawal, 1996). Configurations of columns maintaining one-directed movement of vapor flow (more operable) for $n = 4$ (Fig. 6.13b) and for $n = 5$ were developed in the work (Agrawal, 1999). Columns with separation in one shell and with several vertical partitions (Fig. 6.13c) were introduced for four- and five-component mixtures (Kaibel, 1987; Christiansen, Scogestad, & Lien, 1997a; Agrawal, 1999).

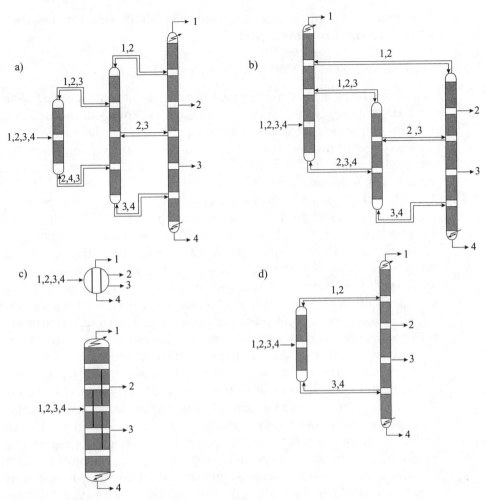

Figure 6.13. Some Petlyuk columns for four-component mixtures: (a) sequential column arrangement; (b) more operable arrangement; (c) with two divided walls; and (d) in two columns (with intermediate split in first column).

Columns with thermal coupling flows, but with sharp intermediate separation between components with average volatilities, were introduced to decrease the number of sections (Cahn et al., 1962; Petlyuk, Platonov, & Avet'an, 1966; Christiansen et al., 1997a; Kaibel, 1987) (Fig. 6.13d). Such sequences require bigger energy expenditures for separation and contain exchange sections between point of withdrawal of products with average volatilities. Flow rates of liquid and vapor in exchange sections are equal.

In Fig. 6.13a ÷ c, the number of components at the exit from each section is smaller by one than at the entrance. It was shown in the work (Christiansen, Scogestad, & Lien, 1997b), that the complete separation of four-component mixtures into pure components in such sequences requires energy expenditures bigger only by 10% ÷ 15% than the separation of two-component mixture.

6.8.   Calculation of Minimum Reflux Mode for Distillation Complexes with Thermal Coupling Flows

6.8.1.   The Columns with Side Withdrawals of Flows

Before examining minimum reflux mode for complexes with branching of flows, we discuss complex columns with side withdrawals of flows. Side products of such columns cannot be pure components at finite reflux, but the number of components in each side product can differ from the number of components in the other side products, in the initial mixture, and in the top and bottom products. In such complex columns in each section, the number of components at the exit from the section is smaller, than at the entrance. The simplest example of separation is: 1 : 1, 2 : 3 (Fig. 6.14). In this case, side product 1,2 is withdrawn above feed. Such splits are sharp. We confine oneself to examining of complex columns with sharp splits. The pseudoproduct of each intermediate section of the column with side withdrawals of products is the sum of all the products above (below) the section under consideration, if this section itself is located above (below) feed. For such splits, all the pseudoproduct points of the intermediate sections are located at the boundary elements of concentration simplex. Therefore, the structure of trajectory bundles for the intermediate sections does not differ from the structure of trajectory bundles for the top or bottom sections at sharp separation.

Figure 6.14 shows trajectories of the intermediate section for separation 1 : 1, 2 : 3 at different modes. Pseudoproduct points x'_{D1} ($D' = D_1 + D$) is located at side 1-2, and joining of the intermediate and bottom sections in the mode of minimum reflux goes on in the same way as for the simple column at indirect split. Trajectory of the intermediate section r_1 tears off from side 1-2 in point S_{r1}, and point of side product x_{D1} can coincide with point S_{r1} (Fig. 6.14a) or lie at segment $1 - S_{r1}$ (Fig. 6.14b). The first of these two modes is optimal because the best separation between top and side products (the mode of the best separation) is achieved at this mode. Zones of constant concentrations in the top and intermediate sections arise in point $S_{r1} \equiv N_{r2}^+$. Therefore, in the mode of minimum reflux in the intermediate section, there are two zones of constant concentrations. At the reflux bigger than minimum, point S_{r1} moves to vertex 2 and at $R = \infty$ this point reaches it (i.e., at $R = \infty$, pure component 2 can be obtained in the infinite column as a side product). Therefore, for the columns with side withdrawals of the products, the mode of the best separation under minimum reflux corresponds to joining of sections in points S_{r1} and N_{r1}^+ of the trajectory bundle of the intermediate section (at sharp separation) or in its vicinity (at quasisharp separation). The trajectory of the column with a side product at minimum reflux at best separation may be described as follows:

$$x_B^{(1)} \rightarrow q S_{s1}^{(2)} \rightarrow x_f^{(3)} \Leftarrow\Downarrow x_{f-1}^{(3)} \xleftarrow[x'_{D1}]{} q S_{r1}^{(2)} \Rightarrow\Uparrow N_{r2}^{+(2)} \rightarrow x_D^{(1)}$$

$$\underset{3}{\overset{1,2}{\operatorname{Reg}_B}} \quad \underset{2,3}{\overset{1}{\operatorname{Reg}_{s1}^t}} \quad \underset{2,3}{\overset{1}{\operatorname{Reg}_{sep,s}^{min,R}}} \quad \underset{1,2}{\overset{}{N_{r1}^+}} \quad \underset{1,2}{\overset{3}{\operatorname{Reg}_{r1}^t}} \quad \underset{1,2}{\overset{3}{\operatorname{Reg}_{r1}^t}} \quad \underset{1}{\overset{2}{\operatorname{Reg}_D}} .$$

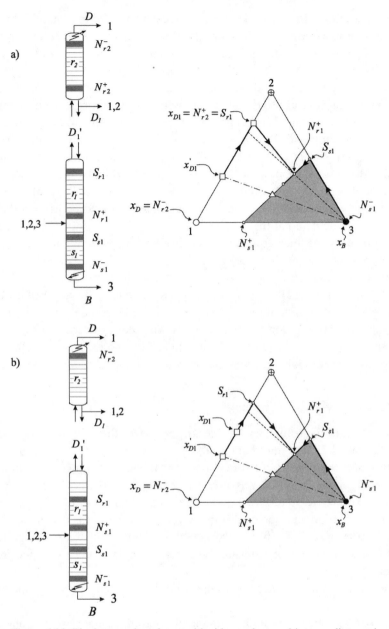

Figure 6.14. Pinch zones in column with side product and intermediate and stripping section trajectories for the ideal ternary mixture: (a) side product composition equal to tear-off point composition; and (b) side product composition unequal to tear-off point composition. Attraction region of point N_{s1}^- (Reg_{att}^3) is shaded.

We now discuss the general algorithm of calculation of minimum reflux mode for the column with several side withdrawals located above and below feed cross-section at sharp separation in each section and at the best separation between products.

Minimum reflux mode is determined by the conditions of joining of trajectories of two sections adjacent to the feed cross-section. Therefore, the interconnected parameters $(L/V)_r^{\min}$ and $(V/L)_s^{\min}$ are determined initially for these two sections. Compositions in points x'_D and x'_B are calculated preliminarily for these sections at set requirements to compositions of all the products at the conditions of sharp or quasisharp separation in each section. Minimum reflux mode is calculated in the same way as for the simple column that separates initial raw materials into products of compositions x'_D and x'_B. Liquid and vapor flow rates for the other sections are calculated at the obtained values of $(L/V)_r^{\min}$ and $(V/L)_s^{\min}$ with the help of material balance equations (strictly speaking, with the help of equations of material and thermal balance).

Trajectories of each section can be calculated at the composition of the pseudo-product of this section. Part of section trajectory between the points of its product and pseudoproduct is fictitious as is seen at Fig. 6.14 (segment $x_{D1} - x'_{D1}$).

Minimum reflux mode for the columns with side withdrawals of flows was discussed in the works (Sugie & Benjamin, 1970; Kohler, Kuen, & Blass, 1994; Rooks, Malone, & Doherty, 1996; Bausa, Watzdorf, & Marquardt, 1997).

6.8.2. The Columns with Side Strippings

We now turn to the columns with side strippings (for the columns with side recti-fiers, the calculation of minimum reflux mode is carried out the same way).

Minimum reflux mode for the columns with side strippings at separation of three-component mixtures was investigated in the works (Glinos & Malone, 1985; Fidkowski & Krolikowski, 1987), and at separation of multicomponent mixtures using Underwood equations system, this mode was investigated in the work (Carlberg & Westerberg, 1989a). Splits without distributed components in each two-section column entering distillation complex were examined in these works.

We shortly examine the general algorithm for multicomponent nonideal mix-tures.

Figure 6.15 shows the simple example of separation of a four-component mix-ture into four pure components in a column with side strippings. As for the columns with side withdrawals of the products, the calculation of minimum reflux mode should be started with determination of the conditions of joining of trajectories of two sections adjacent to feed cross-section. For section r_1, the pseudoproduct equals the sum of top and two side products. The minimum reflux mode for the first two-section column is calculated the same way it is done for the correspond-ing simple column with split $1,2,3 : 4$ (indirect split). In a more general case, when the bottom product contains more than one product component, the intermediate split will be in this column.

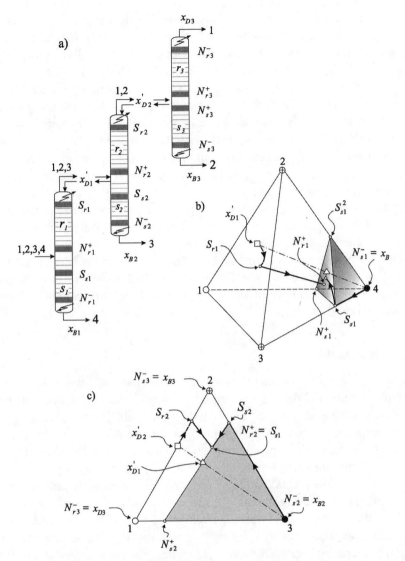

Figure 6.15. Column with two strippings as three two-section columns (a), section trajectories of the first two-section column (b), and section trajectories of the second two-section column (c). Dotted line, an imaginary part of section trajectories between pseudoproduct points and tear-off points. Attraction regions Reg_{att}^4 and Reg_{att}^3 are shaded.

Figure 6.15b shows the trajectories of sections r_1 and s_1 (part of the trajectory from point x'_{D1} to tear-off point S_{r1} is fictitious). The flow rate of vapor V_{r1} and its composition (equilibrium composition in point S_{r1}), and also the flow rate and composition of liquid (flow rate L_{r1} and composition in point S_{r1}), are becoming clear after the determination of parameter $(L/V)_{r1}^{\min}$. It is accepted here that trajectories of sections r_1, r_2, and s_2 join in point S_{r1}. At such optimum joining, the maximum concentration of component 3 in section r_1 is achieved (in point

x'_{D1}, the concentration of this component is smaller than in point S_{r1}). This avoids repeated mixing (remixing) of flows that takes place while using the sequence of the simple columns. As for the columns with side withdrawals of products, such joining of sections maintains the best separation at set expenditures of energy or the smallest expenditures of energy at set requirements to the quality of the separation products. The difference between vapor and liquid flows of section r_1 is the feed of the second two-section column consisting of sections r_2 and s_2 (in contrast to that for a simple column the liquid part of the feed is negative). Figure 6.15c shows trajectories of sections r_2 and s_2 in the mode of minimum reflux (part of the trajectory from point x'_{D2} to tear-off point S_{r2} is fictitious, there is no leap of concentrations in the feed cross-section).

In the example under consideration, in the second column as in the first one there is indirect split 1,2 : 3, but if product of section s_2 contains more than one product component, then there is an intermediate split.

The last two-section column containing sections r_3 and s_3 is calculated in the same way as the second one. The described algorithm is a general one: it embraces columns with any number of side strippings and with any number of product components in each product.

6.8.3. The Petlyuk Columns

The main difference between the minimum reflux mode calculation algorithms for Petlyuk columns and those for columns with side sections is the necessity to take into account the availability of distributed components. The complex in Fig. 6.13d is an exception. Therefore, this complex can be calculated the same way as the columns with side strippings, beginning with the first two-sections column, whose feed is the mixture being separated and whose top and bottom pseudoproducts are the feeds of the second and third columns correspondingly, and then passing to two other two-section columns. The mode in one of these columns is the control one, and the second column works at a reflux bigger than minimum. As well as for the columns with side sections, for Petlyuk columns, the smallest expenditures of energy for separation are achieved at the joining of section trajectories of the first and following columns in cross-sections S_{r1} and S_{s1} without remixing of flows.

The availability of distributed components is of considerable importance for other types of Petlyuk columns. The minimum reflux mode calculation for such columns was examined for three-component mixtures in the work (Fidkowski & Krolikowski, 1986) and for multicomponent mixtures with several distributed components on the basis of the Underwood equation system in the work (Carlberg & Westerberg, 1989b) and also in a number of other works (Cerda & Westerberg, 1981; Glinos & Malone, 1988; Nikolaides & Malone, 1988; Christiansen & Scogestad, 1997). The availability of distributed components, first, leads to the necessity to use for the calculation of minimum reflux mode the corresponding algorithm, and, second, it creates an additional degree of freedom of designing in the corresponding two-section columns of the complex. For example, for the complex at Fig. 6.12c the additional degree of freedom is the ratio between the withdrawals flow rates of pseudoproducts in the first two-section column, and for the complex in Fig. 6.13a, it is this ratio in the first three two-section columns. The preferable split is thermodynamically optimal for two-section columns themselves with distributed components, but such split leads to nonbalancing of vapor and liquid flows in the next columns of the complex (i.e., to the modes of the reflux bigger than minimum in separate columns). It was shown in the work (Christiansen & Scogestad, 1997) for the complex at Fig. 6.12d that preferable split in the first column and separation leading to balancing of flows in the next two columns maintain close to each other expenditures of energy for separation, but the preferable split leads to smaller expenditures. Therefore, it is possible to use the preferable split as optimal. This calculates withdrawals flow rates of pseudoproducts and minimum flows of vapor and liquid in the first column of the complex (and for the complex in Fig. 6.13a also in the other columns with distributed components). Compositions of liquid and vapor at the ends of the first column should correspond to trajectory tear-off points x_r^t (S_r) and $x_s^t(S_s)$ from boundary elements of concentration simplex of the mixture under separation. The transition to the subsequent columns in the course of separation and, finally, to the last product column is carried out after that. This transition is realized in the same way as it is for the columns with side strippings. The calculation of section trajectories at the preferable split in the minimum reflux mode is carried out most easily, compared with the calculation of sections trajectories for other splits, because in this case the minimum value of parameter L/V does not have to be found by means of scanning (it is equal to the ratio of flows in feed cross-section at sharp reversible distillation and it is defined at Eq. [5.6], where K_j is the phase equilibrium coefficient in the feed point of the component absent in top product). Section trajectories in this case should be calculated in the direction from column ends to the feed cross-section using the method "tray by tray" (Fig. 5.6a).

We examined above Petlyuk columns with preferable split in each column – $1,2\ldots n-1:2,3\ldots n$. Along with such sequence one can use in practice sequences, where each product contains several components. The example of such separation given in the work (Amminudin et al., 2001) is the separation of the mixture of light hydrocarbons consisting of nine components into three products: propane fraction, butane fraction, and pentane fraction. In these case, the split of the following type is used in the first column: $1,2,\ldots k,\ldots l:k, k+1,\ldots l,\ldots n$ (i.e., components $k, k+1,\ldots l$ are distributed ones). So far, we examined only splits with one distributed component or with $(n-2)$ distributed components (the preferable split). The split $1,2\ldots k,\ldots l:k, k+1,\ldots l,\ldots n$ has more than one and less than $(n-2)$ distributed components. The main difficulty in the calculation of minimum reflux mode for such splits consists of the fact that distribution coefficients of the distributed components cannot be arbitrary. In order that sections trajectories in minimum reflux mode join each other product points should belong to some regions at the boundary elements of concentration simplex. In the general case, the boundaries of these regions are unknown. However, for zeotropic mixtures, separation product compositions can be determined at the set requirements to the quality of the products with the help of the Underwood equation system

(Amminudin et al., 2001). In spite of the fact that real zeotropic mixtures with wide interval of boiling deviate considerably from the conditions at which the Underwood equation system (Underwood, 1948) is rigorous (the constancy of relative volatilities of the components and the constancy of molar flows of liquid and vapor), calculation separation product compositions in the general case get inside regions for which the joining of section trajectories is feasible. Along with this, the value of minimum reflux number obtained with the help of Underwood equations system should not be used. Instead of that, it is necessary to apply the general algorithm of calculation of minimum reflux mode described in Section 5.6 for splits with one distributed component.

6.9. Distillation Trajectories in Complexes of Heteroazeotropic and Heteroextractive Distillation

We examine separation of the mixtures, concentration space of which contains region of existence of two liquid phases and points of heteroazeotropes. It is considerably easier to separate such mixtures into pure components because one can use for separation the combination of distillation columns and decanters (i.e., heteroazeotropic and heteroextractive complexes). Such complexes are widely used now for separation of binary azeotropic mixtures (e.g., of ethanol and water) and of mixtures that form a tangential azeotrope (e.g., acetic acid and water), adding an entrainer that forms two liquid phases with one or both components of the separated azeotropic mixture. In a number of cases, the initial mixture itself contains a component that forms two liquid phases with one or several components of this mixture. Such a component is an autoentrainer, and it is the easiest to separate the mixture under consideration with the help of heteroazeotropic or heteroextractive complex. The example can be the mixture of acetone, butanol, and water, where butanol is autoentrainer. First, heteroazeotropic distillation of the mixture of ethanol and water with the help of benzene as an entrainer was offered in the work (Young, 1902) in the form of a periodical process and then in the form of a continuous process in the work (Kubierschky, 1915).

General regularities of location of residue curves in the region of existence of two liquid phases, if change of total composition of liquid at open evaporation is shown, remain the same as for homogeneous mixtures (Haasze, 1950; Storonkin & Smirnova, 1963).

Therefore, residue curves intersect the boundary of the region of existence of two liquid phases without any leaps and twists.

This uses the previously stated general theory of section trajectory bundles in columns with one and two feeds for analysis of mixtures separation in the complexes of heteroazeotropic and heteroextractive distillation.

Specific peculiarities of mixtures with two liquid phases are the following ones: (1) the point of heteroazeotrope x_{Haz} can only be the unstable node N^- or the saddle S; and (2) the points of equilibrium vapor y for points of liquid x in the regions of existence of two liquid phases Reg_{L1-L2} should lie in the concentration triangle at some line called the vapor line, in the concentration tetrahedron at the vapor surface etc. (if $x \in \text{Reg}_{L1-L2}$ then $y \in \text{Reg}_{vap}$).

We examine the most typical splits and separation sequences for various types of three-component mixtures (Fig. 6.16 shows examples of heteroazeotropic distillation; Fig. 6.17 shows examples of heteroextractive distillation).

6.9.1. Heteroazeotropic Distillation

Figure 6.16a shows separate usage of distillation column and decanter, when top product of the column, close in composition to ternary heteroazeotrope–unstable node, is directed after cooling to decanter. The example is separation of the mixture ethanol(2)-water(3) using toluene(1) as entrainer (Pilhofer, 1983). At such sequence, the structure and evolution of section trajectory bundles remain the same as at separation of homogeneous mixtures, when one of the products is azeotrope – unstable node – and the second product is pure component (see Chapter 5). The difference from homogeneous mixtures consists of the fact that the point of vapor composition from column y_D should lie in the region of existence of two liquid phases at the vapor line. The distillation trajectory looks as follows:

$$x_B \;\rightarrow\; S_s \;\rightarrow\; x_f \;\Leftarrow\!\Downarrow\; x_{f-1} \;\rightarrow\; y_D \approx N_{Haz}$$
$$\text{Reg}_B \quad \text{Reg}_s^t \quad \text{Reg}_{sep,s}^{sh,R} \quad \text{Reg}_{att} \quad \text{Reg}_{vap}$$

Figure 6.16b shows joint usage of a distillation column and a decanter, when one of two liquid phases is brought in to the reflux of the column from the decanter or some amount of the second phase is added to the first phase. The example is separation of the mixture isopropanol(2)-water(3) using benzene(1) as an entrainer (Bril et al., 1977). Figure 6.16c shows another variant of distillation column for this separation, with one bottom section.

In these cases, one of two liquid phases is brought into the more minimum reflux of the column from the decanter, therefore the necessary number of trays is finite. In other cases, the reflux with one phase is not sufficient for the separation, which makes necessary reflux with mixture of both phases.

In contrast to distillation of homogeneous mixtures, it is not expedient for the heteroazeotropic complex to carry out calculation of minimum reflux mode before calculation of the necessary number of trays. It is offered in Chapter 7 to carry out at the beginning calculation of the necessary number of trays at reflux with one phase and then, only if this reflux is not sufficient, to determine necessary flow rate of reflux with both phases.

The important peculiarity of the sequence at Fig. 6.16b consists in the fact that column section between the cross-section of feed input and the top of the column by its nature is not the top but the intermediate section of the column with two feeds (see Section 6.3.). This leads to important peculiarities of the trajectory bundle ($\text{Reg}_{w,int}^R$) of this section.

As we saw before, the location and direction of trajectories of intermediate sections differ from those of trajectories of the top and bottom sections. In

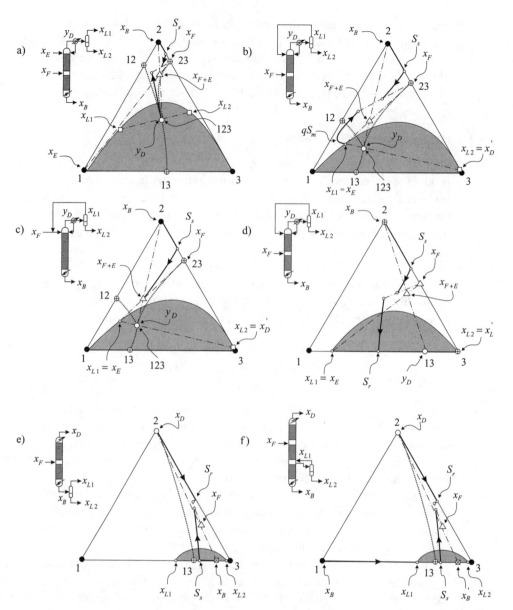

Figure 6.16. Trajectories of heteroazeotropic distillation: (a) distillate from azeocolumn to decanter for separation toluene(1)-ethanol(2)-water(3) mixture; (b) distillate from azeocolumn to decanter and a recycle stream of the entrainer from decanter to azeocolumn for separation benzene(1)-isopropanol(2)-water(3) mixture; (c) distillate from azeostripping to decanter and a recycle stream of the entrainer from decanter to azeostripping for separation benzene(1)-isopropanol(2)-water(3) mixture; (d) distillate from azeocolumn to decanter and a recycle stream of the entrainer from decanter to azeocolumn for separation acetic acid(1)-n-butyl acetate(2)-water(3) mixture; (e) bottom from azeocolumn to decanter for separation butanol(1)-acetone(2)-water(3) mixture; and (f) side product from azeocolumn to decanter for separation butanol(1)-acetone(2)-water(3) mixture. Regions of two liquid phases Reg_{L1-L2} are shaded.

a)

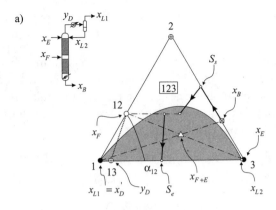

b)

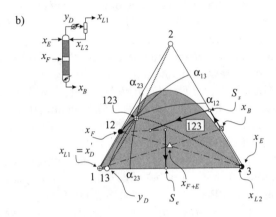

Figure 6.17. Trajectories of heteroextractive distillation: (a) distillate from azeocolumn to decanter and a stream of the entrainer from decanter and additional entrainer to azeocolumn for separation vinyl acetate(1)-methanol(2)-water(3) mixture; (b) distillate from azeocolumn to decanter and a stream of the entrainer from decanter and additional entrainer to azeocolumn for separation chloroform(1)-acetone(2)-water(3) mixture. Regions of two liquid phases Reg_{L1-L2} are shaded.

particular, the trajectories of the intermediate sections can intersect boundaries between distillation regions. Calculations for the sequence at Fig. 6.16b showed that the trajectory of the intermediate section is really located in two distillation regions: one part of the trajectory is located in the region Reg_1^∞ containing isopropanol and the second part in the region Reg_2^∞ containing benzene.

In the stationary points of the trajectory bundle of the intermediate section, the liquid–vapor tie-lines should be directed to pseudoproduct point of this section that is, in the given case, the point of water phase from decanter $x_{L2} \equiv x_D'$. Such quasistationary point is point qS_m, where the calculated trajectory of the intermediate section changes its direction sharply, and the compositions at neighboring trays are very close to each other (quasizones of constant concentrations), and stationary point N_m^+ that coincides with the point of ternary heteroazeotrope and stationary point N_m^- that coincides with the point of binary azeotrope benzene(1)-isopropanol(2). Point S_m is located at reversible distillation trajectory of the intermediate section joining mentioned points N_m^- and N_m^+. Its location at this trajectory is determined by the value of parameter L/V in the intermediate section. At $L/V = 1$ (the mode of infinite reflux), point S_m coincides with point N_m^-. The distillation trajectory looks as follows:

$$x_B \;\to\; S_s \;\to\; x_f \;\Leftarrow\!\Downarrow\; x_{f-1} \;\to\; qS_m \;\to\; y_D \approx N_{Haz} \Leftrightarrow \updownarrow \; x_{L2}$$
$$\mathrm{Reg}_B \quad \mathrm{Reg}_s^t \quad \mathrm{Reg}_{sep,s}^{sh,R} \quad \mathrm{Reg}_{att} \quad \mathrm{Reg}_{w,int}^R \quad \mathrm{Reg}_{vap} \quad \mathrm{Reg}_{L1-L2}^{bound}$$

Comparison of the calculated trajectories of a two-section column (Fig. 6.16b) with those of one-section column (Fig. 6.16c) shows a sharp difference between them. Trajectory of the one-section column lies completely in the distillation region Reg_1^∞ containing isopropanol. This trajectory does not have points of sharp twist that are quasizones of constant concentrations:

$$x_B \;\to\; S_s \;\to\; x_f \;\to\; y_D \approx N_{Haz} \Leftrightarrow \updownarrow \; x_{L2}$$
$$\mathrm{Reg}_B \quad \mathrm{Reg}_s^t \quad \mathrm{Reg}_{sep,s}^{sh,R} \quad \mathrm{Reg}_{vap} \quad \mathrm{Reg}_{L1-L2}^{bound}$$

We note that, in some cases, complex of heteroazeotropic distillation can be used even if the ternary azeotrope–unstable node is located outside the region of existence of two liquid phases (i.e., it is a homoazeotrope). In such cases, composition of the mixture in decanter can be shifted into the region of existence of two liquid phases by adding entrainer into the decanter.

We examined above mixtures with ternary heteroazeotrope. Figure 6.16d shows calculated trajectory for the mixture with binary heteroazeotrope–unstable node. The example is separation of the mixture acetic acid(2)-water(3) with butyl acetate(1) as an entrainer (Othmer, 1978; Bril et al., 1985). Calculated trajectory in homogeneous region from the point of bottom product x_B (acetic acid) passes along the side acetic acid(2)-water(3), then tears off from it in point $S_s = x_s^t$, intersects the boundary of the region of existence of two liquid phases $\mathrm{Reg}_{L1-L2}^{bound}$, and passes through this region to the point of heteroazeotrope $y_D \approx x_{Haz}$ (vapor compositions at trays in this region Reg_{L1-L2} lie at vapor line Reg_{vap}). We note that the bottom section trajectory tear-off from the side acetic acid(2)-water(3) is feasible only in the segment $\alpha_{13} - 3$ because the structural condition of bottom section trajectory tear-off is valid only in region Reg_{ord}^{132} (absent at the side component–butyl acetate(1) becomes the lightest component).

Figure 6.16e shows separate usage of a distillation column and a decanter at the bottom product when binary heteroazeotrope is saddle. The example can be separation of the mixture butanol(1)-acetone(2)-water(3) (Pucci, Mihitenho, & Asselineau, 1986). Sections trajectories do not differ from trajectories at separation of homogeneous mixture of the same type. Figure 6.16f shows joint usage of the distillation column and decanter for the same mixture. The decanter is installed at the side product. Water is withdrawn from the decanter, and the organic phase is returned into the column. The bottom product of the column is butanol.

Heteroazeotropic distillation cannot be used for separation of some types of mixtures having the region of existence of two liquid phases Reg_{L1-L2}, but heteroextractive distillation can be applied.

6.9.2.   Heteroextractive Distillation

At heteroextractive distillation, one of the phases in the decanter plays the role of entrainer, and the section above the input of the mixture being separated plays the role of extractive section. So that the section above the feed cross-section can function as extractive one, the pseudoproduct point of this section should be located outside the concentration triangle (see Section 6.4). For that, it is necessary to bring it into the top cross-section of the column, besides that phase from the decanter which is entrainer, an additional amount of this entrainer. Trajectory tear-off of the section under consideration from the side of the concentration triangle in the points x_e^t, where the component absent at this side is average in volatility, is feasible only in this case. Such a process was described in the works (Polyakova et al., 1977) for the mixture vinyl acetate(1)-methanol(2)-water(3) (Fig. 6.17a) and for the mixture cloroform(1)-acetone(2)-water(3) (Fig. 6.17b) and (Wahnschafft, 1997) for the mixture methanol(1)-isopropyl acetate(2)-water(3). In these examples, water is the entrainer. The mixture chloroform(1)-acetone(2)-water(3) with ternary saddle heteroazeotrope was investigated for the first time in Reinders & De Minjers (1940).

Types of the mixtures at Fig. 6.17a,b are different, but for heteroextractive distillation it is substantial that in both cases there is binary heteroazeotrope and that at the side, where this binary heteroazeotrope is located (Reg_{L1-L2}), is intermediate section trajectory tear-off segment (Reg_e^t) (i.e., the segment where absent at this side component is average in volatility). For sharp heteroextractive distillation it is also necessary that at the side where the bottom product point is located there is the bottom section trajectory tear-off segment (Reg_s^t) (i.e., the segment where it is the lightest component that is absent at this side). The distillation trajectory looks as follows:

$$x_B \;\to\; S_s \;\to\; x_f \;\Leftarrow\!\Downarrow\; x_{f-1} \;\to\; S_e \;\to y_D \approx N_{Haz} \Leftrightarrow\!\Updownarrow \quad x_{L1}$$
$$\text{Reg}_B \quad \text{Reg}_s^t \quad \text{Reg}_{sep,s}^{sh,R} \quad \text{Reg}_{sep,e}^{sh,R} \quad \text{Reg}_e^t \quad \text{Reg}_{vap} \quad\quad \text{Reg}_{L1-L2}^{bound}$$

We note that at heteroextractive distillation, the intermediate section trajectory tear-off segment Reg_e^t can also be located outside the segment of the existence of two liquid phases Reg_{L1-L2}. In this case, one should use a three-section column, into top cross-section of which the reflux is brought in (i.e., the condensate of vapor from the column that is close by composition to binary heteroazeotrope). One of the phases from the decanter that plays the role of entrainer and additional entrainer is brought in below in cross-section between top and extractive sections. At such arrangement of the column, top section trajectory passes along the side from the intermediate section trajectory tear-off point x_e^t to the region of the existence of two liquid phases Reg_{L1-L2}.

The above-examined examples of various sequences and splits of three-component mixtures with the help of heteroazeotropic and heteroextractive distillation allow to make a general conclusion about the usage of these distillation complexes not only for three-component mixtures, but also for mixtures with bigger number of components.

Heteroazeotropic distillation can be effectively used in the following cases: (1) inside concentration simplex there is a heteroazeotrope–unstable node $x_{Haz} = N^-$, and the top product point y_D is close to it (the examples are Fig. 6.16a,b,c); (2) at the boundary element of the concentration simplex there is a heteroazeotrope–unstable node $x_{Haz} = N^-$, and top product point y_D is close to it (the example is Fig. 6.16d); and (3) at the boundary element of concentration simplex there is a heteroazeotrope–saddle $x_{Haz} = S$, the bottom product point x_B is located at this boundary element in the bottom product region Reg_B (the example is Fig. 6.16e,f).

Heteroextractive distillation can be effectively used if at the boundary element of concentration simplex there is a heteroazeotrope–unstable node $x_{Haz} = N^-$ (the example is Fig. 6.17b) or saddle $x_{Haz} = S$ (the example is Fig. 6.17a), and at the same boundary element there is the intermediate section trajectory tear-off region Reg_e^t.

In all the above-mentioned cases of application of heteroazeotropic or heteroextractive distillation, the second product point of distillation column (of the second product, obtained at that column end, where there is no decanter) should belong to the possible product region at the corresponding boundary element of concentration simplex (of bottom product Reg_B at Fig. 6.16a ÷ d and at Fig. 6.17a,b or of top product Reg_D at Fig. 6.16e,f).

6.10. Conclusion

The theory of trajectory tear-off from boundary elements of concentration simplex and the theory of joining of section trajectory bundles find possible product composition regions Reg_D and Reg_B and to calculate the minimum reflux mode not only for simple columns, but also for complex columns and complexes.

For intermediate sections of columns with side products, with side sections, and of Petlyuk columns location of the stationary points of separatrix trajectory bundles (regions $Reg_{sep,int}^{min,R}$) is the same as for simple columns, product compositions of which coincide with pseudoproduct compositions of these intermediate sections (possible product regions Reg_D and Reg_B of simple columns and possible pseudoproduct regions Reg_D and Reg_B of intermediate sections coincide). This extends the use of methods of minimum reflux mode calculation worked out for the simple columns to the previously mentioned complex columns and complexes.

The location of intermediate sections trajectories of columns with two feeds, including those at extractive, heteroazeotropic, and heteroextractive distillation, has fundamental distinctions from that of section trajectories of the simple columns. At sharp extractive or heteroextractive distillation, pseudoproduct point $x'_{D,e}$ of the intermediate section should be located at the continuation of the boundary element, to which components of top product and of entrainer belong. If this condition is valid, the whole trajectory bundle of the intermediate section including trajectory tear-off point x_e^t from the mentioned boundary element is located in the region $Reg_{w,e}^R$ where the top product components are more volatile and the entrainer components are less volatile than the rest of components. The trajectory tear-off point of the intermediate section is the stable node ($x_e^t = N^+$). The conditions of intermediate section trajectory tear-off in different points of trajectory tear-off region Reg_e^t allow to determine limit modes of extractive distillation for each mixture – the mode of minimum flow rate of the entrainer E_{min}, and for the set flow rate of entrainer – to determine the modes of minimum and maximum reflux number (R_{min} and R_{max}).

6.11. Questions

1. Let at sides 1-2 and 2-3 of the concentration triangle be a region of order of components Reg_{ord}^{132}. Is it possible to use extractive distillation to separate this ternary mixture? Which component can be the entrainer? The top product?

2. Let at sides 1-2 and 2-3 of the concentration triangle be a region of order of components Reg_{ord}^{231}. Which process can be used to separate this mixture? How should the components be distributed among the products? Which component should be the entrainer?

3. In the concentration tetrahedron and its faces, there is a region of order of components Reg_{ord}^{1234}. Which sharp extractive splits are feasible for this mixture?

4. In the concentration tetrahedron and at its edges 1-2 and 3-4, there is a region of order of components Reg_{ord}^{1234}. Is it possible to use a "Petlyuk column" to separate this mixture into components? What minimum number of sections should this column have?

5. Write the trajectory of an extractive distillation column at separation of a four-component mixture at minimum reflux, bottom feeding being the control one.

6. Write the trajectories of heteroazeotropic and heteroextractive distillation; indicate the components present and absent in the product points and in trajectory tear-off points.

References

Agrawal, R. (1996). Synthesis of Distillation Column Configurations for a Multicomponent Separation. *Ind. Eng. Chem. Res.*, 35, 1059–71.

Agrawal, R. (1999). More Operable Fully Thermally Coupled Distillation Column Configurations for Multicomponent Distillation. *Trans IChemE.*, 77, Part A, 543–53.

Agrawal, R., & Fidkowski, Z. T. (1998). More Operable Arrangements of Fully Thermally Coupled Distillation Columns. *AIChE J.*, 44, 2565–81.

Agrawal, R., Woodward, D. W., & Modi, A. K. (1997). Coproduction of High Purity Products Using Thermally-Linked Columns. In *Distillation and Absorption Conference*. Maastricht, pp. 511–520.

Amminudin, K. A., Smith, R., Thong, D. Y.-C., & Towler, G. P. (2001). Design and Optimization of Fully Thermally Coupled Distillation Columns. Part 1: Preliminary Design and Optimization Methodology. *Trans. IChemE.*, 79, Part A, 701–15.

Barnes, F. G., Hanson, D. N., & King, C. J. (1972). Calculation of Minimum Reflux for Distillation Columns with Multiple Feeds. *Ind. Eng. Chem. Process Des. Dev.*, 11, 136–47.

Bauer, M. H., & Stichlmair, J. (1995). Synthesis and Optimization of Distillation Sequences for the Separation of Azeotropic Mixtures. *Comput. Chem. Eng.*, 19, 515–20.

Bausa, J., Watzdorf, R. V., & Marquardt, W. (1997). Targeting Sidestream Compositions in Multicomponent Nonideal Distillation. In *Distillation and Absorption Conference*, Maastricht, pp. 735–44.

Becker, H., Godorr, S., & Kreis, H. (2001). Partitioned Distillation Columns – Why, When & How. *Chemical Engineering*, www.che.com.January.

Benedict, M., & Rubin, L. C. (1945). Extractive and Azeotropic Distillation. 1. Theoretical Aspects. *Am. Inst. Chem. Eng.*, 41, 353–70.

Bril, Z. A., Mozzhukhin, A. S., Pershina, L. A., & Serafimov, L. A. (1985). Combined Theoretical and Experimental Design Method for Heteroazeotropic Rectification. *Theor. Found. Chem. Eng.*, 19, 449–54.

Bril, Z. A., Mozzhukhin, A. S., Petlyuk, F. B., & Serafimov, L. A. (1977). Investigations of Optimal Conditions of Heteroazeotropic Rectification. *Theor. Found. Chem. Eng.*, 11, 675–81.

Brugma, A. J. (1942). U. S. Patent No. 2,295,256, September 8.

Cahn, R. P., Di Micelli, E., & Di Micelli, A. G. (1962). U. S. Patent No. 3,058,8., October 16.

Carlberg, N. A., & Westerberg, A. W. (1989a). Temperature (Heat Diagrams for Complex Columns): 2. Underwood's Method for Side Strippers and Enrichers. *Ind. Eng. Chem. Res.*, 28, 1379–86.

Carlberg, N. A., & Westerberg, A. W. (1989b). Temperature (Heat Diagrams for Complex Columns): 3. Underwood's Method for the Petlyuk Configuration. *Ind. Eng. Chem. Res.*, 28, 1386–97.

Cerda, J., & Westerberg, A. W. (1981). Shortcut Methods for Complex Distillation Columns: 1. Minimum Reflux. *Ind. Eng. Chem. Process Des. Dev.*, 20, 546–57.

Christiansen, A. C., & Scogestad, S. (1997). *Energy Savings in Complex Distillation Arrangements: Importance of Using the Preferred Separation*. In AIChE annual meeting, paper 199D, Los Angeles.

Christiansen, A. C., Scogestad, S., & Lien, K. (1997a). Complex Distillation Arrangement: Extending the Petlyuk Ideas. *Comput. Chem. Eng.*, 21, S237–S240.

Christiansen, A. C., Scogestad, S., & Lien, K. (1997b). *Partitioned Petlyuk Arrangements for Quaternary Separations*. IChemE Symp Series No. 142, 745.

Dhole, V. R., & Linnhoff, B. (1993). Distillation Column Targets. *Comput. Chem. Eng.*, 17, 549–60.

Doherty, M. F., & Malone, M. F. (2001). Conceptual Design of Distillation Systems. New York: McGraw-Hill.

Drew, J. W. (1979). Solvent Recovery. In P. A. Schweitzer (Ed.), *Handbook of Separation Techniques for Chemical Engineers*. New York: McGraw-Hill.

Emmrich, G., Gehrke, H., & Ranke, U. (2001, June). Working with an Extractive Distillation Process. Petrochemicals. http://www.eptq.com/eptq/articles/articles_temp.

Ennenbach, F., Kolbe, B., & Ranke, U. (2000, September). Divided-Wall Columns (A Novel Distillation Concept. Process Heating/Fluid Flow. http://www.eptq.com/eptq/articles/articles_temp.

Fidkowski, Z., & Krolikowski, L. (1986). Thermally Coupled System of Distillation Columns: Optimization Procedure. *AIChE J.*, 32, 537–46.

Fidkowski, Z., & Krolikowski, L. (1987). Minimum Energy Requirements of Thermally Coupled Distillation Systems. *AIChE J.*, 33, 643–53.

Glanz, S., & Stichlmair, J. (1997). Minimum Energy Demand of Distillation Columns with Multiple Feeds. *Chem. Eng. Technol.*, 20, 93–103.

Glinos, K., & Malone, M. F. (1985). Minimum Vapor Flows in a Distillation Column with a Side Stream – Stripper. *Ind. Eng. Chem. Process Des. Dev.*, 24, 1087–90.

Glinos, K., & Malone, M. F. (1988). Optimality Regions for Complex Column Alternatives in Distillation Systems. *Chem. Eng. Res. Des.*, 66, 229–40.

Haasze, R. (1950). Verdampfungsgleichgewichte von Mehrstoffgemischen: 6. Ternare Systeme mit Mischungslucke. *Z., Naturforschung*, 5a, 109–24 (Germ.).

Happe, J., Cornell, P. W., & Eastman, D. (1946). Extractive Distillation of C4-Hydrocarbons Using Furfurol. *AIChE J.*, 4, 189–214.

Hoffman, E. G. (1964). *Azeotropic and Extractive Distillation*. New York: Wiley.

Kaibel, G. (1987). Distillation Columns with Vertical Partitions. *Chem. Eng. Technol.*, 10, 92–8.

Kiva, V. N., Timofeev, V. S., Vizhesinghe, A. D. M. C., & Chyue, Vu Tam (1983). *The Separation of Binary Azeotropic Mixtures with a Low-Boiling Entrainer.* Theses of 5th Distillation Conference in USSR, Severodonezk (Rus.).

Knapp, J. P., & Doherty, M. F. (1990). Thermal Integration of Homogeneous Azeotropic Distillation Sequences. *AIChE J.*, 36, 969–84.

Knapp, J. P., & Doherty, M. F. (1992). A New Pressure-Swing Distillation Process for Separating Homogeneous Azeotropic Mixtures. *Ind. Eng. Chem. Res.*, 31, 346–57.

Knapp, J. P., & Doherty, M. F. (1994). Minimum Entrainer Flows for Extractive Distillation: A Bifurcation Theoretic Approach. *AIChE J.*, 40, 243–68.

Knight, J. R., & Doherty, M. F. (1989). Optimal Design and Synthesis of Homogeneous Azeotropic Distillation Sequences. *Ind. Eng. Chem. Res.*, 28, 564–72.

Kogan, V. B. (1971). *Azeotropic and Extractive Distillation*. Leningrad: Khimiya (Rus.).

Kohler, J., Kuen, T., & Blass, E. (1994). Minimum Energy Demand for Distillations with Distributed Components and Sideproduct Withdrawals. *Chem. Eng. Sci.*, 49, 3325–30.

Kubierschky, K. (1915). Verfahren zur Gewinnung von Hochprozentigem, bezw. Absolutem Alkohol – Wassergemischen in Unterbrochenem Betriebe. German Patent 287, 897.

Levy, S. G., & Doherty, M. F. (1986). Design and Synthesis of Homogeneous Azeotropic Distillation. 4. Minimum Reflux Calculations for Multiple-Feed Columns. *Ind. Eng. Chem. Fundam.*, 25, 269–79.

Nikolaides, J. P., & Malone, M. F. (1987). Approximate Design of Multiple Feed/Side-Stream Distillation Systems. *Ind. Eng. Chem. Res.*, 26, 1839–45.

Nikolaides, J. P., & Malone, M. F. (1988). Approximate Design and Optimization of Thermally Coupled Distillation with Prefractionation. *Ind. Eng. Chem. Res.*, 27, 811–18.

Othmer, D. F. (1978). Azeotropic and Extractive Distillation. In *Kirk-Othmer Enciclopedia of Chemical Technology* (pp. 352–377). New York: John Wiley.

Parkinson, G. (2000). Drip and Drop in Column Internals. *Chemical Engineering*, 107(7), 27–31.

Parkinson, G., Kamiya, T., D'Aquino, R., & Ondrey, G. (1999). The Divide in Distillation. *Chemical Engineering*, 106(4), 32–5.

Petlyuk, F. B. (1984). Necessary Condition of Exhaustion of Components at Distillation of Azeotropic Mixtures in Simple and Complex Columns. In *The Calculation Researches of Separation for Refining and Chemical Industry* (pp. 3–22). Moscow: Zniiteneftechim (Rus.).

Petlyuk, F. B., & Danilov, R. Yu. (1999). Sharp Distillation of Azeotropic Mixtures in a Two-Feed Column. *Theor. Found. Chem. Eng.*, 33, 233–42.

Petlyuk, F. B., Platonov, V. M., & Avet'an, V. S. (1966). The Optimal Distillation Flowsheets for Separating Multicomponent Mixtures. *Chem. Industry*, (11), 865–869 (Rus.).

Petlyuk, F. B., Platonov, V. M., & Slavinskii, D. M. (1965). Thermodynamical Optimal Method for Separating of Multicomponent Mixtures. *Int. Chem. Eng.*, 5(2), 309–17.

Pilhofer, T. (1983). Energiesparende Alternativen zur Rektifikation bei der Ruckgewinnung Organischer Stoffe aus Losungen. *Verfahrenstechnik*, 17, 547–9 (Germ.).

Platonov, V. M., Petlyuk, F. B., & Zhvanetskiy, I. B. (1970). Patent USSR No. 292,339, October 23 (Rus.).

Polyakova, E. B., Pavlenko, T. G., Kalabuhova, N. P., Timofeev, V. S., & Serafimov, L. A. (1977). A Research of Distillation of Heterogeneous Mixture Methanol-Vinil Acetate-Water. In *Physical-Chemical Foundation of Distillation*, Moscow. *MIChM*, pp. 160–173 (Rus.).

Pucci, A., Mihitenho, P., & Asselineau, L. (1986). Three-Phase Distillation. Simulation and Application to the Separation of Fermentation Products. *Chem. Eng. Sci.*, 41, 485–94.

Reinders, W., & De Minjers, C. H. (1940). Vapour-Liquid Equilibria in Ternary Sistems. The Sistem Acetone-Cloroform-Benzene. *Rec. Trav. Chim. Pays-Bas*, 59, 392–406.

Rooks, R. E., Malone, M. F., & Doherty, M. F. (1996). Geometric Design Method for Side-Stream Distillation Columns. *Ind. Eng. Chem. Res.*, 35, 3653–64.

Rooks, R. E., Malone, M. F., & Doherty, M. F. (1996). A Geometric Design Method for Sidestream Distillation Column. *Ind. Eng. Chem. Res.*, 35, 3653–64.

Sargent, R. W. S., & Gaminibandara, K. (1976). Optimum Design of Plate Distillation Columns. In *Optimization in Action*. L. W. C. Dixon, pp. 267–314.

Schreinemakers, F. A. H. (1901). Dampfdrucke ternarer Gemische. *Z. Phys. Chem.*, 36, 413–49 (Germ.).

Schultz, M. A., Stewart, D. G., Harris, J. M., Rosenblum, S. P., Shakur, M. S., & O'Brien, D. E. (2002). Reduce Costs with Dividing-Wall Columns. *CEP*, 98(5), 64–71.

Smith, R., & Linnhoff, B. (1988). The Design of Separators in the Context of Overall Processes. *Trans IChemE*, 66, Part A, 195–214.

Stichlmair, J. G., & Fair, J. R. (1998). *Distillation: Principles and Practice*. New York: John Wiley.

Stichlmair, J., Offers, H., & Potthoff, R. W. (1993). Minimum Reflux and Minimum Reboil in Ternary Distillation. *Ind. Eng. Chem. Res.*, 32, 2438–45.

Storonkin, A. V., & Smirnova, N. A. (1963). Certain Aspects of the Thermodynamics of Multicomponent Heterogeneous Systems. 4. Shapes of the Distillation Curves of Ternary Solutions. *J. Phys. Chem.*, 37, 601–7 (Rus.).

Sugie, H., & Benjamin, C. Y. L. (1970). On the Determination of Minimum Reflux Ratio for a Multicomponent Distillation Column with Any Number of Side-Cut Streams. *Chem. Eng. Sci.*, 25, 1837–46.

Terranova, B. E., & Westerberg, A. W. (1989). Temperature (Heat Diagrams for Complex Columns): 1. Intercooled/Interheated Distillation Columns. *Ind. Eng. Chem. Res.*, 28, 1374–79.

Underwood, A. J. V. (1948). Fractional Distillation of Multicomponent Mixtures. *Chem. Eng. Prog.*, 44, 603–14.

Wahnschafft, O. M. (1997). *Advanced Distillation Synthesis Techniques for Non-ideal Mixtures Are Making Headway in Industrial Applications.* Presented at the Distillation and Absorption Conference, Maastricht, pp. 613–23.

Wahnschafft, O. M., Kohler, J., & Westerberg, A. W. (1994). Homogeneous Azeotropic Distillation: Analysis of Separation Feasibility and Consequences for Entrainer Selection and Column Design. *Comput. Chem. Eng.*, 18, S31–S35.

Wahnschafft, O. M., & Westerberg, A. W. (1993). The Product Composition Regions of Azeotropic Distillation Columns. 2. Separability in Two-Feed Columns and Entrainer Selection. *Ind. Eng. Chem. Res.*, 32, 1108–20.

Watkins, R. N. (1979). *Petroleum Refinery Distillation.* Houston, TX: Gulf Publishing Company.

Wright, R. O. (1949). U. S. Patent No. 2,471,134, May 24.

Young, S. (1902). The Preparation of Absolute Alcohol from Strong Spirit. *J. Chem. Sos.*, 81, 707–17.

7

Trajectories of the Finite Columns and Their Design Calculation

7.1. Introduction

In this chapter, we turn from infinite columns to real finite columns. On the basis of the analysis performed previously for infinite columns, we determine the regularities of location of finite columns trajectories in the concentration simplex and, in particular, the regularities of joining of finite column section trajectories. This will allow us to develop simple and reliable algorithms of distillation design calculation.

Designing distillation columns is particularly important because of the great expenditures of energy for mixture separation. Simplified empirical methods were first used for designing. The method of Underwood–Fenske–Gilliland (calculation of minimum reflux number according to Underwood method [1948], calculation of minimum trays number according to Fenske method [1932], and usage of the empirical correlation of Gilliland [1940] for the transition from the infinite number of trays and from the infinite reflux to their finite calculation values) is among them. A big number of other approximate methods that could produce mistakes of unforeseen magnitude, especially for nonideal mixtures, were also introduced.

Therefore, two rigorous in the limits of conception of theoretical step of separation (Sorel, 1893) algorithms of distillation columns calculation – Lewis and Matheson method (1932) and Thiele and Geddes method (1933) – were introduced already in the 1930s. The first of them is based on traywise grouping of the equations describing distillation process (phase equilibrium equations, equations of material and thermal balance), the second one is based on componentwise grouping of these equations. Both methods presuppose a fixed number of trays in the column, and for this reason they are not design ones. Both methods are iterative and do not guarantee the solution of the task (i.e., ensuring of equation system validity with sufficient precision). These methods were widely adopted in practice and developed in various modifications only with the advent of computers in the end of the 1950s. The componentwise methods are most widely practised in modern program systems for simulating calculation of chemical engineering units.

Among these methods, the modification that was named "inside-out" (Russel, 1983) is the most widely used.

But the existing methods of distillation calculation (simulating methods) are poorly adapted to designing. They do not answer essential questions, such as: (1) is the set split feasible? (2) which minimum reflux number is necessary to ensure the set split? and (3) which numbers of the trays in column sections are sufficient to ensure the set split?

Therefore, the answers to these questions are being looked for "in the dark," setting tray numbers in column sections at random.

The second problem is the convergence of iterative process that is not guaranteed but depends on numerous parameters of calculating process, such as those set by the designer assumed in the initial approximation (estimated) profiles of vapor and liquid flow rates, profiles of temperature, and components concentrations at theoretical trays. It is usually necessary to make numerous calculations at different parameters.

If, finally, the designer obtains the result acceptable to his opinion then he does not know how far he is from the optimal parameters of the column and of the mode in it. The search for parameters in such conditions turns into a nearly hopeless task, especially in conditions of time deficit usual to the designing process, and the application of mathematical optimization methods also does not lead to the achievement of the goal because of the difficulties caused by the availability of "local" extremums and discrete variables. This leads to the fact that in practice the task of designing is not solved optimally (i.e., the expenditures for the separation unwarrantably grow).

This situation is explained by the fact that the above-described algorithms do not take into consideration the regularities of location of distillation trajectory bundles in the concentration space.

The geometric theory of distillation suggests a new approach to the task of its designing. This approach ensures guaranteed obtaining of optimal design parameters without any participation of the designer in the calculation process.

The geometric approach based on calculation of reversible distillation trajectories, on linearization of separatrix distillation trajectory bundles, and on realization of calculations by means of the method "tray by tray" required the development of the theory of joining of sections trajectory bundles at reflux bigger than minimum for any splits (Petlyuk & Danilov, 2001a, 2001b). Similar approach was applied before for two splits simplest in calculating aspect – the direct and the indirect ones (Julka & Doherty, 1990), (Doherty & Melone, 2001). But the algorithms useful for the direct and the indirect splits cannot be used for any intermediate ones and ones with a distributed component. Limitations of these algorithms are of fundamental nature (i.e., they are conditioned by the peculiar structure of section trajectory bundles for the direct and the indirect splits). This fact is discussed later in this present chapter.

The geometric approach develops methods of conceptual design calculation of simple and complex distillation columns (i.e., methods of determination of optimal values of the main process parameters, of the numbers of theoretical trays at different reflux numbers). Conceptual design calculation can precede simulating of the process. In this case, the found values of reflux numbers, trays numbers, profiles of liquid and vapor flows, and temperature and component concentrations at the trays of the column are used as initial ones for the realization of simulating calculation by means of known methods at a small number of iterations. At the other approach, the conceptual design calculation used without the simulating calculation.

In complex columns and distillation complexes, geometric conditions of joining of section trajectories are similar to those for simple columns.

Therefore, the algorithm of design calculation of simple columns is a basis for the algorithms of design calculation of any complex distillation columns and distillation complexes.

In connection with it, we examine in detail conditions of the joining of section trajectories and the algorithms of design calculation of simple columns at various splits, and then on this basis we examine these questions for complex columns.

7.2. Distillation Trajectories of Finite Columns: Possible Compositions in Feed Cross Section

7.2.1. Location of Section Trajectories

The main difference between distillation trajectories of finite columns and those of infinite columns consists of the fact that finite columns trajectories do not pass through stationary points $S_r^1, S_s^1, S_r^2, S_s^2 \ldots$ From this, it follows that, in particular, absolutely sharp separation at which sections trajectories should pass through points S_r^1 and S_s^1 is not feasible in finite columns. Also from this, it follows that the parameter $(L/V)_r$ for finite columns cannot have any value at which section trajectories should pass through points S_r^2 and S_s^2.

Section trajectories at quasisharp and nonsharp separation and at reflux bigger than minimum are examined below. At quasisharp separation, each product of the column contains, besides the product components itself, small amounts of impurity components, mostly of the key nonproduct component. The purpose of separation is to obtain in each product a prescribed set of product components at a prescribed summary concentration of impurity components.

For example, for the split $1, 2, \ldots k : k+1, k+2 \ldots n$, components $1, 2 \ldots k$ are product ones for the top product and impurity ones for the bottom product and vice versa for components $k+1, k+2 \ldots n$; key components are k and $k + 1$. Along with that, the heavy key component $(k + 1)$ is a main impurity one in the top product and the light key component k is a main impurity one in the bottom product. The remaining impurity components are contained in the products in small amounts.

If a component is distributed, it is product one for both products of the column. For example, for the split $1, 2 \ldots k, k+1 : k+1, k+2, \ldots n$ components $1, 2 \ldots k, k+1$ are product ones for the top product and components $k+2, \ldots n$ are impurity ones, components $k+1, k+2 \ldots n$ are product ones for the bottom product and components $1, 2 \ldots k$ are impurity ones; key components are k and $k + 2$.

In Chapter 5, we examined in detail the structure and evolution of section trajectory bundles for various sharp splits. In this section, we examine in detail the peculiarities of location of section trajectories at quasisharp and nonsharp separation and at given reflux R bigger than minimum. Along with that, we are interested in the location of trajectories with respect to separatrix sharp split bundles (regions) $\mathrm{Reg}_{sep,r}^{sh,R}(S_r^1 - S_r^2 - N_r^+)$ and $\mathrm{Reg}_{sep,s}^{sh,r}(S_s^1 - S_s^2 - N_s^+)$, to the boundary elements of these bundles $S_r^2 - N_r^+, S_r^1 - S_r^2, S_s^2 - N_s^+, S_s^1 - S_s^2$ and to the product boundary elements of the concentration simplex (i.e., with respect to the boundary elements in the vicinity of which the product points are located). We note that at quasisharp separation stationary points S_r^1 and S_s^1 are absent inside the concentration simplex (they are located outside it close to the product boundary elements), but there are separatrix bundles $\mathrm{Reg}_{sep,r}^{sh,R}(S_r^1 - S_r^2 - N_r^+)$ and $\mathrm{Reg}_{sep,s}^{sh,R}(S_s^1 - S_s^2 - N_s^+)$. These separatrix bundles isolate the working trajectory bundles $\mathrm{Reg}_{w,r}^R$ and $\mathrm{Reg}_{w,s}^R$ to which section trajectory belongs from other bundles of dimensionality $n - 1$.

At quasisharp distillation and at reflux bigger than minimum composition points at the first trays above x_{f-1} and below x_f feed cross-section are very close to these separatrix bundles. In their turn, these separatrix bundles nearly coincide with the separatrix bundles at sharp separation. While decreasing, the sharpness of separation the compositions at the first trays above and below the feed cross-section move away from the trajectory separatrix bundles deep into the working bundles. Therefore, at quasisharp separation, the part of the trajectory of each section passes in the small vicinity of separatrix bundle of this section $S^1 - S^2 - N^+$ and, at nonsharp separation, the whole trajectory of the section passes far from the separatrix bundle. The mentioned regularities are of great importance for the development of the general algorithm of design calculation. This algorithm should include calculation "tray by tray." Any other algorithms, in particular, those based on componentwise grouping of distillation equations, do not take into consideration the structure of trajectory bundles. The choice of the initial point of the calculation and its direction plays the key role in the calculation by method "tray by tray." The calculation from one of the ends of the column is efficient only at direct or indirect splits, because at these splits there is one impurity component in one of the products, which sets the composition of this product with high precision. Besides that, the structure of section trajectory bundles promotes the execution of calculation from this product.

In general, at intermediate splits and splits with a distributed component, the calculation from one of the ends of the column for such splits encounters large difficulties. Determination of possible compositions in the feed cross-section of the column is of great importance for overcoming these difficulties. To estimate correctly the limits of change of component concentrations at the trays above and below feed cross-section, this limits have to be determined at sharp separation ($[x_{f-1}]^{sh}$ and $[x_f]^{sh}$).

At minimum reflux for the splits without distributed components, there is only one composition point at the first tray above the feed cross-section x_{f-1} and only

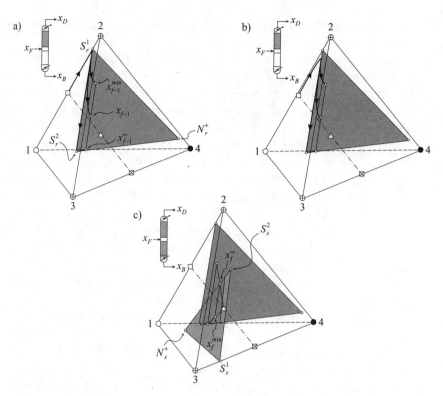

Figure 7.1. Rectifying trajectories at sharp intermediate split and different x_{f-1} ($K_1 > K_2 > K_3$) (a); rectifying trajectories at quasisharp intermediate split and different x_{f-1} (b); section bundles (separatrix sharp spilt regions $\text{Reg}_{sep,r}^{R}{}^{3,4}$ and $\text{Reg}_{sep,s}^{R}{}^{1,2}$) (c). Separatrix sharp split section regions are shaded.

one composition point at the first tray below the feed cross-section x_f. These points lie at the separatrix min-reflux regions of sections trajectories $\text{Reg}_{sep,r}^{\min,R}(S_r^2 - N_r^+)$ and $\text{Reg}_{sep,s}^{\min,R}(S_s^2 - N_s^+)$, respectively.

At reflux bigger than minimum, the sizes of working trajectory bundles of the sections increase and, at the condition of validity of material balance in feed cross-section (Eq. [5.18]), points x_{f-1} and x_f can be located not only at these separatrix regions, but also inside separatrix sharp split regions $\text{Reg}_{sep,r}^{sh,R}(S_r^1 - S_r^2 - N_r^+)$ and $\text{Reg}_{sep,s}^{sh,R}(S_s^1 - S_s^2 - N_s^+)$ (Fig. 7.1a,b). Figure 7.1a,b for clarity's sake shows only the working trajectories and the separatrix sharp split region of the top section at the sharp and quasisharp intermediate split $\text{Reg}_{sep,r}^{sh,R}{}^{3,4}_{1,2}$ of four-component mixture.

The intermediate split is chosen as illustrating the most general case of separating multicomponent mixtures. To determine the set of composition points at the first trays above and below the feed cross-section at *given* reflux larger than minimum, we examine first sharp separation and linear regions $\text{Reg}_{sep,r}^{sh,R}(S_r^1 - S_r^2 - N_r^+)$ and $\text{Reg}_{sep,s}^{sh,R}(S_s^1 - S_s^2 - N_s^+)$. We designate possible compositions in the feed cross-section at these assumptions $(x_{f-1})_{lin}^{sh}$ and $(x_f)_{lin}^{sh}$.

Because the dimensionality of region $\text{Reg}_{sep,r}^{sh,R}(S_r^1 - S_r^2 - N_r^+)$ is larger by one than that of its boundary element $\text{Reg}_{sep,r}^{min,R}(S_r^2 - N_r^+)$, the dimensionality of the set of points $\{x_{f-1}\}_{lin}^{sh}$ at given reflux larger than minimum should be larger by one than the dimensionality at minimum reflux. As far as at minimum reflux, the dimensionality of $\{x_{f-1}\}_{lin}^{sh}$ is equal to zero then at reflux larger than minimum the dimensionality of $\{x_{f-1}\}_{lin}^{sh}$ is equal to one; that is, the set of points $\{x_{f-1}\}_{lin}^{sh}$ is a segment lying in linear region $\text{Reg}_{sep,r}^{sh,R}(S_r^1 - S_r^2 - N_r^+)$. Similarly, the set of points $\{x_f\}_{lin}^{sh}$ is a segment lying in linear region $\text{Reg}_{sep,s}^{sh,R}(S_s^1 - S_s^2 - N_s^+)$. There is a correspondence between each point of segment $[x_{f-1}]_{lin}^{sh}$ and certain point of segment $[x_f]_{lin}^{sh}$ that is connected with the first one by Eq. (5.18). The greater the length of segments $[x_{f-1}]_{lin}^{sh}$ and $[x_f]_{lin}^{sh}$, the greater the reflux number.

One of the ends of segment $[x_{f-1}]_{lin}^{sh}$ that we designate $(x_{f-1}^\infty)_{lin}^{sh}$ should lie at linear boundary element $\text{Reg}_{sep,r}^{min,R}(S_r^2 - N_r^+)_{lin}$. The section trajectory starting in point $(x_{f-1}^\infty)_{lin}^{sh}$ should pass through two stationary points S_r^2 and S_r^1, that is, for point $(x_{f-1}^\infty)_{lin}^{sh}$ the top section is infinite not only at sharp, but also at quasisharp separation (Fig. 7.1b). The other end of the segment $[x_{f-1}]_{lin}^{sh}$ that we designate $(x_{f-1}^{min})_{lin}^{sh}$ should lie inside region $\text{Reg}_{sep,r}^{sh,R}(S_r^1 - S_r^2 - N_r^+)_{lin}$, the farthest possible from boundary element $\text{Reg}_{sep,r}^{min,R}(S_r^2 - N_r^+)_{lin}$. The section trajectory starting in point $(x_{f-1}^{min})_{lin}^{sh}$ should pass only through stationary point S_r^1, the farthest one from stationary point S_r^2. Therefore, in this case, at quasisharp separation the top section has the smallest number of trays.

Similarly, for the bottom section, the ends of segment $[x_f]_{lin}^{sh}$ are $(x_f^\infty)_{lin}^{sh}$ and $(x_f^{min})_{lin}^{sh}$. Point $(x_f^{min})_{lin}^{sh}$ corresponds to point $(x_{f-1}^\infty)_{lin}^{sh}$ and point $(x_f^\infty)_{lin}^{sh}$ corresponds to point $(x_{f-1}^{min})_{lin}^{sh}$ (Fig. 7.1c). In the first case at quasisharp separation, there is an infinite number of trays in the top section and the smallest number in the bottom section, and vice versa in the second case.

The smallest summary number of trays of two sections at quasisharp separation corresponds to some middle location of points $(x_{f-1})_{lin}^{sh}$ and $(x_f)_{lin}^{sh}$. Such compositions in the feed cross-section are optimal.

7.2.2. Possible Compositions in Feed Cross Section

The coordinates of segments $[x_{f-1}]_{lin}^{sh}$ and $[x_f]_{lin}^{sh}$ can be determined from purely geometric considerations from the known coordinates of the stationary points and of point x_F.

While solving this task, we act the way we did when we determined points x_{f-1} and x_f in the mode of minimum reflux (see Section 5.6).

It follows from the condition of material balance in the feed cross-section (Eq. [5.18]) that segments $[x_{f-1}]_{lin}^{sh}$ and $[x_f]_{lin}^{sh}$ should be parallel to each other and to the line of intersection of surfaces or hypersurfaces $\text{Reg}_{sep,r}^{sh,R}(S_r^1 - S_r^2 - N_r^+)_{lin}$ and $\text{Reg}_{sep,s}^{sh,R}(S_s^1 - S_s^2 - N_s^+)_{lin}$. We examine points $(x_{f-1}^{min})_{lin}^{sh}$, $(x_f^\infty)_{lin}^{sh}$ and x_F for which Eq. (5.18) should be valid. Therefore, straight line $(x_{f-1}^{min})_{lin}^{sh} - (x_f^\infty)_{lin}^{sh} - x_F$ should be the intersection line for linear manifolds $\text{Reg}_{sep,r}^{sh,R} \equiv (S_r^1 - S_r^2 - N_r^+)_{lin}$ and

$(x_F - S_s^2 - N_s^+)_{lin}$. The equation of this straight line can be found by means of solving the system of equations describing the mentioned manifolds by coordinates of the stationary points entering into them and of point x_F. Then the coordinates of points $(x_{f-1}^{min})_{lin}^{sh}$ and $(x_f^\infty)_{lin}^{sh}$ at this straight line are determined from Eq. (5.18). The coordinates of points $(x_f^{min})_{lin}^{sh}$ and $(x_{f-1}^\infty)_{lin}^{sh}$ are determined in the same way (Fig. 7.1c).

Possible composition segments at the first trays above and below the feed cross-section in the real columns $[x_{f-1}]$ and $[x_f]$ located at vicinity of curvilinear separatrix manifolds $\text{Reg}_{sep,r}^{sh,R}(S_r^1 - S_r^2 - N_r^+)$ and $\text{Reg}_{sep,s}^{sh,R}(S_s^1 - S_s^2 - N_s^+)$, or inside the working trajectory bundles $\text{Reg}_{w,r}^R$ and $\text{Reg}_{w,s}^R$, correspond to segments $[x_{f-1}]_{lin}^{sh}$ and $[x_f]_{lin}^{sh}$.

At the direct split, $S_r^1 \equiv N_r^-$ and $S_s^2 \equiv N_s^+$ (Fig. 7.2a). Therefore, at the direct split, segment $[x_{f-1}]$ is located inside working trajectory bundle of the top section

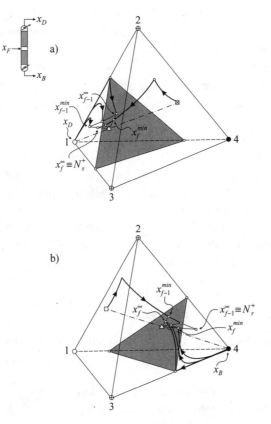

Figure 7.2. Section regions at given reflux $\text{Reg}_{w,r}^R = \underset{1}{\text{Reg}_{att}^1}, \underset{1,2,3}{\text{Reg}_{sep,s}^{sh,R}}$ and $\underset{2,3,4}{\text{Reg}_{w,s}^R} = \underset{4}{\text{Reg}_{att}^4}, \underset{1,2,3}{\text{Reg}_{sep,r}^{sh,R}}$ and possible composition segments $[x_{f-1}]$ and $[x_f]$ in feed cross-section: (a) direct split, (b) indirect split. Separatrix sharp split section regions are shaded.

$\text{Reg}^{R}_{w,r} = \text{Reg}^{1}_{att}\overset{2,3,4}{(N^{-}_{r} - S^{2}_{r} - N^{+}_{r})}$, and segment $[x_{f}]$ is located at separatrix sharp split region $\text{Reg}^{sh,R}_{sep,s}$ (separatrix $\overset{1}{S^{1}_{s}} - N^{+}_{s}$). Similarly, at the indirect split $S^{1}_{s} \equiv N^{-}_{s}$ and $S^{2}_{r} \equiv N^{+}_{r}$ (Fig. 7.2b). Therefore, at the indirect split, segment $[x_{f}]$ is located inside the working trajectory bundle of the bottom section $\text{Reg}^{R}_{w,s} = \overset{1,2,3}{\text{Reg}^{4}_{att}}\overset{1,2,3}{(N^{-}_{s} - }$

$S^{2}_{s} - N^{+}_{s})$ and segment $[x_{f-1}]$ is located at separatrix sharp split region $\text{Reg}^{sh,R}_{sep,r}$ (separatrix $\overset{1,2,3}{S^{1}_{r} - N^{+}_{r}}$).

For splits with one distributed component, the summary dimensionality of separatrix sharp split regions $\text{Reg}^{sh,R}_{sep,r}(S^{1}_{r} - S^{2}_{r} - N^{+}_{r})$ and $\text{Reg}^{sh,R}_{sep,s}(S^{1}_{s} - S^{2}_{s} - N^{+}_{s})$ is smaller by one than that for splits without distributed components (see Chapter 5). This leads to the decrease by one of the dimensionality of the set of intersection points of these separatrix bundles and, correspondingly, to the decrease of the dimensionality of sets of the points $\{x_{f-1}\}^{sh}_{lin}$ and $\{x_{f}\}^{sh}_{lin}$. Therefore, these sets of points at reflux larger than minimum have zero dimensionality; that is for split $1,2 : 2,3,4$, they are the following points: point $(x_{f-1})^{sh}_{lin}$ is located in separatrix region $\text{Reg}^{R}_{sep,r}\overset{3,4}{(S^{1}_{r} - S^{2}_{r} - N^{+}_{r})_{lin}}$, and point $(x_{f})^{sh}_{lin}$, is located in separatrix region $\overset{1}{\text{Reg}^{R}_{sep,s}}\overset{1,2}{(S^{1}_{s} - S^{2}_{s} - N^{+}_{s})_{lin}}$ (Fig. 7.3).

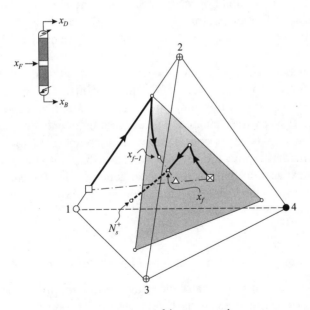

Figure 7.3. Section regions $\overset{3,4}{\text{Reg}^{R}_{sep,r}}$ and $\overset{1}{\text{Reg}^{R}_{sep,s}}$ and possible composition points (x_{f-1}) and (x_{f}) in feed cross-section for the split $1,2 : 2,3,4$ with a distributed component. Rectifying separatrix sharp split region is shaded.

7.3.   Design Calculation of Two-Section Columns

Purity of the products is the set (specified) parameter at designing, and number of trays in each section n_r and n_s and reflux number L/D are the parameters that have to be determined.

Knowledge of the general regularities of location of separatrix trajectory bundles of the sections and of possible composition segments in the feed cross-section develops reliable and fast algorithms of calculation of the necessary tray numbers. These algorithms include (1) the determination of coordinates of the stationary points of sections trajectory bundles at different values of parameter $(L/V)_r$ (i.e., calculation of reversible distillation trajectories for the set product points at sharp separation); (2) the obtaining of linear equation systems describing planes and hyperplanes $S_r^2 - N_r^+$, $S_r^1 - S_r^2 - N_r^+$, $S_s^2 - N_s^+$, and $S_s^1 - S_s^2 - N_s^+$ by the coordinates of the stationary points through which these planes and hyperplanes pass (we remember that the number of the stationary points of bundles $\mathrm{Reg}_{sep,r}^{min,R}(S_r^2 - N_r^+)$ and $\mathrm{Reg}_{sep,s}^{min,R}(S_s^2 - N_s^+)$ is equal to the number of components absent in the top and bottom products, respectively, and the number of the stationary points of bundles $\mathrm{Reg}_{sep,r}^{sh,R}(S_r^1 - S_r^2 - N_r^+)$ and $\mathrm{Reg}_{sep,s}^{sh,R}(S_s^1 - S_s^2 - N_s^+)$ is larger by one); (3) obtaining linear equation systems describing planes and hyperplanes $x_F - S_r^2 - N_r^+$ and $x_F - S_s^2 - N_s^+$ by the coordinates of point x_F and of the corresponding stationary points; (4) the determination of the coordinates of points $(x_{f-1}^{min})_{lin}^{sh}$, $(x_{f-1}^{\infty})_{lin}^{sh}$, $(x_f^{min})_{lin}^{sh}$, and $(x_f^{\infty})_{lin}^{sh}$ with the help of the algorithm described in Section 7.2; and (5) the calculation of sections trajectories by method "tray by tray" and the determination of necessary tray numbers n_r and n_s in the sections at the set value of parameter $(L/V)_r([L/V]_r > [L/V]_r^{min})$ and at different coordinates of points $(x_{f-1})_{lin}^{sh}$ and $(x_f)_{lin}^{sh}$ at segments $[x_{f-1}]_{lin}^{sh}$ and $[x_f]_{lin}^{sh}$.

The first four items of this algorithm are of general nature and do not depend on the split. But the efficiency of the choice of the initial point and of the direction of calculation by method "tray by tray" depends to a great extent on the accepted split. In some cases, it is easy to calculate the whole column in one direction (the direct and the indirect splits). It is considerably more complicated to perform calculation at intermediate splits and at splits with one distributed component. It is shown in the next section that for these most general splits the calculation of each section trajectory should be performed from the end of the column. We examine all the listed cases.

7.3.1.   Direct and Indirect Splits of Mixtures with Any Number of Components

At direct split $1 : 2, 3 \ldots n$ and at set small concentrations of impurities in products $(1 - \eta_D)$ and $(1 - \eta_B)$, one can quite precisely set the bottom product composition $x_B : x_{B1} = 1 - \eta_B$, $x_{B2} = x_{F2}/(1 - D/F) - D/F(1 - \eta_D)(1 - D/F)$, $Bx_{Bi} = Fx_{Fi}$ for $i = 3, 4 \ldots n$. Components are arranged in the order of decreasing phase equilibrium coefficients. For split $1 : 2,3,4$, it is important that component 1 is the lightest one in all the points of both section trajectories (both trajectories are located in the sharp split region Reg_{sh} for the direct split). It is also important that component 2 is the second in the value of phase equilibrium coefficient in point $S_r^1 \equiv N_r^-$ (i.e., in vertex 1). Calculation of the bottom section is carried out by method "tray by tray" from point x_B to a chosen point x_f within segment $[x_f^{\min}, x_f^\infty]$. This calculation is stable, because after an abrupt change of direction of the trajectory under calculation in the vicinity of point S_s^1 it is attracted to separatrix $S_s^1 - N_s^+$ and to node $N_s^+ \equiv x_f^\infty$ (see Fig. 7.2a). The composition in point x_{f-1} located inside trajectory bundle of the top section is determined from the material balance in the feed cross-section by the compositions in points x_F and x_f. Then the calculation of the top section is performed from point x_{f-1} to point $x_D \approx N_r^-$ (i.e., until the condition $x_{1j} \geq \eta_D$ is valid). The calculation of the top section is also stable because point x_{f-1} is located in the region of attraction Reg_{att} of node N_r^-, and the calculation trajectory is attracted to this node. For azeotropic mixtures at more minimum reflux like at minimum reflux (see Fig. 5.26b) is a set of attraction regions Reg_{att}. The working region Reg_{att} is determined by composition x_{f-1}. The column's trajectory at direct split may be put as follows:

$$
\begin{array}{ccccccc}
x_B & \rightarrow & qS_s^1 & \rightarrow & x_f & \Leftarrow\Downarrow & x_{f-1} & \rightarrow & qS_r & \rightarrow & x_D \\
\text{Reg}_B & & \text{Reg}_s^t & & [x_f] \in \text{Reg}_{sep,s}^{sh,R} & & [x_{f-1}] \in \text{Reg}_{att} & & \text{Reg}_r^t & & \text{Reg}_D
\end{array}
$$

To determine x_f^{opt} at which $(n_r + n_s)$ is minimum, it is necessary to perform several calculations of the column at different points x_f at segment $[x_f^{\min}, x_f^\infty]$. This algorithm was introduced in the work (Julka, 1993). The similar algorithm can also be used at indirect separation, but calculation should be executed top-down from point x_D.

At small set concentrations of impurities $(1 - \eta_D)$ and $(1 - \eta_B)$ to determine value $n_{\min} = (n_r + n_s)_{\min}$, one does not have to make iterations by product compositions x_B and x_D (i.e., for chosen points x_f and x_{f-1} calculation "tray by tray" is executed once).

However, if we want to achieve full satisfaction of the distillation equation system and to obtain precise product compositions x_B and x_D, it is necessary to execute iterations by these compositions (i.e., to take into consideration the fact that at the direct split not only the second component is an impurity one in the top product). These iterations become more necessary the larger the set value of $(1 - \eta_D)$ at the direct split or the set value of $(1 - \eta_B)$ at the indirect split.

The simplest organization of the iteration process is "simple iteration," when the found composition x_D or x_B from the previous iteration is used to determine composition x_B or x_D for the following iteration at the direct or indirect separation, respectively. Besides "simple iteration," one can also use other more complicated but more reliable and faster methods.

7.3.2. Intermediate Splits of Mixtures with Any Number of Components

The number of intermediate splits $1, 2, \ldots k : k + 1, k + 2 \ldots n$ equals $(n - 3)$, while at any value of n there is only one direct split and one indirect split. Therefore, at $n > 4$, intermediate splits prevail.

Along with that, the creation of design calculation algorithms for intermediate splits is a considerably more complicated task than that for direct and indirect splits. For intermediate splits, it is impossible to conduct design calculation from one end of the column to the other one.

The determination of the end coordinates of possible composition segments in the feed cross-section $[x_{f-1}]_{lin}^{sh}$ and $[x_f]_{lin}^{sh}$ at the assumption about linearity of the separatrix sharp split regions of the section $\text{Reg}_{sep}^{sh,R} \equiv S^1 - S^2 - N^+$ and about sharp separation is one of the most important elements of new algorithms of design calculation for intermediate splits (see Section 7.3.1).

The above-mentioned assumptions are valid entirely only under the condition of absense of impurity components in the separation products (i.e., absence of components $k+1, k+2\ldots n$ in the top product and of components $1, 2\ldots k$ in the bottom product) and also at $\alpha_i = const$ and $L_r = const$, $V_r = const$, $L_s = const$, $V_s = const$.

For real mixtures, at a set content of impurity components in separation products, the ends of segments $[x_{f-1}]$ and $[x_f]$ shift relatively to the ends of segments $[x_{f-1}]_{lin}^{sh}$ and $[x_f]_{lin}^{sh}$. This shifting takes place for several reasons:

1. For real mixtures, separatrix trajectory bundles $\text{Reg}_{bound}^{sh} \equiv S^1 - S^2 - N^+$ at sharp separation are curvilinear (calculations prove that this curvature is not big for any mixture).
2. At the decrease of sharpness of separation (at the increase of the set content of impurity components in separation products), segments $[x_{f-1}]$ and $[x_f]$ become longer and move from separatrix sharp split regions of the section $\text{Reg}_{sep}^{sh,R} \equiv S^1 - S^2 - N^+$ deep into the working bundles Reg_w^R.

Therefore, the new algorithms of design calculation include the correction of compositions in the feed cross-section x_{f-1} and x_f compared with compositions at segments $[x_{f-1}]_{lin}^{sh}$ and $[x_f]_{lin}^{sh}$.

The correction is realized on the basis of execution of the series of trial calculations of section trajectories by method "tray by tray" in the direction from the feed cross-section to the ends of the column or vice versa.

Theoretic analysis proved that the algorithms employing calculation from the feed cross-section (Petlyuk & Danilov, 2001a) can be applied only at $n = 4$, while algorithms that use calculation from the ends of column can be applied at any n.

At the design calculation, the following summary concentrations of impurity components in separation products are set (specified):

$$x_D^{imp} = x_{D,k+1} + x_{D,k+2} + \cdots + x_{D,n} \tag{7.1}$$
$$x_B^{imp} = x_{B,1} + x_{B,2} + \cdots + x_{B,k} \tag{7.2}$$

The distribution of non-key impurity components in the separation products $(x_{D,k+2}, x_{D,k+3} \ldots x_{D,n}, x_{B,1}, x_{B,2} \ldots x_{B,k+1})$ and the numbers of trays in column sections N_r and N_s are unknown.

Different summary concentrations of impurity components at the trays adjacent to the feed cross-section correspond to different ratios of tray numbers in column sections N_r/N_s:

$$x_{f-1}^{imp} = x_{f-1,\,k+1} + x_{f-1,\,k+2} + \cdots + x_{f-1,\,n} \tag{7.3}$$

$$x_f^{imp} = x_{f,1} + x_{f,2} + \cdots + x_{f,k} \tag{7.4}$$

In spite of the fact that points x_{f-1} and $(x_{f-1})_{lin}^{sh}$ and points x_f and $(x_f)_{lin}^{sh}$ that correspond to them in material balance do not coincide, the values of x_{f-1}^{imp} can be accepted the same as in any points of segment $[x_{f-1}]_{lin}^{sh}$, and the values of x_f^{imp} can be accepted the same as in the points of segment $[x_f]_{lin}^{sh}$ that correspond to points of segment $[x_{f-1}]_{lin}^{sh}$ in material balance.

Therefore, while executing a series of trial calculations of section trajectories, some point $(x_{f-1})_{lin}^{sh}$ at segment $[x_{f-1}]_{lin}^{sh}$ and point $(x_f)_{lin}^{sh}$ at segment $[x_f]_{lin}^{sh}$, corresponding to it in material balance, are chosen preliminarily.

The values of x_{f-1}^{imp} and x_f^{imp} are determined and fixed for these points, and trial calculations of section trajectories from column ends are realized further up to reaching at some section tray of these values of x_{f-1}^{imp} and x_f^{imp}. Therefore, the fractional numbers N_r and N_s are determined as a result of the trial calculation itself.

Trial calculations of the top section are realized at different little concentrations of non-key impurity components in the top product $x_{D,k+2}, x_{D,k+3} \ldots x_{D,n}$, and trial calculations of the bottom section are realized at different little concentrations of non-key impurity components in the bottom product $x_{B,1}, x_{B,2} \ldots x_{B,k-1}$.

Design calculation comes to the search of the concentrations of these components that would ensure validity of material balance in the feed cross-section. Such design calculation is realized for the fixed set of points $[(x_{f-1})_{lin}^{sh} - (x_{f-1}^{\infty})_{lin}^{sh}]/[(x_{f-1}^{\min})_{lin}^{sh} - (x_{f-1}^{\infty})_{lin}^{sh}]$, which determines $(N_r/N_s)_{opt}$, at which the total number of column trays is minimum.

The algorithm of design calculation includes a preliminary search for little concentrations of non-key impurity components in separation products and subsequently more precise definition of these concentrations.

The little concentrations $x_{D,k+2}^{\circ}, x_{D,k+3}^{\circ} \ldots x_{D,n}^{\circ}$ at which trajectory of top section comes to point x_{f-1}°, where the summary concentrations of the non-key impurity components of the top product $[x_{f-1}^{\circ}]_{nkey}^{imp}$ are the same as in fixed set of point $[(x_{f-1})_{lin}^{sh} - (x_{f-1}^{\infty})_{lin}^{sh}]/[(x_{f-1}^{\min})_{lin}^{sh} - (x_{f-1}^{\infty})_{lin}^{sh}]$; that is, $[x_{f-1}^{\circ}]_{nkey}^{imp} = [(x_{f-1})_{lin}^{sh}]_{nkey}^{imp}$ are determined at preliminary search. The top section trajectory at quasisharp separation after preliminary search may be presented as follows: $x_D^{\circ} \rightarrow qS_r^1 \rightarrow qS_r^2 \rightarrow x_{f-1}^{\circ}$.
$\qquad\qquad\qquad\qquad\qquad\qquad\qquad\qquad\qquad\quad \text{Reg}_D \quad\quad \text{Reg}_r'$
$\text{Reg}_r' \qquad \text{Reg}_{sep,r}^{qsh,R}$

The little concentrations $x_{B,1}^{\circ}, x_{B,2}^{\circ} \ldots x_{B,k-1}^{\circ}$ at which trajectory of bottom section comes to point x_f°, where the concentrations of the non-key impurity components of the bottom product are the same as in point $(x_f)_{lin}^{sh}$ corresponding to point $(x_{f-1})_{lin}^{sh}$ in material balance are determined in the same way; that is, $[x_f^{\circ}]_{nkey}^{imp} = [(x_f)_{lin}^{sh}]_{nkey}^{imp}$. The bottom section trajectory at quasisharp separation after preliminary search may be presented as follows: $x_B^{\circ} \rightarrow qS_s^1 \rightarrow qS_s^2 \rightarrow x_f^{\circ}$.
$\qquad\qquad\qquad\qquad\qquad\qquad\quad \text{Reg}_B \quad\quad \text{Reg}_s' \quad\quad \text{Reg}_s' \quad\quad \text{Reg}_{sep,s}^{qsh,R}$

For the above-described conditions of validity of trial calculations, the concentrations of the non-key impurity components in the feed cross-section are monotonously increasing functions of the concentrations of these components in the separation products. Besides that, in the vicinity of points x_{f-1}° and x_f°, the concentration of each non-key impurity component is a linear function of the concentrations of all non-key impurity components in the corresponding separation product.

Therefore, no calculation difficulties arise at the stage of preliminary search under consideration.

We note, nevertheless, that points x_{f-1}° and x_f° do not meet the conditions of material balance in the feed cross-section. Therefore, the following step of the algorithm of determination of the concentrations of the non-key impurity components in the separation products is necessary, as it will ensure the material balance.

Validity of this specifying step becomes easier by the fact that sought for points x_{f-1} and x_f are sufficiently close to already found points x_{f-1}° and x_f°.

Therefore, at the specifying step, one may accept that the concentration of each non-key impurity component in the feed cross-section is a linear function of the little concentrations of all the non-key impurity components in the corresponding separation product:

$$x_{f-1,i} \approx a_i + a_{i,k+2}x_{D,k+2} + a_{i,k+3}x_{D,k+3} + \cdots + a_{i,n}x_{D,n} \tag{7.5}$$

$$x_{f,i} \approx b_i + b_{i,1}x_{B,1} + b_{i,2}x_{B,2} + \cdots + b_{i,k-1}x_{B,k-1} \tag{7.6}$$

To determine coefficients $a_i, a_{i,k+2}, a_{i,k+3}, \ldots a_{i,n}$ and $b_i, b_{i,1}, b_{i,2}, \ldots b_{i,k-1}$, increments are given to the concentrations $\Delta x_{D,i}$ or $\Delta x_{B,i}$, while the concentrations of the rest of the components in the separation products $x_{D,j}^{\circ}$ or $x_{B,j}^{\circ}$ are fixed, and the calculation of section trajectories is carried out.

Then, using Eqs. (7.5) and (7.6), the system of equations for discrepancies of material balance in the feed cross-section is solved for all non-key impurity components in both separation products (in the described algorithm, the validity of material balance at non-key impurity components leads to balance validity at all components):

$$L_r \left(a_i + a_{i,k+2}x_{D,k+2} + a_{i,k+3}x_{D,k+3} + \cdots + a_{i,n}x_{D,n}\right) + L_F x_{F,i}$$
$$- L_s \left(b_i + b_{i,1}x_{B,1} + b_{i,2}x_{B,2} + \cdots + b_{i,k-1}x_{B,k-1}\right) = 0$$
$$(i = 1, 2, \ldots, k-1, k+2, k+3, \ldots, n) \tag{7.7}$$

By means of solution of Eq. (7.7), we determine more precise values of little concentrations of non-key impurity components in the separation products $x_{D,k+2}^1, x_{D,k+3}^1, \ldots, x_{D,n}^1, x_{B,1}^1, x_{B,2}^1, \ldots, x_{B,k-1}^1$, ensuring smaller values of discrepancies of material balance in the feed cross-section than the preliminarily found concentrations $x_{D,k+2}^{\circ}, x_{D,k+3}^{\circ}, \ldots, x_{D,n}^{\circ}, x_{B,1}^{\circ}, x_{B,2}^{\circ}, \ldots, x_{B,k-1}^{\circ}$.

To obtain solutions with a set precision, the above-described specifying step should be taken a few times, because Eqs. (7.5) and (7.6) are rough linear and become more and more precise, while obtained points x_{f-1}^{calc} and x_f^{calc} come nearer to the sought-for points x_{f-1} and x_f. The column trajectory at quasisharp separation, with the calculation finished, may be presented as follows:

$$
\begin{array}{cccccc}
x_D & \to & qS_r^1 & \to & qS_r^2 & \to & x_{f-1} & \Rightarrow\Downarrow & x_f & \leftarrow & qS_s^2 \\
\text{Reg}_D & & \text{Reg}_r^t & & \text{Reg}_r^t & & [x_{f-1}]\in\text{Reg}_{sep,r}^{qsh,R} & & [x_f]\in\text{Reg}_{sep,s}^{qsh,R} & & \text{Reg}_s^t
\end{array}
$$

$$
\begin{array}{cccc}
& \leftarrow & qS_s^1 & \leftarrow & x_B & . \\
& & \text{Reg}_s^t & & \text{Reg}_B
\end{array}
$$

To entirely solve the design task, the described algorithm is applied for the fixed set of points $(x_{f-1})_{lin}^{sh}$ at segment $[x_{f-1}]_{lin}^{sh}$ and for the fixed set of excess reflux factors $\sigma = R/R_{min}$.

Therefore, the above-described algorithm includes the following steps:

1. The stationary points of section trajectory bundles $S_r^1, S_r^2, S_r^3 \ldots N_r^+$ and $S_s^1, S_s^2, S_s^3 \ldots N_s^+$ are determined for the set value $\sigma = R/R_{min}$. We note that for azeotropic mixtures stationary points can be located not only at the corresponding boundary elements of the concentrations simplex, but also at the α-manifolds (the example of location of point $S_s^{2(\alpha)}$ at an α-surface is given in Fig. 5.28b).

2. The ends of segments $[x_{f-1}]_{lin}^{sh}$ and $[x_f]_{lin}^{sh}$ are determined for the set value of $\sigma = R/R_{min}$.

3. Preliminary values of little concentrations of impurity non-key components in separation products $x_{D,k+2}^\circ, x_{D,k+3}^\circ, \ldots, x_{D,n}^\circ, x_{B,1}^\circ, x_{B,2}^\circ, \ldots, x_{B,k-1}^\circ$ are determined for the set value of $\sigma = R/R_{min}$ and for the set point $(x_{f-1})_{lin}^{sh}$ at segment $[x_{f-1}]_{lin}^{sh}$.

4. Coefficients $a_{i,j}$ and $b_{i,j}$ of Eqs. (7.5) and (7.6) are determined for the set value of $\sigma = R/R_{min}$ and for set point $(x_{f-1})_{lin}^{sh}$.

5. The system of equations for componentwise discrepancies of material balance in the feed cross-section (Eq. [7.7]) is solved, determining specified values of little concentrations of the impurity non-key components in the separation products, for the set value of $\sigma = R/R_{min}$ and the set point $(x_{f-1})_{lin}^{sh}$.

6. Steps 4 and 5 are repeated to ensure the set precision of componentwise material balances in the feed cross-section.

7. Steps 3, 4, and 5 are repeated for different points $(x_{f-1})_{lin}^{sh}$ at segment $[x_{f-1}]_{lin}^{sh}$, determining point opt $(x_{f-1})_{lin}^{sh}$.

8. Points $1 \div 7$ are repeated for different values of $\sigma = R/R_{min}$.

The above-described algorithm can be somewhat modified for more precise determination of the value R_{min} and the coordinates of the segments $[x_{f-1}]_{lin}$ and $[x_f]_{lin}$ at nonsharp separation. The modified algorithm should take into consideration that at nonsharp separation in the mode of minimum reflux each product contains only key non-product components as impurity ones (i.e., for split $1, 2 \ldots k : k+1, k+2 \ldots n$, the top product contains impurity component $k + 1$, and the bottom product contains impurity component k):

$$x_{D,k+1} = 1 - \eta_D \tag{7.8}$$

$$x_{B,k} = 1 - \eta_B \tag{7.9}$$

Given the product compositions at nonsharp separation, one can determine stationary points of section trajectory bundles $S_r^2, S_r^3 \dots N_r^+$ and $S_s^2, S_s^3 \dots N_s^+$ for any values R. Points S_r^2 and S_s^2 will belong to the same boundary elements of the concentration simplex as at sharp separation, but they will be shifted relatively to points S_r^2 and S_s^2 for sharp separation. Apparently, this shift will be the bigger the smaller is the purity of products η_D and η_B.

Points $S_r^3 \dots N_r^+$ and $S_s^3 \dots N_s^+$ will be located inside the concentration simplex at some distance from those of its boundary elements to which they belong at sharp separation. This distance is the bigger the smaller is purity of products η_D and η_B. Points S_r^1 and S_s^1 are located outside the concentration simplex at some distance from its boundary elements, which they belong to at sharp separation.

Coordinates of these points cannot be determined precisely for a nonideal mixture because mathematic models used to describe phase equilibrium are determined only inside the concentration simplex. However, there is no necessity of that because only composition in the rest of the stationary points is significant for determination of the value $R_{\min}$ and coordinates of the ends of segments $[x_{f-1}]_{\text{lin}}$ and $[x_f]_{\text{lin}}$. The most rigorous variant of the algorithm requires that the composition in all stationary points, besides points S_r^1 and S_s^1, are determined for nonsharp separation. In other respects, the modified algorithm of design calculation does not differ from the one described above, taking into consideration the fact that segments $[x_{f-1}]_{\text{lin}}$, $[x_f]_{\text{lin}}$ and points $(x_{f-1})_{\text{lin}}$, $(x_f)_{\text{lin}}$ are determined instead of segments $[x_{f-1}]_{\text{lin}}^{sh}$, $[x_f]_{\text{lin}}^{sh}$ and points $(x_{f-1})_{\text{lin}}^{sh}$, $(x_f)_{\text{lin}}^{sh}$.

We note, however, that the main algorithm described above can be used in the majority of cases. It will be used in the examples given below. The modified algorithm is necessary only for the modes close to the mode of minimum reflux.

We now illustrate the algorithm described at the example of an ideal four-component mixture ($K_1 > K_2 > K_3 > K_4$) for split $1,2 : 3,4$. Feed composition $x_{F,i}$ and purity of products η_D and η_B are set.

Component 4 is the non-key impurity component in the top product, and component 1 is the non-key impurity component in the bottom product. Trial calculations of the top section are carried out until the summary concentration of components 3 and 4, which is equal to the concentration of these components in a chosen point $(x_{f-1})_{\text{lin}}^{sh}$ of segment $[x_{f-1}]_{\text{lin}}^{sh}$, is achieved at some tray. Similarly, trial calculations of the bottom section are performed until the summary concentration of components 1 and 2, which is equal to the concentration of these components in a chosen point $(x_f)_{\text{lin}}^{sh}$, is achieved at some tray.

The little concentrations of non-key impurity components in products $x_{D,4}$ and $x_{B,1}$ are determined during the search process with the help of the described algorithm, and the concentrations of the rest of the components in the products are

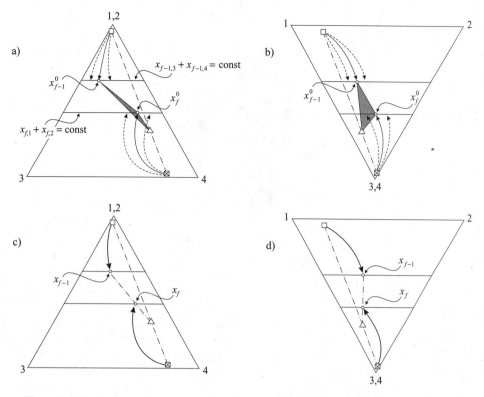

Figure 7.4. Normal projections to edges 1-2 (a,c) and 3-4 (b,d) of section trajectories at nonsharp intermediate split 1,2:3,4 of ideal mixture at different x_{D4} and x_{B1} (components 1 and 4 – non-key impurity components in bottom and overhead products, respectively) illustrating algorithm of design calculation. Dotted lines with arrows, trial trajectories; thin lines with arrows, trajectories at $x_{D,4}^{\circ}$, x_{f-1}°, $x_{B,1}^{\circ}$ and x_f°; thick lines with arrows, calculated trajectories at zero disbalancement in feed cross-section (i.e., at $x_{D,4}$, x_{f-1}, $x_{B,1}$ and x_f; area of shaded triangle is proportional to disbalancement in feed cross-section).

determined from the conditions of material balance, according to the set values of $x_{F,i}$, η_D, and η_B :

$$x_{D,1} = \frac{F}{D}x_{F,1} \qquad\qquad x_{B,2} = (1 - \eta_B) - x_{B,1}$$

$$x_{D,2} = \frac{F}{D}x_{F,2} - \frac{B}{D}(1 - \eta_B) \qquad x_{B,3} = \frac{F}{B}x_{F,3} - \frac{D}{B}(1 - \eta_D) \qquad (7.10)$$

$$x_{D,3} = (1 - \eta_D) - x_{D,4} \qquad x_{B,4} = \frac{F}{B}x_{F,4}$$

The preliminary determination of little concentrations $x_{D,4}^{\circ}$ and $x_{B,1}^{\circ}$ is shown in Fig. 7.4a,b, and their specified $x_{D,4}$ and $x_{B,1}$ determination is shown in Fig. 7.4c,d.

Figure 7.5 shows the application of this algorithm for equimolar zeotropic mixture pentane(1)-hexane(2)-heptane(3)-octane(4) at different products purities $\eta_D = \eta_B = 0,999; 0,99; 0,95; 0,9$. The values $L/V = 1.3(L/V)_{min}^{sh}$ and $[(x_{f-1})_{lin}^{sh} - (x_{f-1}^{\infty})_{lin}^{sh}]/[(x_{f-1}^{min})_{lin}^{sh} - (x_{f-1}^{\infty})_{lin}^{sh}] = 0,3$ were accepted as set parameters of design calculation. One can make out from this figure that, at a sufficiently

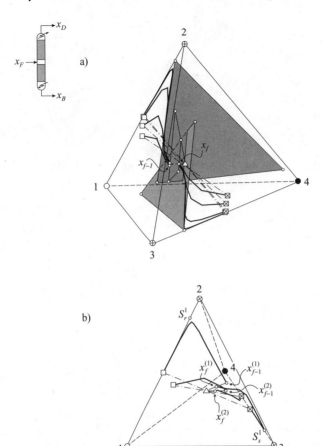

Figure 7.5. Section trajectories at quasisharp and nonsharp intermediate split for the equimolar pentane(1)-hexane(2)-heptane(3)-octane(4) mixture for $L/V = 1.3(L/V)^{sh}_{min}$, $(L/V)^{sh}_{min} = 0.614$, $x_{f-1} = x^{\infty}_{f-1} + 0.3\Delta x^{sh}_{f-1}$, $\eta_D = \eta_B = 0.999; 0.99; 0.95; 0.90$, separatrix sharp split section regions $\text{Reg}^{R\,3,4}_{sep,r}$ and $\text{Reg}^{R\,1,2}_{sep,s}$ are shaded (a); normal proection to face 1-2-3, $(1) - \eta_D = \eta_B = 0.999$, $(2) - \eta_D = \eta_B = 0.9$ (b).

sharp separation (at $\eta_D = \eta_B = 0.999$), calculation trajectories of sections come close to points S^1_r and S^1_s, and at nonsharp separation they are far from them. Correspondingly, at sharp separation the necessary tray number is big, and at nonsharp separation it is small.

The other projection at Fig. 7.5b shows two of these trajectories at $\eta = 0.999$ and $\eta = 0.9$. This projection makes it quite clear that at different purity values the points x_{f-1} and x_f are different: with a decrease of purity, these points move away from separatrix surfaces for sharp split $\text{Reg}^{R\,3,4}_{sep,r}$ and $\text{Reg}^{R\,1,2}_{sep,s}$, and they go deep into working trajectory bundles of the sections $\text{Reg}^{R\,3,4}_{w,r}$ and $\text{Reg}^{R\,3,4\,1,2}_{w,s}$. Table 7.1

Table 7.1. Product concentration of non-key impurity
components

$\eta_D = \eta_B$, mol. fraction	x_{D4}, mol. fraction	x_{B1}, mol. fraction
0.999	$6 \cdot 10^{-10}$	$3 \cdot 10^{-6}$
0.990	$5 \cdot 10^{-7}$	$2 \cdot 10^{-4}$
0.950	$1 \cdot 10^{-4}$	$4.4 \cdot 10^{-3}$
0.900	$1.2 \cdot 10^{-3}$	$1.71 \cdot 10^{-2}$

shows the little concentrations of non-key impurity components in the products $x_{D,4}$ and $x_{B,1}$ found for this example.

Figure 7.6 shows section trajectories for different points $(x_{f-1})^{sh}_{lin}$ at segment $[x_{f-1}]^{sh}_{lin}$ for this example at purities $\eta_D = \eta_B = 0.99$. Figure 7.7a shows how the tray number in the sections and in the whole column depends on these points. Similar dependences are shown at Fig. 7.7b for purity $\eta_D = \eta_B = 0.9$. A decrease of purity of products, as it was written above, leads to an expansion of segments $[x_{f-1}]$ and $[x_f]$. Therefore, the last points in Fig. 7.7 exceed the limits of segments $[x_{f-1}]^{sh}_{lin}$ and $[x_f]^{sh}_{lin}$.

Figure 7.8a shows the calculation trajectory of the sections at $\eta_D = \eta_B = 0.99$ and $[(x_{f-1})^{sh}_{lin} - (x^{\infty}_{f-1})^{sh}_{lin}]/[(x^{min}_{f-1})^{sh}_{lin} - (x^{\infty}_{f-1})^{sh}_{lin}] = 0.3$ for two values $(L/V)_r$, one of which is close to minimum. One can make out from this figure that at a mode

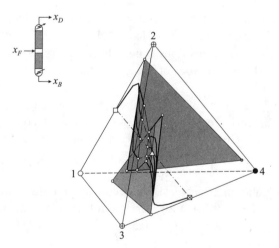

Figure 7.6. Section trajectories at quasisharp intermediate split for the equimolar pentane(1)-hexane(2)-heptane(3)-octane(4) mixture for $L/V = 1.3(L/V)^{sh}_{min}$, $(L/V)^{sh}_{min} = 0.614$, $\eta_D = \eta_B = 0.99$ at different points $(x_{f-1})^{sh}_{lin}$ at segment $[x_{f-1}]^{sh}_{lin}$. Separatrix sharp split section regions $\text{Reg}^R_{sep,r}$ and $\text{Reg}^R_{sep,s}$ are shaded.

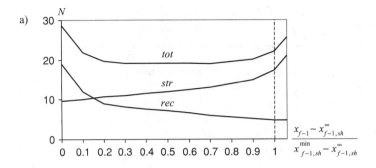

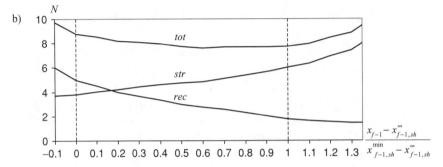

Figure 7.7. Number of trays in each column section at quasisharp intermediate split for the equimolar pentane(1)-hexane(2)-heptane(3)-octane(4) mixture for $L/V = 1.3(L/V)^{sh}_{min}$, $(L/V)^{sh}_{min} = 0.614$, $\eta_D = \eta_B = 0.99$ (a), and $\eta_D = \eta_B = 0.90$ (b) at different points $(x_{f-1})^{sh}_{lin}$ at segment $[x_{f-1}]^{sh}_{lin}$. *tot*, total, *rec*, rectifying, *str*, stripping.

close to the mode of minimum reflux the trajectories of sections come close to points S^2_r and S^2_s, in the vicinity of which they sharply change their direction. In the points of turn of trajectories compositions at neighboring trays show but little difference, which leads to the big number of trays in sections and in the whole column.

Figure 7.8b shows the dependence of tray number on the value of $(L/V)_r$ for this example. It is typical that an increase in the excess reflux factor leads to a sharp growth in the little concentration of non-key impurity components in the products: $x_{D,4} = 1.0 \cdot 10^{-12}$, $x_{B,1} = 1.4 \cdot 10^{-6}$ at $(L/V)_r = 0.455$, $x_{D,4} = 7.3 \cdot 10^{-7}$, $x_{B,1} = 3.1 \cdot 10^{-4}$ at $(L/V)_r = 0.713$.

Figure 7.9a shows calculation trajectories of sections for equimolar azeotropic mixture acetone(1)- benzene(2)-chloroform(3)-toluene(4) at $\eta_D = \eta_B = 0.99$; $(L/V)_r = 0,778$ and $(x_{f-1})^{sh}_{lin} = (x^\infty_{f-1})^{sh}_{lin} + 0.5[(x^{min}_{f-1})^{sh}_{lin} - (x^\infty_{f-1})^{sh}_{lin}]$. The separatrix trajectory bundle of the stripping section $Reg^R_{ssep,s}$ for this example has a number of characteristic peculiarities that are rendered the clearest by a projection perpendicular to edge 2-4 (Fig. 7.9b) to which the points of the bottom product belong.

The main peculiarity consists of the fact that at the value of parameter $(L/V)_r = 0,778$ the structural conditions of trajectory tear-off are broken in a

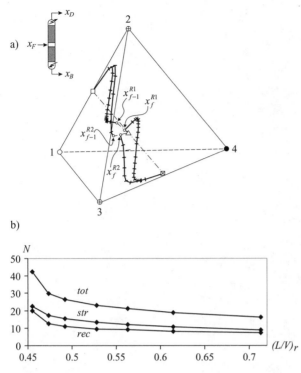

Figure 7.8. Section trajectories at quasisharp intermediate split for the equimolar pentane(1)-hexane(2)-heptane(3)-octane(4) mixture for $L/V = 0.455$ ($R1$) and $L/V = 0.713$ ($R2$), $x_{f-1} = x_{f-1}^{\infty} + 0.3\Delta x_{f-1}^{sh}$, $\eta_D = \eta_B = 0.99$ (a); dependence of trays number on the value of $(L/V)_r$ (b).

stationary point of the bundle located in face 2-3-4: this stationary point is located in component-order region Reg_{ord}^{3124}, where component 1 absent in face 2-3-4 is not the lightest one. Therefore, the stationary point in face 2-3-4 is the stable node $N_s^{+(2)}$ but not the saddle S_s^2.

Saddle point $S_s^{2(\alpha)}$ of the separatrix bundle $\text{Reg}_{sep,s}^{R}{}_{2,4}^{1,3}$ is located at the line of its intersection with α_{13}-surface (see also Fig. 5.28b). This generates in the separatrix bundle the separatrix $S_s^1 - S_s^{2(\alpha)}$ that breaks the whole bundle into two bundles $S_s^1 - S_s^{2(\alpha)} - N_s^{+(1)}$ and $S_s^1 - S_s^{2(\alpha)} - N_s^{+(2)}$. The trajectory of the bottom section should be located in the first of these bundles. This example makes it obvious that at the validity of design calculation it is necessary to examine stationary points S^2 located not only in the boundary elements of the concentration simplex, but also in the α-manifolds (for point x_B located on the edge 2-4 stationary points $S_s^{2(\alpha)}$ lie at different L/V on some line in the α_{13} – surface).

Other peculiarities of the separatrix bundle under consideration include the S-shape course of the trajectories and a big deviation of line $S_s^{2(\alpha)} - N_s^{+(1)}$ from linearity and, correspondingly, a big deviation of point x_f^{∞} from point $x_{f,lin}^{\infty}$. This deviation leads to a decrease in the length of segment $[x_f]$ as compared with that of segment $[x_f]_{lin}^{sh}$. Therefore, in Fig. 7.10, that shows dependence N_r, N_s, and N_{tot}

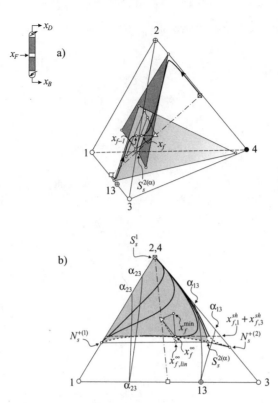

Figure 7.9. (a) Section trajectories at quasisharp intermediate split for the equimolar acetone(1)-benzene (2)-chloroform(3)-toluene(4) mixture ($L/V = 0.778$, $x_{f-1} = x_{f-1}^{\infty} + 0.5\Delta x_{f-1}^{sh}$, products purity – 0.99), separatrix sharp split section regions $\text{Reg}_{sep,r}^{R}\underset{1,3}{\overset{2,4}{}}$ and $\text{Reg}_{sep,s}^{R}\underset{2,4}{\overset{1,3}{}}$ are shaded, (b) stripping bundle a different $x_{B,1}$ (natural projection to edge 2-4) illustrating that the bundle contain the saddle point $S_s^{2(\alpha)}$ on α_{13}-surface if $(L/V)_s < K_1$ and $K_3 > K_1$ in point $N_s^{+(2)}$ on face 2-3-4. Working part of stripping separatrix sharp bundle $S_s^1 - S_s^{2(\alpha)} - N_s^{+(1)} \in \text{Reg}_{sep,s}^{R}\underset{2,4}{\overset{1,3}{}}$ is shaded.

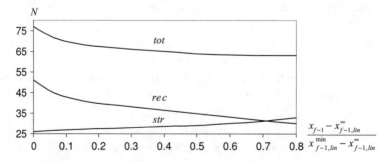

Figure 7.10. Number of trays in each column section and whole column at quasisharp intermediate split for the equimolar acetone(1)-benzene(2)-chloroform(3)-toluene(4) mixture for different x_{f-1} ($L/V = 0.778$, products purity – 0.99).

on parameter $[(x_{f-1})^{sh}_{lin} - (x^{\infty}_{f-1})^{sh}_{lin}]/[(x^{min}_{f-1})^{sh}_{lin} - (x^{\infty}_{f-1})^{sh}_{lin}]$, the maximum value of this parameter is not 1.0, but 0.8.

7.3.3. Splits with a Distributed Component

The number of splits with distributed components $1, 2, \ldots, k-1, k : k, k+1 \ldots n$ equals $(n-2)$. The coefficient of distribution of this component $\beta = \frac{x_{D,k}D}{x_{B,k}B}$ can be set (specified) arbitrarily. At the set value of β, the joining of section trajectories is possible at a unique pair of composition x_{f-1} and x_f. Optimal value of β_{opt} at which energy expenses for separation are minimum.

The existence of a unique pair of composition x_{f-1} and x_f leads to a necessity for a change in the algorithm that would make it different from the algorithm for intermediate splits without distributed components. The new algorithm includes the following steps:

1. The stationary points of trajectory bundles of the sections are determined for the set value of $\sigma = R/R_{min}$ in the same way as in the algorithm for splits without distributed components.

2. Points $(x_{f-1})^{sh}_{lin}$ and $(x_f)^{sh}_{lin}$, are determined for the set value $\sigma = R/R_{min}$. Coefficients of equations describing two straight lines of intersection of linear manifolds $x_F - S^1_r - S^2_r - N^+_r$ and $S^1_s - S^2_s - N^+_s$ and of linear manifolds $x_F - S^1_s - S^2_s - N^+_s$ and $S^1_r - S^2_r - N^+_r$ are determined for this purpose. Points $(x_{f-1})^{sh}_{lin}$ and $(x_f)^{sh}_{lin}$ meeting condition of material balance in the feed cross-section are found at these straight lines.

3. Preliminary values of little concentrations of impurity non-key components in separation products $x^{\circ}_{D,k+2}, x^{\circ}_{D,k+3}, \ldots x^{\circ}_{D,n}, x^{\circ}_{B,1}, x^{\circ}_{B,2}, \ldots x^{\circ}_{B,k-2}$ and preliminary values of number of trays N°_r and N°_s are determined for the set value of $\sigma = R/R_{min}$. For this purpose, the concentrations and tray numbers in each section are varied and trial calculations of sections from the ends of the column to tray composition points, the distance from which to point $(x_{f-1})^{sh}_{lin}$ or $(x_f)^{sh}_{lin}$ is minimal, are realized. At $x^{\circ}_{D,k+2}, x^{\circ}_{D,k+3}, \ldots x^{\circ}_{D,n}$ and N°_r, the trajectory of the top section is finished in point x°_{f-1}, the distance from which to point $(x_{f-1})^{sh}_{lin}$ is minimum of trial section calculations. This step of algorithm differs from the corresponding step of the algorithm for intermediate splits by the fact that tray numbers in the sections are independent variables during the process of search, but they are not determined during the process of calculation of section trajectories. A similar search is carried out for the bottom section. Section trajectories at quasisharp split after preliminary calculation may be put as follows:
$$\begin{array}{ccc} x^{\circ}_D & \to qS^1_r \to & x^{\circ}_{f-1} \\ \text{Reg}_D & \text{Reg}^t_r & \text{Reg}^{qsh,R}_{sep,r} \end{array}$$
and
$$\begin{array}{ccc} x^{\circ}_B & \to qS^1_s \to & x^{\circ}_f \\ \text{Reg}_B & \text{Reg}^t_s & \text{Reg}^{qsh,R\cdot}_{sep,s} \end{array}$$

Because for points x°_{f-1} and x°_f the material balance in the feed cross-section is not valid, and further more precise definition of little concentrations of non-key impurity components in separation products and of tray numbers in the section of the column is required.

4. The system of the equations for the componentwise discrepancies of the material balance in the feed cross-section is solved for the set value of $\sigma = R/R_{\min}$, determining the more precise values of little concentrations of the non-key impurity components in the separation products and of tray numbers in the column sections. The difference of that from the corresponding step of the algorithm for intermediate splits consists of the fact that tray numbers in the sections are included into the number of independent variables besides the concentrations of the non-key impurity components in the separation products. In accordance with that, it is accepted that the concentration of each component in the feed cross-section is a linear function not only of the little concentrations of the non-key impurity components in the corresponding product, but also of the tray numbers in the corresponding section:

$$x_{f-1,i} \approx a_i + a_{i,k+2}x_{D,k+2} + a_{i,k+3}x_{D,k+3} + \cdots + a_{i,n}x_{D,n} + a_{i,N_r} N_r \quad (7.11)$$

$$x_{f,i} \approx b_i + b_{i,1}x_{B,1} + b_{i,2}x_{B,2} + \cdots + b_{i,k-2}x_{B,k-2} + b_{i,N_s} N_s \quad (7.12)$$

The coefficients of these equations are determined in the same way as in the algorithm for intermediate splits without distributed components.

The system of linear equations of material balance in the feed cross-section looks as follows:

$$L_r(a_i + a_{i,k+2}x_{D,k+2} + a_{i,k+3}x_{D,k+3} + \cdots + a_{i,n}x_{D,n} + a_{i,N_r} N_r) + L_F x_{F,i}$$

$$- L_s(b_i + b_{i,1}x_{B,1} + b_{i,2}x_{B,2} + \cdots + b_{i,k-2}x_{B,k-2} + b_{i,N_s} N_s) = 0 \quad (7.13)$$

$$(i = 1, 2, \ldots, k-2, k+2, k+3, \ldots, n)$$

5. Specifying Step 4 is to be taken a few times to ensure the set precision. The column trajectory at quasisharp separation, with the calculation finished, may be presented as follows:

$$x_D \;\rightarrow\; q S_r^1 \;\rightarrow\; x_{f-1} \;\Rightarrow\Downarrow\; x_f \;\leftarrow\; q S_s^1 \;\leftarrow\; x_B$$
$$\mathrm{Reg}_D \quad \mathrm{Reg}_r^t \quad \mathrm{Reg}_{sep,r}^{qsh,R} \quad \mathrm{Reg}_{sep,s}^{qsh,R} \quad \mathrm{Reg}_s^t \quad \mathrm{Reg}_B$$

6. Steps $1 \div 5$ are repeated for different values $\sigma = R/R_{\min}$.

The described algorithm for nonsharp separation and for the modes that are close to the mode of minimum reflux can be made more rigorous in the same way as it was described above for the intermediate splits without distributed components. The content of impurity component $k+1$ in the top product and the content of impurity component $k-1$ in the bottom product should be considered while determining points $S_r^2, S_r^3 \ldots N_r^+$ and $S_s^2, S_s^3 \ldots N_s^+$ and by the modified algorithm.

Let's examine split $1,2 : 2,3,4$ $(K_1 > K_2 > K_3 > K_4)$ as an example. The summary concentration of impurity components 3 and 4 is set in the top product, and the concentration of component 1 is set in the bottom product. Besides that, the distribution coefficient β of component 2 or its concentration in the top product $x_{D,2}$ is set.

The concentration of the only non-key impurity component in the top product, that is, component 4, and of tray numbers in the sections are determined with the help of the described algorithm, and the concentrations of the rest of the components in the products are determined from the conditions of material balance according to the set values $x_{F,i}$, η_D, η_B, and β:

$$x_{D,1} = x_{F,1}\frac{F}{D} \qquad\qquad x_{B,1} = 1 - \eta_B$$

$$x_{D,2} = x_{F,2}\frac{F\beta}{D(1+\beta)} \qquad x_{B,2} = x_{F,2}\frac{F}{B(1+\beta)} \qquad (7.14)$$

$$x_{D,3} = (1 - \eta_D) - x_{D,4} \qquad x_{B,3} = x_{F,3}\frac{F}{B} - (1-\eta_D)\frac{D}{B}$$

$$x_{B,4} = x_{F,4}\frac{F}{B}$$

Figure 7.11 shows separatrix trajectory bundles $\operatorname{Reg}^{R}_{sep,r} \overset{3,4}{\underset{1,2}{}}$ and $\operatorname{Reg}^{R}_{sep,s} \overset{1}{\underset{2,3,4}{}}$ of the sections and calculation trajectories for equimolar mixture pentane(1)-hexane

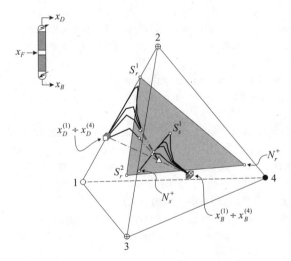

Figure 7.11. Section trajectories at quasisharp split with the distributed component 2 for the equimolar pentane(1)-hexane(2)-heptane(3)-octane(4) mixture $((L/V)_r = 0.5$, $x_{D,2} = 0.336$, product purity $-$ 0.999, 0.99, 0.98, 0.95); rectifying separatrix sharp split region $\operatorname{Reg}^{R}_{sep,r} \overset{3,4}{\underset{1,2}{}}$ is shaded.

Table 7.2. Product concentration of non-key impurity component and number of trays

$\eta_D = \eta_B$, mol. fraction	x_{D4}, mol. fraction	N_r	N_s	N_{tot}
0.999	$7.7 \cdot 10^{-8}$	7.06	14.29	21.35
0.990	$3.7 \cdot 10^{-5}$	3.68	7.63	11.31
0.980	$2.4 \cdot 10^{-4}$	2.69	5.68	8.37
0.950	$1.8 \cdot 10^{-3}$	1.75	3.22	4.97

(2)-heptane(3)-octane(4) at $(L/V)_r = 0.5$, $x_{D,2} = 0.336$ and at several values of purity of the products $\eta_D = \eta_B = 0.999, 0.99, 0.98$, and 0.95. Table 7.2 gives the values of $x_{D,4}$, N_r, N_s, and N_{tot} obtained as a result of calculation for this example.

7.3.4. Splits with Several Distributed Components: Preferred Split

We examine in conclusion splits with several distributed components $-1, 2, \ldots, k_1^{dist}, k_2^{dist}, \ldots, k_m^{dist} : k_1^{dist}, k_2^{dist}, \ldots, k_m^{dist}, \ldots n$.

Components $k_1^{dist}, k_2^{dist}, \ldots, k_m^{dist}$ are distributed ones, component $(k_1^{dist} - 1)$ is the light key one, component $(k_m^{dist} + 1)$ is the heavy key one, components $(k_m^{dist} + 2), \ldots, n$ are non-key impurity ones in the top product, and components $1, 2 \ldots, (k_1^{dist} - 2)$ are non-key impurity ones in the bottom product. For the splits under consideration, one can arbitrarily set the distribution coefficient for only one of the distributed components β ($\min \beta < \beta < \max \beta$). Distribution coefficients of the rest of the distributed components are some unknown functions of β. Therefore, the algorithm of design calculation for splits with several distributed components includes the search for distribution coefficients of these components.

The preferred split, for which $k_1^{dist} = 2$ and $k_m^{dist} = (n - 1)$, is an exclusion. Components 1 and n are the key ones in this case, and non-key impurity components are absent. Only one distribution coefficient for one component can be chosen. Distribution coefficients for other components should be determined by phase equilibrium coefficients of all the components in point x_F.

Points x_D and x_B should lie in the straight line passing through the liquid–vapor tie-line $x_F \to y_F$. The compositions of separation products can be determined from these conditions.

Design calculation for the set value of $\sigma = R/R_{\min}$ comes to the determination of tray numbers in the sections of the column N_r and N_s at which section trajectories are joined (i.e., the componentwise material balance in the feed cross-section is valid). The distillation trajectory may be put as follows:

$$x_D \quad \to \quad qS_r^1 \quad \to \quad x_{f-1} \quad \Rightarrow\Downarrow \quad x_f \quad \leftarrow \quad qS_s^1 \quad \leftarrow \quad x_B \ .$$
$$\text{Reg}_D \quad \text{Reg}_r^t \quad \text{Reg}_{rev} \qquad \text{Reg}_{rev} \quad \text{Reg}_s^t \quad \text{Reg}_B$$

Table 7.3. Comparison of the algorithms of design and simulation

Item	Design	Simulation
1	N_r and N_s should not be set.	N_r and N_s should be set.
2	Possibility of the chosen split for azeotropic mixtures is determined preliminarily during the process of calculation.	Possibility of the chosen split is not determined.
3	N_r and N_s are determined during the process of calculation.	Possibility of obtaining of products of a set purity at set N_r and N_s is not determined.
4	The solution of the task is always achieved.	The solution of the task can be unachieved, even if the chosen split and the set purity of products are feasible, for the reason of absence of convergence of iteration process.
5	Interference of the user in the process of calculation is not required.	A change in the estimated profiles of temperatures, compositions, and flow rates and other user-defined parameters of calculation process can be required to solve the task.
6	Minimum input information is required to solve the task.	Large amounts of input information can be required to solve the task.
7	Designed separation process is optimal in expenditures.	It is necessary to carry out a big volume of calculations to design the separation process optimal in expenditures.
8	See item 4.	No information about the reasons why the task was not solved is available.
9	The user can obtain the entire information (part of it visual) about the peculiarities of designed separation process.	The entire information about the peculiarities of projected separation process is not available.
10	No participation of qualified users is required to perform calculations.	Participation of qualified users is required to carry out calculations.
11	Conceptual designing is short.	Conceptual designing is long.

7.3.5. Advantages of New Design Algorithms

The described new algorithms of design calculation for different splits have great advantages compared with the known simulation algorithms. These advantages are listed in Table 7.3.

7.4. Design Calculation of Extractive Distillation Columns

The general approach to design calculation of extractive distillation columns is similar to the approach applied for two-section columns. We use our notions about the structure of intermediate section trajectory bundles (see Sections 6.4 ÷ 6.6), about possible compositions at the trays adjacent to the feed cross-section from above and below (see Section 7.2), and about possible directions of calculation of section trajectories (see Section 7.3). In contrast to two-section columns, for columns of extractive distillation, we have an additional degree of freedom of designing (the entrainer rate) and an additional cross-section of joining of section trajectories (the cross-section of input of the entrainer).

A number of works paid great attention to the questions of optimal designing of extractive distillation columns for separation of binary azeotropic mixtures (Levy & Doherty, 1986; Knight & Doherty, 1989; Knapp & Doherty, 1990; Knapp & Doherty, 1992; Wahnschafft & Westerberg, 1993; Knapp & Doherty, 1994; Bauer & Stichlmair, 1995; Rooks, Malone, & Doherty, 1996; 1993). The region of possible mode parameters of extractive distillation process, limited by minimum rate of the entrainer and by limits of changing of reflux number between minimum and maximum values, was investigated. Some heuristic rules were introduced for the choice of rate of the entrainer and the reflux number.

The development of the theory of intermediate section trajectory tear-off from boundary elements of concentration simplex (Petlyuk, 1984; Petlyuk & Danilov, 1999) expanded the application sphere of extractive distillation process to multicomponent mixtures.

This is especially important for the solution of the task of separation flowsheet synthesis of multicomponent azeotropic mixtures (see Chapter 8) because this, in many cases, uses autoextractive distillation (i.e., to exclude the application of entrainers).

The product purity is set (specified) at designing, while the tray numbers in the three sections of the column n_r, n_m, and n_s, the ratio between the flow rate of the entrainer, and the flow rate of the initial mixture E/F (it is frequently convenient to use, instead of this parameter, the ratio between the flow rate of the entrainer and the flow rate of top product E/D) and the reflux number in top section L/D (it is frequently convenient to use instead of this parameter the ratio of flow rates of liquid and vapor in the intermediate section $(L/V)_m$) are main design parameters that have to be determined.

The main part of the algorithm is the calculation of section trajectories and the determination of optimal tray numbers in the sections at set two mode parameters. This includes in the the same main stages as at the calculation of two-section columns: the calculation of reversible distillation trajectories of the three sections; the obtaining of linear equation systems for separatrix bundles of the three sections and for the manifolds, including the boundary elements of these bundles and point x_F; the determination of coordinates of possible composition segments in the feed cross-section and in that of input of the entrainer; the calculation of section trajectories by method "tray by tray."

As in the two-section column at reflux larger than minimum, in the column of extractive distillation at $(L/V)_m > (L/V)_m^{\min}$ there are possible composition segments at the trays adjacent to the feed cross-section from above and below $[x_{f-1}]_{lin}^{sh}$ and $[x_f]_{lin}^{sh}$ (Fig. 7.12). Same as for the two-section columns, the coordinates of the ends of these segments can be determined from purely geometric considerations from the known coordinates of the stationary point and point x_F (see Section 7.2).

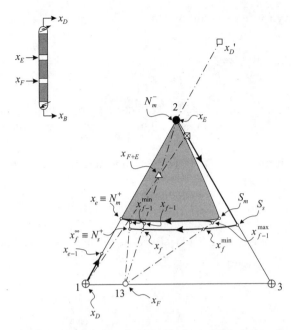

Figure 7.12. Calculated section trajectories for acetone (1)-water(2)-methanol(3) extractive distillation. $E/F = 2.03$, $R_r = 7$, x_F(0.6, 0.0, 0.4), the region of intermediate section trajectories $\mathrm{Reg}^R_{w,e}$ is shaded.

7.4.1. Three-Component Azeotropic Mixtures

In Fig. 7.12, $x_f^\infty = N_s^+$ and coordinates of point x_{f-1}^{min} tied with it by the conditions of material balance in the feed cross-section are determined with the help of Eq. (5.18). Point x_f^{min} is located at line $S_s - N_s^+(\mathrm{Reg}^{qsh,R}_{ssep,s})$, and point x_{f-1}^{max} is located at line $S_m - N_m^+(\mathrm{Reg}^{qsh,R}_{ssep,e})$. This determines coordinates of points $(x_{f-1}^{max})_{lin}^{sh}$ and $(x_f^{min})_{lin}^{sh}$. Along with that segment, $[x_{f-1}]_{lin}^{sh}$ can be located outside the working trajectory bundle of the intermediate section $N_m^- - S_m - N_m^+$ in Reg_{att} of point N_m^+. The same way as for two-section columns at segments $[x_{f-1}]_{lin}^{sh}$ and $[x_f]_{lin}^{sh}$, there are points x_{f-1}^{opt} and x_f^{opt} interconnected by conditions of material balance for which summary number of trays in the bottom and intermediate sections is minimum.

As far as the joining of trajectories of the top and the intermediate sections is concerned, the composition point at the first tray below the cross-section of the entrainer input x_e should be located quite close to the boundary element of the concentration simplex that contains components of the top product and of the entrainer (in Fig. 7.12 – to side 1-2). Allowable concentration of impurity components in point x_e is determined by the requirements to the purity of the top product. Therefore, the composition in point x_e is not an optimized parameter and the composition in point x_{e-1} at the first tray above the cross-section of entrainer input is determined by the conditions of material balance in this cross-section.

The distillation trajectory for the column under consideration may be presented as follows:

$$x_B \;\; \to \; qS_s \to \quad\quad x_f \quad\quad \Leftarrow\!\Downarrow \quad\quad x_{f-1} \quad\quad \to \; x_e \; \Leftarrow\!\Downarrow \; x_{e-1} \to \; x_D$$

$$\mathrm{Reg}_B \quad \mathrm{Reg}_s^t \quad [x_f] \in \mathrm{Reg}_{sep,s}^{qsh,R} \quad [x_{f-1}] \in \mathrm{Reg}_{sep,e}^{qsh,R} \quad \mathrm{Reg}_e^t \quad \mathrm{Reg}_{att} \quad \mathrm{Reg}_D$$

We now examine the calculation of section trajectories by method "from tray to tray" for various splits of multicomponent mixtures by extractive distillation.

7.4.2. The Multicomponent Mixtures: The Top Product and the Entrainer Are Pure Components ($m_r = 1$, $m_e = 2$)

Section 6.5 shows that the joining of section trajectories in the feed cross-section and in the cross-section of the entrainer input at such split is similar to direct separation in two-section column (Fig. 7.13a). Because the bottom product of the

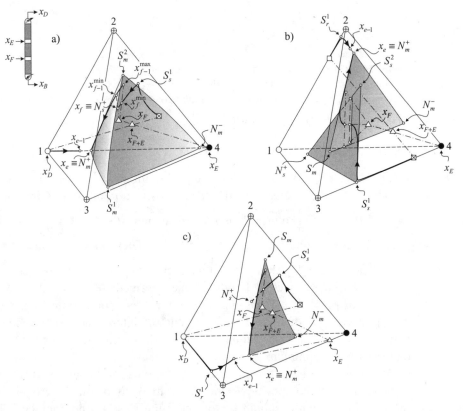

Figure 7.13. Section separatrix sharp split regions for extractive distillation of ideal four-component mixtures ($K_1 > K_2 > K_3 > K_4$): (a) $\mathrm{Reg}_{w,e}^{R}$ [2,3] (component 1 is overhead product, component 4 is entrainer), (b) $\mathrm{Reg}_{w,e}^{R}$ [3] and $\mathrm{Reg}_{sep,s}^{R}$ [1,2] [1,4] (mixture 1,2 is overhead product, component 4 is entrainer), (c) $\mathrm{Reg}_{w,e}^{R}$ [2] [1,2,4] (component 1 is overhead product, mixture 3,4 is entrainer). The separatrix sharp split surfaces ($\mathrm{Reg}_{sep,e}^{sh,R}$ [1,3,4] and $\mathrm{Reg}_{sep,s}^{sh,R}$ [3,4]) are shaded.

column contains only one impurity component (component 1), its composition can be set quite exactly. The calculation of trajectories of the sections is carried out by method "tray by tray" in an upward direction. The trajectory of the bottom section is attracted to separatrix sharp region $S_s^1 - N_s^+(\text{Reg}_{sep,s}^{sh,R})$ and to node N_s^+. The calculation is carried out until the achievement of chosen point x_f at segment $[x_f^{min}, x_f^\infty]$. To determine x_f^{opt}, this calculation is carried out many times, finishing in different points. The composition in point x_{f-1} is determined from the conditions of material balance in the feed cross-section. After that, the calculation of the intermediate section from point x_{f-1} to point x_e is carried out. The trajectory of the intermediate section is attracted to separatrix bundle $S_m^1 - S_m^2 - N_m^+(\text{Reg}_{sep,e}^{sh,R})$ and to node point N_m^+. The composition in point x_{e-1} is determined from the conditions of material balance in the cross-section of entrainer input, and the calculation of the top section from point x_{e-1} to point x_D is carried out. The trajectory of the top section is attracted to node point N_r^- of edge of the concentration simplex $1 - 4(\text{Reg}_{sep,r}^{sh,R})$. The whole calculation is quite stable and it does not require iterations by product composition. The distillation trajectory for the column under consideration may be presented as follows:

$$x_B \;\rightarrow\; qS_s^1 \;\rightarrow\; x_f \;\Leftarrow\!\Downarrow\; x_{f-1} \;\rightarrow\; x_e \;\Leftarrow\!\Downarrow\; x_{e-1} \;\rightarrow\; x_D$$
$$\text{Reg}_B \quad \text{Reg}_s^t \quad [x_f] \in \text{Reg}_{sep,s}^{qsh,R} \quad [x_{f-1}] \in \text{Reg}_{sep,e}^{qsh,R} \quad \text{Reg}_e^t \quad \text{Reg}_{att} \quad \text{Reg}_D$$

The way it is presented here is the same we used for three-component mixtures, but here the number of components in points x_B and qS_s^1 is 3, not 2, and that in points x_f and x_{f-1} is 4, not 3.

7.4.3. The Multicomponent Mixtures: The Top Product Is a Binary Mixture, the Entrainer Is a Pure Component ($m_r = 2$, $m_e > 2$)

According to Section 6.5, in this case the joining of trajectories of the bottom and the intermediate sections is similar to the intermediate split in two-section columns, and the joining of trajectories of the top and the intermediate sections takes place according to the split with one distributed component. Therefore, the calculation of section trajectories should be carried out according to the general algorithm described in Section 7.3 for two-section columns at the intermediate split and at the split with a distributed component (Fig. 7.13b). The distillation trajectory for the column under consideration may be presented as follows:

$$x_B \;\rightarrow\; qS_s^1 \;\rightarrow\; x_f \;\Leftarrow\!\Downarrow\; x_{f-1} \;\rightarrow\; x_e \;\Leftarrow\!\Downarrow\; x_{e-1} \;\leftarrow\; qS_r^1 \;\leftarrow\; x_D$$
$$\text{Reg}_B \quad \text{Reg}_s^t \quad [x_f] \in \text{Reg}_{sep,s}^{qsh,R} \quad [x_{f-1}] \in \text{Reg}_{sep,e}^{qsh,R} \quad \text{Reg}_e^t \quad \text{Reg}_{sep,r}^{qsh,R} \quad \text{Reg}_r^t \quad \text{Reg}_D$$

7.4.4. The Multicomponent Mixtures: The Top Product Is Pure Component, the Entrainer Is a Mixture ($m_r = 1$, $m_e > 2$)

According to Section 6.5, in this case the joining of trajectories of the bottom and intermediate sections is similar to joining of sections of two-section column with a

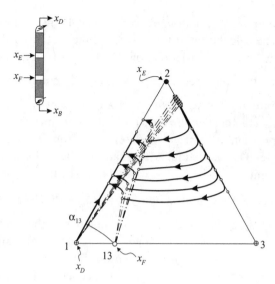

Figure 7.14. Calculated section trajectories for acetone (1)-water(2)-methanol(3) extractive distillation at $(L/V)_m/(L/V)_m^{\min} = 1.3$ and different E/D. The little circles are compositions at trajectories tear-off points and at feeds cross-section.

distributed component. This means that instead of segment $[x_f^{\min}, x_f^{\infty}]$ there is only point x_f (Fig. 7.13c). The distillation trajectory for the column under consideration may be presented as follows:

$$
\begin{array}{ccccccc}
x_B & \to q S_s^1 \to & x_f & \Leftarrow\Downarrow \ x_{f-1} & \to \ x_e & \Leftarrow\Downarrow \ x_{e-1} & \to \ x_D \\
\text{Reg}_B & \text{Reg}_s^t & \text{Reg}_{sep,s}^{qsh,R} & \text{Reg}_{sep,e}^{qsh,R} & \text{Reg}_e^t & \text{Reg}_{att} & \text{Reg}_D
\end{array}.
$$

The task of designing of extractive distillation columns, besides calculation of section trajectories, includes a number of subtasks. These are the same subtasks as for two-section columns and additional subtasks of determination of minimum entrainer flow rate and of choice of design entrainer flow rate. Optimal designing of extractive or autoextractive distillation includes optimization by two parameters – by entrainer flow rate and by reflux number. Figure 7.14 shows influence of entrainer flow rate on section trajectories at fixed value of parameter $\sigma = (L/V)_m/K_j^t$ (as is shown in Section 6.4 $(L/V)_m^{\min} = K_j^t$).

The entrainer flow rate influences expenditures for separation not only in extractive distillation column itself, but also in the column of the entrainer recovery. In the case of separation of a multicomponent azeotropic mixture in an autoextractive distillation column (see Chapter 8), the intermediate columns can be located between this column and the column of autoentrainer recovery. In this case, the flow rate of the entrainer also influences expenditures for separation in the intermediate columns. In connection with the aforesaid, the necessity arises to carry out design calculation at several values of parameter E/D and, for each value of E/D, at several values of parameter L/D.

The general algorithm of design calculation includes the following steps:

1. Calculation of minimum flow rate of the entrainer or autoentrainer $(E/D)_{min}$ (see Section 6.6).
2. Determination of the average calculated value of the rate of the entrainer or autoentrainer $(E/D)_{mean}= 3(E/D)_{min}$ according to the heuristic rule (Knapp & Doherty, 1994) and of a number of other calculated values of E/D in the set interval.
3. Calculation of minimum reflux number $(L/D)_{min}$ at $(E/D)_{mean}$ (see Section 6.5).
4. Determination of the average calculated value of reflux number $(L/D)_{mean}= 1,5(L/D)_{min}$ according to heuristic rule (Knight & Doherty, 1989) and of a number of other calculated values of L/D in the set interval.
5. Calculation of necessary trays number at their optimal distribution in sections at different set values of parameters E/D and L/D and choice of design values of these parameters.

We note that there is but a slight dependence between the values of $(L/D)_{min}$ and E/D. This allows us to confine ourselves to the single-stage calculation of $(L/D)_{min}$ and to use the same series of values of L/D at different values of E/D.

The final choice of design and mode parameters is carried out taking into consideration expenditures on separation not only in the column of autoextractive distillation, but also in that of the entrainer or autoentrainer recovery and in the intermediate columns, if there are any.

7.5. Design Calculation of "Petlyuk Columns" and of Columns with Side Sections

7.5.1. Design Calculation of "Petlyuk Columns"

Imperfections of application of simulating software for the purposes of designing distillation complexes sharply grow compared with those of two-section columns because of the dramatic increase in the number of degrees of freedom of designing. Therefore, the application of algorithms of design calculation to distillation complexes is of especially great importance. In connection with this, a number of simplified methods of design calculation of Petlyuk columns that should precede rigorous simulation were developed. Simplified methods were based on limitations and assumptions such as ternary mixtures, sharp separation, liquid feeding, constant molar flows, and constant relative volatilities of components (Fonyo, Scabo, & Foldes, 1974; Tedder & Rudd, 1978; Nikolaides & Malone, 1987; Cerda & Westerberg, 1981; Triantafyllou & Smith, 1992). A great number of assumptions and empirical ratios used in these simplified methods do not meet set requirements

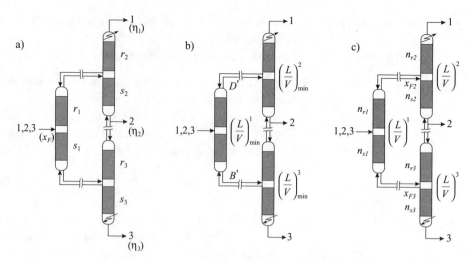

Figure 7.15. Calculation of Petlyuk columns: (a) specifications (in brackets), (b) calculation of minimum reflux and reboil ratios and product distribution in two-section columns, and (c) calculation of design variables.

to the quality of the products after transition to rigorous simulation and lead to nonoptimal design solution. Rigorous models and methods of mathematical optimization are used in the work (Dunnebier & Pentelides, 1999), but its application comes across the difficulties that were mentioned in Section 7.1.

A new simplified method was introduced in the work (Amminudin et al., 2001). This method is based on preliminary determination of product compositions meeting set requirements to their quality with the help of solution of Underwood equation system (1948) for the mode of minimum reflux. The number of trays in sections is determined by means of calculation by method "tray by tray" from the ends of each two-section column entering into the distillation complex at the assumption about constancy of molar flows and about the above-mentioned product compositions. The joining of section trajectories is carried out in a simplified way. Optimal design solution is obtained by means of minimization of objective function, including energy and capital expenses. Minimization is carried out by reflux and vapor ratios in each two-section column entering into distillation complex. This method does not have many of the imperfections of other simplified methods. It embraces Petlyuk columns, products of which contain not only one, but also several components. This method besides a number of assumptions does not take into consideration the general regularities of location of distillation trajectories, which can lead to "nonjoining" of the calculation trajectories starting at the ends of two-section columns (see Section 7.3).

We state below a rigorous method of design calculation of Petlyuk columns and of columns with side sections based on the design calculation of two-section columns described in Section 7.3.

At design of Petlyuk columns, as of distillation columns and complexes of other types, the purity of separation products is a set (specified) parameter, while the tray numbers for all sections n_{r1}, n_{s1}, n_{r2}, n_{s2}, n_{r3}, and n_{s3} (Fig. 7.15) and reflux and vapor ratios in each section are calculated variables (it is convenient to use instead of them ratios of rates of liquid and vapor $(L/V)^{r1}$, $(L/V)^{s1}$, $(L/V)^{r2}$, $(L/V)^{s2}$, $(L/V)^{r3}$, $(L/V)^{s3}$. Instead of parameter $(L/V)^{s1}$ for the first column (prefractionator), it is convenient to use the distribution coefficient of flows D'_1/B'_1, where $D'_1 = D_2 + B_2$, $B'_1 = D_3 + B_3$, and $B_2 + D_3 = S$ is the side product of the main column.

It is expedient to determine the value of optimal distribution coefficient D'_1/B'_1 at the stage of calculation of minimum reflux mode (see Section 6.8). In particular, at separation of a three-component mixture, the optimal value of coefficient D'_1/B'_1 corresponds to the preferable separation (i.e., can be found most easily, see Section 7.3.4). In the general case at separation of a multicomponent mixture, the optimal value of D'_1/B'_1 should be preliminary determined by means of the solution of Underwood equation system in the case of sharp separation of a multicomponent mixture with several distributed components. Then the optimal value of D'_1/B'_1 should be specified by means of variation of pseudoproduct composition and minium reflux calculations (see Section 5.6)

The parameters $(L/V)^{s2}$ and $(L/V)^{s3}$ can be determined from the condition of material and heat balances in the second and third two-section columns at chosen values of $(L/V)^{r2}$ and $(L/V)^{r3}$. Therefore, after exclusion of the above-mentioned mode parameters, three mode parameters $(L/V)^{r1}$, $(L/V)^{r2}$, and $(L/V)^{r3}$ remain.

Because minimum values of these parameters were determined before at the stage of calculation of the mode of minimum reflux (see Section 6.8), design their values should be chosen reasoning from economic considerations taking into account energy and capital expenditures. This choice is similar to that of optimal reflux excess coefficient for two-section columns. Along with that, the equality of vapor flow rates in the second and third columns in the cross-section of output of side product is taken into consideration.

The main purpose of design calculation is to determine necessary tray numbers for all sections at fixed values of mode parameters. At design calculation, one takes into consideration the equality of compositions at the tray of output of the side product obtained at the calculation of the second and third columns. Each two-section column entering into a Petlyuk column is calculated with the help of algorithms described before for two-section columns. The algorithm of calculation for splits with a distributed component is used for the first column, the algorithms for the direct and the indirect splits are used for the second, and the third columns at separation of a three-component mixture, respectively. At separation of multicomponent mixtures, the algorithms for intermediate separation are used.

The calculation of the first column is carried out at set compositions of its pseudoproducts $x_{D'}$ and $x_{B'}$ and at compositions at the ends of the column $x_{up} = S_{r1}$ and $x_{low} = S_{s1}$, corresponding to the maximum concentration in the trays of the component intermediate by bubble temperature at separation of three-component mixture (see Section 6.8.3) or heavy key component of the side product in point x_{up} and light key one in point x_{low} at separation of multicomponent mixture (for split $1, 2 \ldots k, k + 1, \ldots l : k + 1, \ldots l, l + 1, \ldots n$ component $(k + 1)$ is a light key component of the side product and l is a heavy key one).

Purity of intermediate pseudoproducts D_1' and B_1' are important variables at the calculation of tray numbers. In separating three-component mixture, it is necessary to determine the concentration of heavy component in top pseudoproduct D_1' and of light component in bottom pseudoproduct B_1'. These concentrations are determinated by the means of series of calculations of three columns of distillation complex from the condition of equality of compositions at the tray of output of the side product obtained at calculation of the second and third columns for given purity of the side product. Concentrations of key components of the side product in pseudoproducts D_1' and B_1' at separation of multicomponent mixture are determined in the same way.

The general algorithm of design calculation of Petlyuk columns includes the following stages:

1. Determination of optimal distribution coefficient D_1'/B_1', pseudoproduct compositions of the first column, and calculation of the mode of minimum reflux in first column (a) from the preferable split in the case of separation of a three-component mixture (see Section 6.8) or (b) by means of the solution of the Underwood equation system and up-to-date calculation in the case of sharp separation of a multicomponent mixture with several distributed components.
2. Calculation of tray numbers of three two-section columns beginning with the first column by algorithms of design calculation of two-section columns at fixed parameters L/V and fixed purity of pseudoproducts D_1' and B_1' of the first column (see Section 7.3).
3. Determination of purity of pseudoproducts D_1' and B_1' reasoning from the equality of compositions in the cross-section of output of the side product obtained at calculation of the second and third columns for given purity of the side product by means of the series of calculations at different purities of pseudoproducts.
4. Calculation of tray numbers at different values of parameters $(L/V)^{r1}$, $(L/V)^{r2}$, $(L/V)^{r3}$ and choice of design values of these parameters.

7.5.2. Design Calculation of Columns with Side Sections

Columns of this type with side strippings were used for petroleum refining for many decades. Design of petroleum refining units is based on usage of the existing experience and on simulation trials. The procedure of designing is nonsystematic. As a rule, distillation columns are designed at the beginning and then the system of heat exchangers used for heating of petroleum and for recuperation of the heat withdrawn from the distillation columns is designed. The procedure of designing is described in old works (Nelson, 1936; Packie, 1941; Watkins, 1979). This procedure includes correlations based on empirical data about tray numbers, reflux, sharpness of separation, and amount of water vapor for stripping. It was proposed in a number of works to decompose the column with side strippings into a sequence of columns with indirect split (Hengstebeck, 1961; Glinos & Malone, 1985; Carlberg & Westerberg, 1989; Liebmann, Dhole, & Jobson, 1998). In the latter work, it is proposed, besides that, to design distillation columns and the system of heat exchangers simultaneously using pinch analysis (Dhole & Linnhoff, 1993). The calculation of each two-section column is carried out with the help of simplified methods that examine the following modifications: (1) absence of thermal coupling between two-section columns (absence of liquid flows between two-section columns), (2) its availability, (3) steam stripping, and (4) stripping with the help of a steam-heated reboiler. This method considerably decreases expenditures of energy for separation (approximately by 20%).

We propose a method of design calculation based on decomposition of the column with side strippings into a system of two-section columns, on rigorous design calculation of each two-section column and on simultaneous design of distillation columns and of system of heat exchange.

The main task of designing columns with side strippings is the determination of necessary number of plates in each section, of optimal thermal duties on pumparounds and of the rates of steam for stripping or thermal duties on reboilers.

The main element of design calculation is the determination of necessary tray numbers in each two-section column at fixed-mode parameters. In each two-section column entering into the column with side strippings, intermediate separation without distributed product components decreases the number of steps of design calculation compared with calculation of Petlyuk columns. Two-section columns entering into the column with side strippings are calculated consecutively from the column to where the initial mixture arrives. Calculation of two-section columns is carried out at set compositions of pseudoproducts that are determined by the set purity of the corresponding products. The composition at the end of the two-section column where it joins the following two-section column corresponds to the maximum concentration of the heavy key component of the following two-section column. If, for example, split $1, 2 \ldots k : k + 1, \ldots l$ is in the following two-section column, then the mentioned key component is component $k + 1$.

Ratios of the flow rates of liquid and vapor in two-section columns are determined by reasoning from the calculated mode of minimum reflux and specified set of values of reflux excess coefficients, taking into consideration the output of heat by pumparounds decreasing the vapor flow rate passing from one two-section column into another. The amount of heat withdrawn by pumparounds is determined through application of the pinch method (Liebmann et al., 1998).

The same method is used to choose the method of stripping: with the help of steam, reboiler, or a combination of steam and reboiler. Application of reboilers compared with steam stripping, on the one hand, requires additional investment cost and heat expenditures, but on the other hand, it leads to a decrease of vapor flow in the column (i.e., to a decrease of necessary diameter and to an increase of the temperature of stripping product used for heating the crude oil). Besides that, the application of reboilers is inadmissible for heavy products because the resulting increase of the product's temperature may lead to its chemical decomposition. In the case when application of reboilers is admissible, the choice between reboilers,

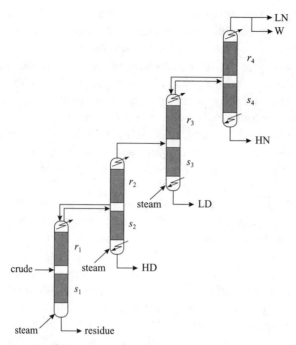

Figure 7.16. Calculation of crude oil column with side strippings using pinch analysis. LN, light naphtha; HN, heavy naphtha; LD, light diesel; HD, heavy diesel.

steam stripping, and a mixed variant is determined from economic considerations by means of construction of "temperature-enthalpy" curves for heat sources and heat sinks (pinch method).

The choice between the absence and availability of liquid flow from one two-section column into the other and the determination of optimal flow rate are performed using the same method. Such optimization simultaneously concerns the distillation column and the system of heat exchangers. As a result of such optimization in different two-section columns entering into the column with side strippings, it turns out to be profitable to use the different modification (reboilers or steam stripping, availability or absence of liquid flow between columns) (Fig. 7.16).

Special requirements to the refining property make necessary a preliminary step to the design procedure: the determination of the set of pseudocomponents specified by their normal bubble temperatures (at 1 atmosphere) in each product, and the determination of admissible concentration of impurity components in each product.

The general algorithm of design calculation of the column with side strippings includes the following stages (the algorithm for columns with side rectifiers includes similar stages):

1. Determination of the sets of product pseudocomponents and of admissible concentrations of impurity pseudocomponents in each product.
2. Calculation of the mode of minimum reflux in the first two-section column.

3. Calculation of necessary tray numbers in the sections of the first two-section column at the specified set of excess reflux coefficients.
4. Choice of the following parameters for the first two-section column: (1) excess coefficient of reflux, (2) the way of stripping, and (3) duty on pumparound.
5. Repetition of items $2 \div 4$ for the rest of two-section columns.

7.6. Determination of Necessary Tray Numbers at Heteroazeotropic and Heteroextractive Distillation

For homogeneous mixtures, all steps of conceptual design calculation from the choice of the split until determination of all main design and mode parameters can be carried out almost without any participation of the designer. For heterogeneous mixtures, the situation is more complicated. The structure of the field of equilibrium coefficients liquid–liquid (regions of the existence of two liquid phases) is superimposed on the structure of the field of equilibrium coefficients liquid–vapor (distillation regions, component-order regions). This leads to a great variety of possible conditions of separation at heteroazeotropic and heteroextractive distillation of various mixtures. Various configurations of the columns (one, two, or three sections), and various sequences of columns and decantors, are possible.

It is necessary to solve the questions of optimal arrangement and of optimal sequence for each particular mixture by means of conduction of calculation investigations. At this stage, there is a necessity for participation of a specialist possessing software for calculation of necessary tray numbers. This software should guarantee the determination of necessary tray numbers in sections, the determination of the best conditions of refluxing (reflux with one or two phases at their optimal correlation) and the determination of the best number of sections.

Questions of optimal designing of heteroazeotropic distillation units were discussed in the works (Bril et al., 1974; Bril et al., 1975; Bril et al., 1977; Bril et al., 1985; Ryan & Doherty, 1989; Pham & Doherty, 1990a; Pham & Doherty, 1990b; Pham & Doherty, 1990c).

The preliminary stage of design – collection of experimental data on phase equilibrium liquid–vapour (VLE) and liquid–liquid (LLE) for binary and ternary constituents of the mixture under separation and creation of adequate model on this basis – is of great importance for heteroazeotropic mixtures.

Works of a number of investigators have shown that usage of data only on binary VLE does not create an adequate model that would satisfactorily describe the conditions at heteroazeotropic or heteroextractive distillation. The same can be said of the usage of data only on ternary LLE.

To adequately describe three-phase equilibrium liquid–liquid–vapor (VLLE), with the help of such models of solution as NRTL and UNIQUAC, it is necessary to find parameters of these models by means of simultaneous processing of experimental data on binary VLE and ternary LLE.

The following preliminary stage of designing is analysis of the structure of concentration space (i.e., the analysis of location in concentration simplex of node and saddle stationary points) of regions of the existence of two liquid phases, of liquid–liquid tie-lines for heteroazeotropes, of component-order regions, and of residue curves. The most interesting splits in the main distillation column with a decantor are chosen, taking into consideration this analysis (see Section 6.9).

The following central stage of designing is design calculation of the main column with a decantor for a chosen split (i.e., the determination of the necessary number of trays and sections and of the best conditions of refluxing).

The method "tray by tray" in the direction upward along the column with posterior calculation of decantor at corresponding organization of iterations is the fastest and the most reliable method to solve the set task. Such choice of calculation methods is conditioned by the fact that, for the most interesting splits discussed in Section 6.9, the column complete with the decantor is an analog of a simple column working at direct split (see Section 7.3). Besides stability of such calculation, its additional advantage at heteroazeotropic distillation is in the possibility to carry out the determination of the necessary number of sections in the column (one or two) and of the necessary number of phases brought in into reflux (one or two) in the process of calculation itself "tray by tray." It is supposed that the choice of one section and refluxing by one phase are preferable. Therefore, the calculation is begun, taking into consideration these preferable conditions. If during the calculation it turns out that at these conditions the required result of separation cannot be achieved, they are rejected.

We examine the main steps of the algorithm. The given data at heteroazeotropic and heteroextractive distillation are concentration of impurity components in the bottom product. This information unambiguously determines the bottom product composition if one component is impurity one (see Section 6.9 Figs. 7.16e,f and 7.17a,b). If two or more components are impurity ones (Fig. 6.16a÷d), then the bottom product composition is set in initial approximation, taking into consideration the ratios of phase equilibrium coefficients of impurity components in the bottom product point. In this case, the bottom product composition is defined more exactly later at iterations.

For the beginning of calculation by method "tray by tray," it is necessary to also determine the estimated composition at the top end of the column y_D. In the cases in Fig. 6.16e,f, this composition is determined by the set concentration of impurities in the top product and, in the other cases in Figs. 6.16 and 6.17, this composition is set at the intersection of vapor line with a certain liquid–liquid tie-line close the heteroazeotropic liquid–liquid tie-line (the distance in the concentration space from the point of vapor going into the decantor to this tie-line is set). For heteroextractive distillation, it is necessary to also set the flow rate of entrainer determining preliminarily its minimum flow rate (see Section 6.6).

Based on these data, the material balance of the column is calculated, the composition of gross-feeding x_{F+E}, compositions, and flow rates of phases L_1 and L_2, and flow rates of vapor and liquid in column sections at refluxing by one phase are determined.

Then the calculation of the bottom section by method "tray by tray" is carried out until obtaining at some tray the vapor composition corresponding to

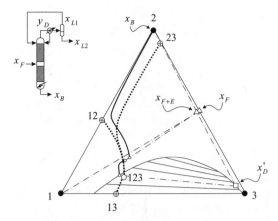

Figure 7.17. Calculated section trajectories for ethyl acetate(1)-ethanol(2)-water(3) heteroazeotropic distillation (ethyl acetate is entrainer).

the before-chosen liquid–liquid tie-line or until the achievement of constant concentrations zone in point N_s^+. In the first case, it is accepted that the preferable conditions (one section and one phase of refluxing) can be preserved, and iterations by product compositions are carried out. In the second case, one proceeds to calculation of the second section accepting $x_f = N_s^+ - \Delta x$, Δx is set beforehand difference of concentrations of one of the components. If at some tray, vapor composition corresponding to chosen before liquid–liquid tie-line is obtained then the number of sections should be equal to two and refluxing should be carried out by one phase. In the opposite case, one gradually increases vapor ratio in the bottom section proceeding to refluxing by two phases and determining optimal vapor ratio and corresponding to it fraction of the second phase in the reflux. In all cases, product compositions are defined more exactly in the process of iterations and the optimal number of trays in each section (optimal x_f) is determined by modifying the value of Δx (see Section 7.3). Figure 7.17 shows calculation trajectories obtained by means of usage of this algorithm.

After design calculation of the main column and of decantor, additional columns entering into the unit are calculated and comparison of different sequences is carried out.

7.7.   Conclusion

The knowledge of the trajectory bundles structure at finite reflux and of their location in the concentration simplex allowed a new class of design calculation algorithms to be developed that guarantees a full optimal solution of the task without the participation of the user.

At reflux bigger than minimum and at separation without distributed components for all feasible quasisharp splits – the direct, the indirect, and the intermediate ones – possible compositions at the trays adjacent to the feed cross-section from above x_{f-1} and below x_f fill some segments $[x_{f-1}]$ and $[x_f]$ located in the vicinity of separatrix sharp split regions $\text{Reg}_{sep,r}^{sh,R}(S_r^1 - S_r^2 - N_r^+)$ and $\text{Reg}_{sep,s}^{sh,R}(S_s^1 - S_s^2 - N_s^+)$. At reflux bigger than minimum and at quasisharp separation with distributed component at the set distribution of this component among the products, there is the only one composition at the first tray above the feed cross-section expressed by point x_{f-1} in the vicinity of separatrix sharp split region $\text{Reg}_{sep,r}^{sh,R}(S_r^1 - S_r^2 - N_r^+)$ and the only one composition at the first tray below the feed cross-section expressed by point x_f in the vicinity of separatrix sharp split region $\text{Reg}_{sep,s}^{sh,R}$ $(S_s^1 - S_s^2 - N_s^+)$. The less sharp is separation the farther from separatrix sharp split trajectory bundles of the sections the composition points in the feed cross-section are located.

Possible compositions at the trays adjacent to the feed cross-section from above and below for any splits at the assumption about the linearity of separatrix trajectory bundles and about sharp separation $[x_{f-1}]_{lin}^{sh}, [x_f]_{lin}^{sh}$ or $(x_{f-1})_{lin}^{sh}, (x_f)_{lin}^{sh}$ can be determined from purely geometric considerations by means of solution of corresponding systems of linear equations.

At calculation by method "tray by tray" from the ends of the column for quasisharp separation, the calculation section trajectories are "attracted" to separatrix line $S^1 - S^2$. At direct and indirect splits, calculation "tray by tray" should be carried out in the upward or downward direction correspondingly.

At intermediate splits and splits with distributed component calculation, "tray by tray" calculation should be carried out from the ends of the column. Design calculation of necessary trays number in each section is carried out for the set of values of the summary concentrations of the impurity components x_{f-1}^{imp}, x_f^{imp} at the first trays above and below the feed cross-section inside possible composition segments $[x_{f-1}]_{lin}^{sh}, [x_f]_{lin}^{sh}$. The distribution of the impurity components in separation products is determined for each value from this set from the condition of joining of section trajectories in the feed cross-section by means of systematic search and execution of trial calculations from the ends of the column. The algorithm of systematic search ensures guaranteed solution of the task. The mentioned set of the values of the summary concentrations of the impurity components is limited by the values in points $(x_{f-1}^\infty)_{lin}^{sh}, (x_{f-1}^{min})_{lin}^{sh}, (x_f^\infty)_{lin}^{sh}, (x_f^{min})_{lin}^{sh}$. That value from this set, at which calculation tray numbers in the column $(N_r + N_s)$ are minimum, is optimal.

Design calculation of three-section column of extractive multicomponent distillation at one-component top product $x_D^{(1)}$ and entrainer $x_E^{(1)}$ can be carried out by method "tray by tray" upward with optimization by flow rates of entrainer and reflux and by distribution of trays among sections. The general design calculation of three-section column of extractive multicomponent distillation includes in the same main stages as at the calculation of two-section column.

Design calculation of "Petlyuk columns" is carried out on the basis of the algorithm of design calculation of two-section columns with optimization by distribution coefficient of pseudoproduct flow rates of the prefractionator, by excess factors of reflux and by distribution of trays among sections in each two-section column.

Design calculation of columns with side strippings is carried out on the basis of the algorithm of design calculation of two-section columns and on the basis of pinch method with optimization by distribution of trays among sections, by distribution of heat loads among pumparounds, and with the choice of the way of stripping in each side stripping.

Optimal cross-sections of joining of previous and posterior two-section columns at calculation by method "tray by tray" of "Petlyuk columns" and of columns with side strippings are cross-sections of previous columns closest by composition to points S^1.

Calculation of necessary tray numbers at heteroazeotropic and heteroextractive distillation can be carried out by method "tray by tray" upward with optimization by number of sections, by number of phases from the decantor for refluxing, and by distribution of trays among the sections.

7.8. Questions

1. In which direction should one calculate the bottom section for split $1,2 : 2,3 \ldots$ n using the method "tray by tray"? What about the top section?

2. Which components are product ones in the top product for split $1,2,3 : 3,4,5$? Which are impurity ones? Which are key ones?

3. What makes impossible the calculation of a two-section column by method "tray by tray" for intermediate split in the direction upward or downward?

4. Where should the composition point x_{f-1} at the first tray above the feed cross-section at quasisharp intermediate separation be located for the necessary tray numbers in this section to be minimum? Infinite?

5. How can one find the coordinates of point S_r^1 at a set composition x_D and set parameter $(L/V)^r$ for split $1,2 : 3,4,5$?

6. Where in the concentration simplex is the composition point in that cross-section of the top section where concentration of light key component at the intermediate split is maximum located?

7. What is the pseudoproduct of the second section above the feed cross-section of the main column with side strippings if its products in the direction downward are D, S_1, S_2, S_3, B?

8. Why can one calculate the column of heteroextractive distillation by method "tray by tray" in the upward direction?

7.9. Exercises with Software

1. Determine the necessary tray numbers and liquid and vapor flow rates in the top and bottom sections of the column at optimal location of the feed tray for separation of mixture of pentane(1)-hexane(2)-heptane(3)-octane(4) of composition 0.2, 0.3, 0.15, 0.35 at split 1: 2,3,4, reflux excess coefficients 1.2 and 1.5, product purities (top and bottom): a) 0.99, 0.98; b) 0.95, 0.99; c) 0.98, 0.98.

2. Same for split 1,2,3:4.

3. Same for split 1,2:3,4.

4. Same for split 1,2:2,3,4 with a) equal and b) optimal shares of the distributed component in the products.

5. Same as in item 4 for split 1,2,3:3,4.

6. State the feasible splits for the equimolar mixture of acetone(1), benzene(2), chloroform(3), and toluene(4). For each split, determine the necessary tray numbers and liquid and vapor flow rates in the top and bottom sections of the column at optimal location of the feed tray and optimal distribution of the component and reflux excess coefficients and product purities as in item 1.

References

Amminudin, K. A., Smith, R., Thong, D. Y.-C., & Towler, G. P. (2001). Design and Optimization of Fully Thermally Coupled Distillation Columns. Part 1: Preliminary Design and Optimization Methodology. *Trans IChemE*, 79, Part A, 701–15.

Bauer, M. H., & Stichlmair, J. (1995). Synthesis and Optimization of Distillation Sequences for the Separation of Azeotropic Mixtures. *Comput. Chem. Eng.*, 19, 515–20.

Bril, Z. A., Mozzhukhin, A. S., Pershina, L. A., & Serafimov, L. A. (1985). Combined Theoretical and Experimental Design Method for Heteroazeotropic Rectification. *Theor. Found. Chem. Eng.*, 19, 449–54.

Bril, Z. A., Mozzhukhin, A. S., Petlyuk, F. B., & Serafimov, L. A. (1974). Simulation of Distillation of Multicomponent Heterogeneous Azeotropic Mixtures. *Theor. Found. Chem. Eng.*, 8, 351–60.

Bril, Z. A., Mozzhukhin, A. S., Petlyuk, F. B., & Serafimov, L. A. (1975). Simulation and Research Heteroazeotropic Distillation. *Theor. Found. Chem. Eng.*, 9, 811–21.

Bril, Z. A., Mozzhukhin, A. S., Petlyuk, F. B., & Serafimov, L. A. (1977). Investigations of Optimal Conditions of Heteroazeotropic Rectification. *Theor. Found. Chem. Eng.*, 11, 675–81.

Carlberg, N. A., & Westerberg, A. W. (1989). Temperature (Heat Diagrams for Complex Columns: 2. Underwood's Method for Side Strippers and Enrichers. *Ind. Eng. Chem. Res.*, 28, 1379–86.

Cerda, J., & Westerberg, A. W. (1981). Shortcut Methods for Complex Distillation Columns: 1. Minimum Reflux. *Ind. Eng. Chem. Process Des. Dev.*, 20, 546–57.

Dhole, V. R., & Linnhoff, B. (1993). Distillation Column Targets. *Comput. Chem. Eng.*, 17, 549–60.

Doherty, M. F. & Melone, M. F. (2001). Conceptual Design of Distillation Systems. NY: McGraw-Hill.

Dunnebier, G., & Pentelides, C. C. (1999). Optimal Design of Thermally Coupled Distillation Columns. *Ind. Eng. Chem. Res.*, 38, 162.

Fenske, M. R. (1932). Fractionation of straight-Run Pennsylvania Gasoline. *Ind. Eng. Chem.* 24, 482–485.

Fonyo, Z., Scabo, J., & Foldes, P. (1974). Study of Thermally Coupled Distillation Columns. *Acta Chim.*, 82, 235–49.

Gilliland, E. R. (1940). Multicomponent Rectification. Optimum Feed-Plate Composition. *Ind. Eng. Chem.*, 32, 918–20.

Glinos, K., & Malone, M. F. (1985). Minimum Vapor Flows in a Distillation Column with a Side Stream – Stripper. *Ind. Eng. Chem. Process Des. Dev.*, 24, 1087–90.

Hengstebeck, R. J. (1961). *Distillation: Principles and Design Procedures*. New York: Reinhold Publishing, pp. 147–9.

Julka, V., & Doherty, M. F. (1990). Geometric Behavior and Minimum Flows for Nonideal Multicomponent Distillation. *Chem. Eng. Sci.*, 45, 1801–22.

Knapp, J. P., & Doherty, M. F. (1990). Thermal Integration of Homogeneous Azeotropic Distillation Sequences. *AIChE J.*, 36, 969–84.

Knapp, J. P., & Doherty, M. F. (1992). A New Pressure-Swing Distillation Process for Separating Homogeneous Azeotropic Mixtures. *Ind. Eng. Chem. Res.*, 31, 346–57.

Knapp, J. P., & Doherty, M. F. (1994). Minimum Entrainer Flows for Extractive Distillation: A Bifurcation Theoretic Approach. *AIChE J.*, 40, 243–68.

Knight, J. R., & Doherty, M. F. (1989). Optimal Design and Synthesis of Homogeneous Azeotropic Distillation Sequences. *Ind. Eng. Chem. Res.*, 28, 564–72.

Levy, S. G., & Doherty, M. F. (1986). Design and Synthesis of Homogeneous Azeotropic Distillation. 4. Minimum Reflux Calculations for Multiple Feed Columns. *Ind. Eng. Chem. Fundam.*, 25, 269–79.

Lewis, W. K., & Matheson, G. L. (1932). Studies in Distillation Design of Rectifying Columns for Natural and Refinery Gasoline. *Ind. Eng. Chem.*, 24(5), 494–498.

Liebmann, K., Dhole, V. R., & Jobson, M. (1998). Integrated Design of a Conventional Crude Oil Distillation Tower Using Pinch Analysis. *Trans IChemE*, 76, Part A, 335–47.

Nelson, W. L. (1936). *Petroleum Refinery Engineering*. New York: McGraw-Hill.

Nikolaides, J. P., & Malone, M. F. (1987). Approximate Design of Multiple Feed/Side-Stream Distillation Systems. *Ind. Eng. Chem. Res.*, 26, 1839–45.

Packie, J. W. (1941). Distillation Equipment in the Oil Refining Industry. *AIChE Trans.*, 27, 51–8.

Petlyuk, F. B. (1984). Necessary Condition of Disappearance of Components at Distillation of Azeotropic Mixtures in Simple and Complex Columns. In *The Calculation Researches of Separation for Refining and Chemical Industry* (pp. 3–22). Moscow: Zniiteneftechim (Rus.).

Petlyuk, F. B., & Danilov, R. Yu. (1999). Sharp Distillation of Azeotropic Mixtures in a Two-Feed Column. *Theor. Found. Chem. Eng.*, 33, 233–42.

Petlyuk, F. B., & Danilov, R. Yu. (2001a). Few-Step Iterative Methods for Distillation Process Design Using the Trajectory Bundle Theory: Algorithm Structure. *Theor. Found. Chem. Eng.*, 35, 224–36.

Petlyuk, F. B., & Danilov, R. Yu. (2001b). Theory of Distillation Trajectory Bundles and its Application to the Optimal Design of Separation Units: Distillation Trajectory Bundles at Finite Reflux. *Trans IChemE*, 79, Part A, 733–46.

Pham, H. N., & Doherty, M. F. (1990a). Design and Synthesis of Azeotropic Distillation: I. Heterogeneous Phase Diagram. *Chem. Eng. Sci.*, 45, 1823–36.

Pham, H. N., & Doherty, M. F. (1990b). Design and Synthesis of Azeotropic Distillation: II. Residue Curve Maps. *Chem. Eng. Sci.*, 45, 1837–43.

Pham, H. N., & Doherty, M. F. (1990c). Design and Synthesis of Azeotropic Distillation: III. Column Sequences. *Chem. Eng. Sci.*, 45, 1845–54.

Rooks, R. E., Malone, M. F., & Doherty, M. F. (1996). Geometric Design Method for Side-Stream Distillation Columns. *Ind. Eng. Chem. Res.*, 35, 3653–64.

Russel, R. A. (1983). A Flexible and Reliable Method Solves Single-Tower and Crude-Distillation Column Problems. *Chem. Eng.*, 90, 53.

Ryan, P. J., & Doherty M. F. (1989). Design/Optimization of Ternary Heterogeneous Azeotropic Distillation Sequences. *AIChE J.*, 35, 1592–601.

Sorel, E. (1893). *La Rectification de l'Alcohol*. Paris: Gauthier-Villars. (French).

Tedder, D. W., & Rudd, D. F. (1978). Parametric Studies in Industrial Distillation. *AIChE J.*, 24, 303–15.

Thiele, E. W., & Geddes, R. L. (1933). Computation of Distillation Apparatus for Hydrocarbon Mixtures. *Ind. Eng. Chem.*, 25, 289–95.

Triantafyllou, C., & Smith, R. (1992). The Design and Optimization of Fully Thermally Coupled Distillation Columns. *Trans IChemE.*, 70, Part A, 118–32.

Underwood, A. J. V. (1948). Fractional Distillation of Multicomponent Mixtures. *Chem. Eng. Prog.*, 44, 603–14.

Wahnschafft, O. M., & Westerberg, A. W. (1993). The Product Composition Regions of Azeotropic Distillation Columns. 2. Separability in Two-Feed Columns and Entrainer Selection. *Ind. Eng. Chem. Res.*, 32, 1108–20.

Watkins, R. N. (1979). *Petroleum Refinery Distillation*. Houston, TX: Gulf Publishing.

8

Synthesis of Separation Flowsheets

8.1. Introduction

Synthesis of a separation flowsheet consists of determining the best sequence of distillation columns and complexes that will ensure the obtaining of the set of products of a set quality from the initial mixture.

Distillation columns and complexes entering into this sequence differ by splits and kinds of complexes. The best sequence is characterized by the smallest summary expenditures on separation (energy expenditures and capital costs, taking into consideration their payback period).

A large number of various research works were dedicated to the task of synthesis. However, these works present nothing more than examination of particular examples of mixtures, and the results of these works prove the complicity of the task of synthesis. The methods used in these works are too laborious to be widely adopted in separation units designing. Therefore, while designing, the separation flowsheet is chosen, as a rule, on the basis of analogy with existing units or some too simple heuristic rules, or comparison of a small number of alternative sequences of columns is carried out. Very often, it so happens that the most "interesting" separation flowsheets remain unexamined. It concerns, first of all, flowsheets that imply the use of distillation complexes and separation flowsheets for azeotropic mixtures. Such practice leads to great excessive expenditures on separation.

For zeotropic mixtures, the main difficulty of the solution of synthesis task consists of the large number of alternative sequences that have to be calculated and compared with each other in terms of expenditures. This number greatly increases when the number of the products into which the mixture should be separated increases. The best sequence (or several sequences with close values of expenditures) depends on the concentrations of the components in the mixture under separation and on the field of phase equilibrium coefficients of the components in the concentration simplex. To ensure the solubility of the task of synthesis for multicomponent zeotropic mixtures, it is necessary to create a program system that would include as main modules programs of automatic design calculation (i.e., calculation without participation of the designer) of simple distillation columns and distillation complexes with branching of flows, a program of calculation of each sequence, a program of estimation of expenditures for each sequence, a program of automatic identification of each feasible sequence, and a program of systematic sorting of these sequences and their selection. The program system should also include limitations on complexity of the distillation complexes used in synthesis that can be installed in accordance with the user's desire (limitation on the number of sections or on the number of columns entering into the complex).

Only the development of the methods of automatic design calculation of simple columns and distillation complexes with branching of flows described in the previous chapters and the increase of performance of computers makes real the creation of such a software product. Traditional methods of design calculation based on calculation investigations with the help of simulation software are of no use when solving a synthesis task.

For azeotropic mixtures, the main difficulty of the solution of the task of synthesis consists not in the multiplicity of feasible sequences of columns and complexes but in the necessity for the determination of feasible splits in each potential column or in the complex. The questions of synthesis of separation flowsheets for azeotropic mixtures were investigated in a great number of works. But these works mainly concern three-component mixtures and splits at infinite reflux. In a small number of works, mixtures with a larger number of components are considered; however, in these works, the discussion is limited to the identification of splits at infinite reflux and linear boundaries between distillation regions Reg^{∞}. Yet, it is important to identify all feasible splits, not only the splits feasible in simple columns at infinite reflux and at linear boundaries between distillation regions. It is important, in particular, to identify the splits feasible in simple columns at finite reflux and curvilinear boundaries between distillation regions and also the splits feasible only in three-section columns of extractive distillation.

The theory of trajectory bundles described in Chapters 5 and 6 ensures the possibility of identification of all feasible splits of multicomponent azeotropic mixtures. The software for synthesis of separation units for multicomponent azeotropic mixtures should include, besides the module of identification of feasible splits, a module of preliminary selection of these splits (i.e., choice of the most "interesting" splits, a module of determination of necessary recycle flow rates, a module of choice of entrainers, and also modules entering into the system of synthesis for zeotropic mixtures).

Besides general questions of synthesis for zeotropic and azeotropic mixtures, we also discuss a particular but very important from the practical point of view task of conceptual designing of units of petroleum refining. The peculiarity of these units is the limitation of separability for the reason of thermolability. Therefore, the most important criterion of efficiency of these units is the degree of recovery of the most valuable ("light") oil products.

8.2. Zeotropic Mixtures

8.2.1. Heuristic Rules of Synthesis

The easiest and the oldest way of synthesis is to use a number of heuristic rules, such as "first isolate the lightest product" (direct separation), "choose such a boundary of first division that the amounts of the top and bottom products should be the closest ones" (the principle of dichotomy) (Harbert, 1957), "choose such a boundary of first division that the separation should be the easiest one" (Thompson & King, 1972), and "the most difficult separation should be the last one." These heuristics obtained from the experience of calculation are supplemented by such evident heuristics as follows: "obnoxious and corrosion-active components should be removed at the beginning" and "it is desirable to obtain end products as distillates" (Hendry, Rudd, & Seader, 1973).

The first group of heuristic rules can be substantiated if some assumptions about the mixture under separation are accepted. Such substantiation was made in the works (Modi & Westerberg, 1992) and (Westerberg & Wahnschafft, 1996) using the Underwood method for calculation of summary vapor flow in the sequence of column in the minimum reflux mode.

The heuristic way of synthesis is very simple and almost does not require calculations; however, its application can lead to nonoptimal sequences of separation and, therefore, to substantial excessive expenditures. Heuristic rules can contradict each other. For example, it is really not too hard to find a mixture for which the first and the second rule recommend different boundaries of division. Such an example is given in the work (Malone et al., 1985) for the mixture with volatilities of the components $\alpha_{AB} = 2.59$ and $\alpha_{BC} = 1.2$ and composition (0.05; 0.05; 0.90).

According to the first rule, it is direct separation that should be in the first column $(A : B,C) \to (B : C)$; according to the second rule, there must be indirect separation $(A,B : C) \to (A : B)$. Calculation check-up proves that summary vapor flow in the two columns is 60% larger at direct separation than at indirect separation in the first column.

Thus, to correctly choose the best sequence of columns, the estimation of expenditures on separation is necessary.

8.2.2. Estimation of the Expenditures on Separation

At the stage of synthesis, it is not necessary to determine all expenditures on separation for each sequence of columns being compared. The number of alternative sequences being great, an aspiration to simplify as much as possible the calculation of these expenditures, not committing errors in estimation of preferability of these sequences, arises.

For estimation of the expenditures, it was proposed in many works to use the summary vapor flow V in all columns of the sequence under consideration, calculated according to the Underwood method for the minimum reflux mode. Such approach presupposes that the mixture is a close-to-ideal one and that expensive utilities of heat and cold are not used in all columns of each sequence. Besides that, such an approach presupposes that only simple columns are used and the number of such columns is minimum (for the separation of an n-component mixture $[n-1]$ columns are used in each column there is separation without distributed components).

Synthesis experiments (Kafarov et al., 1975; Glinos & Malone, 1984) proved that energy expenditures reflected by summary vapor flow V constitute the main share of yearly expenditures, and capital costs do not differ greatly for alternative sequences. Therefore, the flowsheet with smaller summary vapor flow V, as a rule, requires smaller total expenditures on separation. This allows, in the first approximation, to use the summary vapor flow V for estimation of preferability of the sequences at the above-mentioned limitations.

If expensive utilities of heat or cold are used in the columns, for example, at low-temperature separation, then the above-described way of expenditures estimation is not good. In this case, the temperature at which heat should be brought in in the reboilers and withdrawn in the condensers acquires great importance. The total energy expenditures on separation can be estimated by the value of energy of separation that depends on the amount of heat brought in and withdrawn, and on the temperatures at which this heat is brought in or withdrawn (see Chapter 4):

$$W \approx T_0 \sum Q_k/T_k \tag{8.1}$$

where k is number of reboilers or condensers, T_0 is temperature of ambient air, and Q_k is amount of heat brought in or withdrawn.

For a column with one reboiler and one condenser in the case of liquid feeding, instead of Eq. (8.1), we have:

$$W \approx Q T_0 (1/T_{con} - 1/T_{reb}) \tag{8.2}$$

Energy expenditures in the case of the thermodynamically reversible process of separation into pure components (Petlyuk & Platonov, 1964; Petlyuk, Platonov, & Girsanov, 1964):

$$W_{rev} = -RT \sum_{i=1}^{n} x_{Fi} \ln x_{Fi} \tag{8.3}$$

To estimate preferability of different separation flowsheets, one can use thermodynamic efficiency instead of energy:

$$\eta = W_{rev}/W \tag{8.4}$$

Such an approach was used in the works (Petlyuk & Platonov, 1965; Petlyuk, Platonov, & Slavinskii, 1965) and in more detail in Agrawal and Fidkowski (1998, 1999). In the latter works the comparison not only of sequences of simple columns, but also of distillation complexes, was made for three-component mixtures. Capital costs can differ considerably for such separation flowsheets, but in this case it is not important because energy expenditures greatly exceed capital ones.

In other cases, if there is a difference of columns number in alternative sequences, for example, if a flowsheet with prefractionator and flowsheets with minimum number of columns are compared, it is necessary to estimate both kinds of expenditures – energy ones and capital ones. For this purpose, in a number of works simplified methods of design calculation of simple columns and of distillation complexes (method Fenske–Underwood–Gilliland) and simplified models for estimation of the expenditures (Kafarov et al., 1975; Malone et al., 1985; Glinos & Malone, 1988) were used.

Simplified methods of this type are good only for rough estimations because empirical ratios of Gilliland and the assumption about constancy of relative volatilities of components can lead to essential mistakes.

At present, after the development of methods of automatic design calculation (see Chapter 7), the application of simplified methods should be rejected and rigorous methods should be used. To decrease the volume of calculations, some stages can be excluded from these methods (those of iterations and of parameter optimization).

8.2.3. Preferability Regions for Ternary Mixtures

Depending on the composition of the mixture under separation, this or that separation flowsheet is preferable. Many researchers investigated the location of preferability regions in the concentration triangle.

The comparison of sequences with direct and indirect separation in the first column, with three columns, with prefractionator and complex column, and of Petlyuk column (Figs. 8.1 and 6.12c, d) was made in the works (Petlyuk & Platonov, 1965; Petlyuk et al., 1965). The summary flow rate of vapor V and expenditures of energy of separation W were used for estimation of the expenditures.

The mixtures with volatilities $\alpha_{AC} = 1.2, \alpha_{BC} = 1.1$ and with compositions $x_{FA} = x_{FC}$ at different x_{FB} were examined. It turned out for these conditions that, if estimated by summary rate of vapor flow V, the flowsheet with three columns is more preferable than the flowsheets with two columns at $x_{FB} > 0.3$; the flowsheet with prefractionator and complex column is more preferable at any x_{FB}, and Petlyuk column is the most preferable at any x_{FB}. In the case of estimation by energy expenditures, W the flowsheet with three columns and the flowsheet with prefractionator and complex column are the most preferable ones. Petlyuk column yields to these flowsheets because for it all heat is brought in at the highest temperature and is withdrawn at the lowest.

The comparison of various distillation complexes and of ordinary flowsheets is given in Tedder and Rudd (1978). Columns with side strippings and side rectifiers, Petlyuk columns, flowsheet with prefracionator, and also some other feasible configurations of two columns were examined. It was shown, in particular, that Petlyuk columns are preferable at big content of average volatile component.

In the monograph (Stichlmair & Fair, 1998), the preferability regions in concentration triangle for mixture 2.2-dimethylbutane(A)-2-dimethylpentane(B)-hexane(C)(relative volatilities $\alpha_{AC} = 1.887$ and $\alpha_{BC} = 1.329$) were shown. The comparison was made in accordance with summary vapor flow V. The flowsheets with a minimum number of simple columns, with three columns, and with

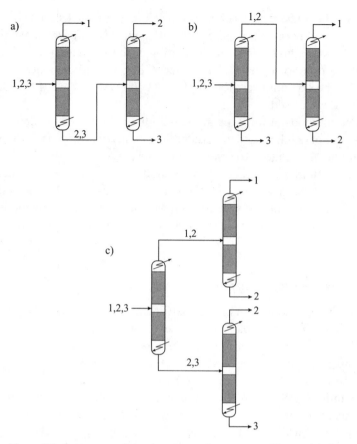

Figure 8.1. Sequences of simple columns for ternary mixtures: (a) the direct sequence, (b) the indirect sequence, and (c) the prefractionator sequence.

prefractionator and complex column were examined. It was shown that for this mixture if the flowsheet with prefractionator and complex column is excluded, in concentration triangle there are three large, approximately equal by area, regions of preferability adjacent to vertexes. At large content of component A, direct split $(A: B,C)$ is preferable in the first column; at large content of component C, indirect split is preferable in the first column $(A,B : C)$; at large content of component B, three-column sequence is preferable $(A,B : B,C) \rightarrow (A : B),(B : C)$. As far as the flowsheet with prefractionator and complex column is concerned, its region of preferability will occupy the bigger part of the area of the concentration triangle.

It was shown in other works (Agrawal & Fidkowski, 1998) that, while estimating the expenditures by summary vapor flow V, the region of preferability of Petlyuk columns occupies the whole area of the concentration triangle.

However, while estimating expenditures by thermodynamic efficiency η (Agrawal & Fidkowski, 1999), as has to be expected, the region of preferability of Petlyuk columns occupies only a small part of the area of the concentration triangle, compared with sequences of simple columns and other distillation complexes.

8.2.4. Systematic Identification of Alternative Sequences

At automatic synthesis of the best sequence from the big number of feasible alternative sequences, one of the tasks is their systematic identification. This task is quite easily solved for the sequences of simple columns without distributed components. In this case, any column of any sequence can be identified by the number of components in its feeding $- I$, by the number of the first of these components $- J$, and by the number of the top product components $- K$ (Kafarov et al., 1975) (i.e., column I, J, K is column $[J, J+1, \ldots, J+K-1 : J+K, \ldots, J+I-1]$).

It is supposed that all components were numbered beforehand in order of decreasing volatility.

The values of parameters I and J should not include impurity components entering into the feeding of the column under consideration. Pseudocomponents (fractions) that have to be obtained as one product can act as components while setting parameters I, J, and K.

The identification of complete set of sequences starts with the first column, for which $I_{(1)} = n$ (total number of components or pseudocomponents of the mixture under separation), $J_{(1)} = 1$, and $K_{(1)} = 1 \div (I_{(1)} - 1)$, i.e. $(1 : 2, \ldots, n)$, $(1,2 : 3, \ldots, n), \ldots, (1,2, \ldots, n-1 : n)$.

Each value $K_{(1)}$ generates the numbers $I_{(2)}$, $J_{(2)}$, $K_{(2)}$ of feasible second columns. For example, at $K_{(1)} = 1$ columns $I_{(2)} = I_{(1)} - 1$, $J_{(2)} = 2$, $K_{(2)} = 2 \div (I_{(1)} - 1)$, i.e., $(2 : 3, \ldots, n)$, $(2,3 : 4, \ldots, n), \ldots, (2,3, \ldots, n-1 : n)$ are feasible, at $K_{(1)} = 2$ columns $I_{(2)} = 2$, $J_{(2)} = 1$, $K_{(2)} = 1$, i.e., $(1 : 2)$ and $I_{(2)} = I_{(1)} - 2$, $J_{(2)} = 3$, $K_{(2)} = 3 \div (I_{(1)} - 1)$, i.e., $(3 : 4, \ldots, n)$, $(3,4 : 5, \ldots n), \ldots, (3,4, \ldots n-1 : n)$ are feasible, etc. It is easy to develop the general algorithm of systematic sorting of all potential columns and their sequences.

The algorithm of systematic identification and sorting of all feasible sequences, including not only simple columns, but also various distillation complexes with branching of flows, is considerably more complicated. This task was solved (Agrawal, 1996; Sargent, 1998). The general name of the approach introduced was given in the latter work: "state-task network." This approach assumes that the main element of any column or distillation complex is a section. The initial mixture, intermediate products, and end products are nodes of the network (states), and the lines joining these nodes show the changes of these states with the help of one section of the distillation column or complex (tasks). We note that section is a main element not only at synthesis of sequences, but also at design calculation (see Chapter 7).

For the top sections, the summary flow (the difference of vapor and liquid flows) is directed upward, and for the bottom sections, it is directed downward. The top sections should get liquid flow from the condenser or from the bottom of another section, and the bottom sections should get vapor flow from the reboiler or from the top of another section.

A sequence of simple columns can be obtained if each top section is provided with a condenser and each bottom section is provided with a reboiler.

A section joining two nodes, including component with intermediate volatility, is called an intermediate section. For example, for the sequence $(1 : 2,3) \rightarrow (2 : 3)$, the section 1,2,3 → 2,3 is intermediate. If for each bottom intermediate section in the sequence of simple columns the reboiler is replaced by vapor flow from the bottom section of the other column, then we get a column with side rectifiers. Similarly, if in each top intermediate section the condenser is replaced by liquid flow from the top section of other column, then we get the column with side strippings.

If a sequence of simple columns has minimum number of columns $(n - 1)$ and a summary number of reboilers and condensers $2(n - 1)$, then the column with side strippings or rectifiers (partially coupled sequence) has a summary number of reboilers and condensers n.

We note that sections can be grouped into columns in various ways. For example, at separation of ternary mixture 1,2,3 in a partially coupled sequence, the sections can be grouped into a column with side rectifier (i.e., sections 1,2,3 → 1, 1,2,3 → 2,3 and 2,3 → 3 can be grouped into the main column and section 2,3 → 2 can be placed in the side column). Another variant of grouping of section is sections 1,2,3 → 1 and 1,2,3 → 2,3 are in one column and sections 2,3 → 2 and 2,3 → 3 are in the other one. From a thermodynamic point of view, both of these variants of grouping of sections are equivalent.

We can turn from partially coupled sequences to completely coupled ones (to Petlyuk columns). For this purpose, the reboilers and condensers connected with the sections, products of which are components with intermediate volatilities, should be excluded. It is possible if each excluded condenser or reboiler is replaced by flow of liquid or vapor from another section. To ensure this flow, it is necessary to supplement the network with an additional node (i.e., with one top and one bottom section). We note that inclusion of additional nodes is equivalent to inclusion of columns with distributed components into the sequence. For example, in the side rectifier 2,3 → 2 condenser can be replaced if network 1,2,3 → 1, 1,2,3 → 2,3, 2,3 → 2, 2,3 → 3 is supplemented with the node 1,2, (i.e., we get a new network 1,2,3 → 1,2, 1,2,3 → 2,3, 1,2 → 1, 1,2 → 2, 2,3 → 2, 2,3 → 3). This network has only one reboiler connected with the section where the heaviest component 3 is obtained and one condenser connected with section where the lightest component 1 is obtained. We have a classical Petlyuk column for ternary mixture separation. Four thermodynamically equivalent groupings of sections into two columns are possible for this column:

1. 1,2,3 → 1,2 and 1,2,3 → 2,3 (1st column), 1,2 → 1; 1,2 → 2; 2,3 → 2 and 2,3 → 3 (2nd column)
2. 1,2,3 → 1,2; 1,2,3 → 2,3 and 2,3 → 3 (1st column), 1,2 → 1; 1,2 → 2 and 2,3 → 2 (2nd column)
3. 1,2,3 → 1,2; 1,2,3 → 2,3 and 1,2 → 1 (1st column), 1,2 → 2; 2,3 → 2 and 2,3 → 3 (2nd column)
4. 1,2,3 → 1,2; 1,2,3 → 2,3, 1,2 → 1 and 2,3 → 3 (1st column), 1,2 → 2 and 2,3 → 2 (2nd column)

We note that for groupings 2 and 3 movement of vapor between columns is unidirectional, which makes easier their practical realization (grouping 3 is shown in Fig. 6.12e).

One can get from completely coupled sequences all the other feasible sequences, including columns with distributed components, if top sections are supplemented with condensers and bottom sections are supplemented with reboilers. For example, the sequence shown in Fig. 6.12d (with prefractionator and complex column) can be obtained in this way.

The general algorithm of synthesis of all feasible sequence has to be started with the sequence containing the maximum number of sections and the maximum number of heat exchangers. Each section of this sequence has at the end one component fewer than at the beginning (i.e., in each top section, the heaviest component disappears and, in each bottom section, the lightest component disappears). Each top section of this sequence has a condenser, and each bottom section has a reboiler. The example of such a sequence is shown in Fig. 8.1c. Then one by one heat exchangers are excluded from this sequence (except the condenser of the section where the lightest component is obtained and the reboiler of the section where the heaviest component is obtained). After that, one by one the nodes are excluded from this sequence and, for each new sequence, one by one heat exchangers are excluded again.

With the increase of the number of components, the total number of feasible sequences grows very quickly. The number of feasible groupings of sections grows even more quickly. Nevertheless, the above-described algorithm identifies all these sequences.

8.2.5. Examples of Synthesis of Separation Flowsheets

For synthesis of separation flowsheets from simple columns, the method of dynamic programming was developed (Kafarov et al., 1975). This method compares systematically all feasible flowsheets at any number of components and to exclude numerous repeated calculations of identical columns entering into various sequences. The main idea of this method consists of the synthesis of sequences step by step, moving from the end of sequence to the beginning (i.e., starting with the smallest groups of components or pseudocomponents [$I = 2$], turning to bigger groups [to $I = 3$, then to $I = 4$, etc.] and obtaining optimum fragments of the sequences). $S_{I,J,K}$ is annual expenditures on separation in column I,J,K, and $F_{I,J}$ is expenditures on complete separation of the group of components or pseudocomponents I,J at optimal sequence for this group. Because column I,J,K in the general case divides group I,J into two smallest groups, we get:

$$F_{I,J} = \min_{K_{I,J}} \left(S_{I,J,K} + F_{K-J+1,J} + F_{I-K+J-1,K+1} \right) \tag{8.5}$$

Applying Eq. (8.5) to the gradually augmenting groups of components or pseudocomponents, one can find optimal values of $K_{I,J}$ for all these groups and the corresponding values of expenditures $F_{I,J}$. As a result, we get the optimal value $K_{n,1}$ for the separation in the first column and minimum expenditures $F_{n,1}$ for the separation of initial mixture:

$$F_{1,J} = 0$$
$$F_{2,J} = S_{2,J,1}$$
$$F_{3,J} = \min_{K_{3,J}} (S_{3,J,K} + F_{K-J+1,J} + F_{3-K,K+1})$$

$$\cdots\cdots\cdots\cdots\cdots\cdots\cdots\cdots\cdots\cdots\cdots\cdots\cdots$$

$$F_{n,1} = \min_{K_{n,1}} (S_{n,1,K} + F_{K1} + F_{n-K,\ K+1})$$

(8.6)

The method of dynamic programming synthesizes the optimal sequence starting from its end. Therefore, expenditures $S_{I,J,K}$ should be determined without calculation of the previous part of the flowsheet. For this purpose, it is necessary to determine the composition of feeding of column I,J,K. It can be done easily, if it is accepted that each product of separation sequence i contains as impurity components only adjacent components $(i-1)$ and $(i+1)$ (i.e., the set permissible concentrations of impurity components):

$$\eta_i^L = \eta_i^{i-1} \tag{8.7a}$$
$$\eta_i^H = \eta_i^{i+1} \tag{8.7b}$$

Where η_i^L and η_i^H are set permissible concentrations of light and heavy impurity components in product i correspondingly, η_i^{i-1} and η_i^{i+1} concentrations of components $(i-1)$ and $(i+1)$ in product i correspondingly.

At this assumption, the amount P_i of the product i can be determined from the system of linear equations of componentwise material balance that has three-diagonal form:

$$f_1 = P_1(1 - \eta_1^H) + P_2\eta_2^L$$
$$f_2 = P_1\eta_1^H + P_2(1 - \eta_2^L - \eta_2^H) + P_3\eta_3^L$$

$$\cdots\cdots\cdots\cdots\cdots\cdots\cdots\cdots$$

$$f_n = P_{n-1}\eta_{n-1}^H + P_n(1 - \eta_n^L)$$

(8.8)

After that, the feeding of column I, J can be determined:

$$f_i^{I,J} = 0 \quad (\text{for } i < J - 1)$$
$$f_i^{I,J} = P_J\eta_J^L \quad (\text{for } i = J - 1)$$
$$f_i^{I,J} = f_J - P_{J-1}\eta_{J-1}^H \quad (\text{for } i = J)$$
$$f_i^{I,J} = f_i \quad (\text{for } J + 1 \leq i \leq J + I - 2)$$
$$f_i^{I,J} = f_{J+I-1} - P_{J+I}\eta_{J+I}^L \quad (\text{for } i = J + I - 1)$$
$$f_i^{I,J} = P_{J+I-1}\eta_{J+I-1}^H \quad (\text{for } i = J + I)$$
$$f_i^{I,J} = 0 \quad (\text{for } i \geq J + I + 1)$$

(8.9)

The above-described algorithm was used for the synthesis of separation flowsheet of mixture of hydrocarbon gases C_3H_8, $i\text{-}C_4H_{10}$, $n\text{-}C_4H_{10}$, $i\text{-}C_5H_{12}$, $n\text{-}C_5H_{12}$, $i\text{-}C_6H_{14}$, and $n\text{-}C_6H_{14}$ $(n = 7)$. At $(n = 7)$, the number of alternative sequences is 132. Composition and flow rate of feed, permissible impurities in the products, and flow rates of the products are given in Table 8.1. The comparison of alternative sequences was made in accordance with the value of annual expenditures. The calculation of the columns were executed in accordance with simplified method Fenske–Underwood–Gilliland. Figure 8.2a shows the graph of dependence of total

Table 8.1. Composition and flowrates of feed and products for separation flowsheet of hydrocarbon gases

Component	Feed		Impurities, % mass		Product rates, Kt/yr						
	Kt/yr	% mass	Light	Heavy	C_3H_8	iC_4H_{10}	nC_4H_{10}	iC_5H_{12}	nC_5H_{12}	iC_6H_{14}	nC_6H_{14}
C_3H_8	61.2	8.6	0	2.5	59.4	1.5	0	0	0	0	0
iC_4H_{10}	69.0	9.7	2.6	1.6	1.8	67.2	1.1	0	0	0	0
nC_4H_{10}	32.5	4.5	1.0	2.0	0	0.3	30.4	0.7	0	0	0
iC_5H_{12}	72.6	10.1	1.6	3.0	0	0	1.0	61.5	1.9	0	0
nC_5H_{12}	193.5	27.0	4.9	3.6	0	0	0	10.4	190.0	7.5	0
iC_6H_{14}	49.0	6.8	4.3	4.8	0	0	0	0	1.6	34.1	1.8
nC_6H_{14}	238.3	33.3	3.0	0	0	0	0	0	0	7.4	236.5
Total	716.1	100.0			61.2	69.0	32.5	72.6	193.5	49.0	238.3

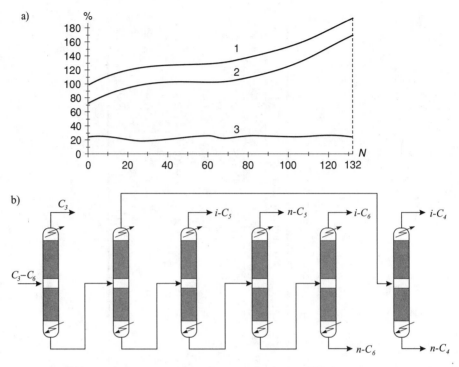

Figure 8.2. (a) The relative expenditures for the 132 simple distillation sequences for given feed composition of 7-component C_3-C_6 hydrocarbon mixture. 1, total expenditures; 2, energy expenditures; 3, capital expenditures. (b) The most economical sequence.

annual expenditures, energy annual expenditures, and capital annual expenditures on the ordinal number of the sequence (all sequences are put in order of increasing total annual expenditures). One can gather from the figure that the energy expenditures are principal (i.e., the sequence that is the best one in total expenditures is also the best from the point of view of energy expenditures, and the worst sequence differs from the best one by 87%). The following sequence is the best: $(1 : 2,3,4,5,6,7) \rightarrow (2,3 : 4,5,6,7) \rightarrow (4 : 5,6,7) \rightarrow (5 : 6,7) \rightarrow (6 : 7) \rightarrow (2 : 3)$ (Fig. 8.2b). For the given example, various heuristic rules give the following worsening of expenditures comparing to the best sequence: "the easiest separation – first one" – by 3%, direct separation – by 10%, the rule of dichotomy – by 33%.

The similar results were also obtained in the work (Malone et al., 1985) for five-component mixture of alcohols: ethanol, i-propanol, n-propanol, i-butanol, and n-butanol with molar composition (0,25; 0,25; 0,35; 0,10; 0,15).

In this case, as earlier, the best sequence (direct separation) in accordance with the value of total expenditures (method Fenske–Underwood–Gilliland was used) coincides with the best sequence in accordance with the value of summary flow rate of vapor V, and the worst one differs from the best one only by 20%.

However, a much bigger difference between the best and the worst sequences (four times) was shown in this work for the mixture obtained at the unit of alkylation of butane for production of i-octane. This is a mixture of five components and pseudocomponents: C_3H_8, $(i\text{-}C_4H_{10} + 1\text{-}C_4H_8)$, $n\text{-}C_4H_{10}$, $i\text{-}C_8H_{18}$, C_9H_{20} (here $[i\text{-}C_4H_{10} + 1\text{-}C_4H_8]$ – one pseudocomponent) with molar composition (0.05; 0.70; 0.10; 0.10; 0.05). In this case, as earlier, the flowsheet of direct separation is the best one in total expenditures and in summary vapor flow rate V.

Besides sequences of simple columns, some types of distillation complexes, each of which can replace two adjacent simple columns, were examined in work (Glinos & Malone, 1988). The following complex columns and distillation complexes were examined: column with side output above the feed cross-section, column with side rectifier, column with side stripping; flowsheet with prefractionator, Petlyuk column; top and side flows from the first column into the second one (Fig. 8.3a),

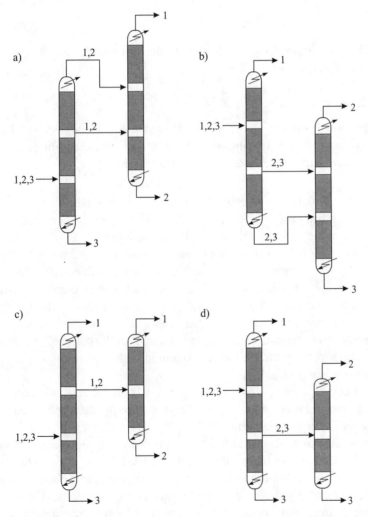

Figure 8.3. Some complex columns for the ternary mixtures: (a) indirect with overhead product and sidestream of first column between columns, (b) direct with bottom product and sidestream of first column between columns, (c) indirect with sidestream of first column between columns, and (d) direct with sidestream of first column between columns.

bottom and side flows from the first column into the second one (Fig. 8.3b), side flow from the top section of the first column into the second one (Fig. 8.3c), and side flow from the bottom section of the first column into the second one (Fig. 8.3d).

The simplified method similar to method Fenske–Underwood–Gilliland was developed for calculation of distillation complexes. The comparison of different sequences was made at the example of the mixture obtained at the unit of alkylation. The most interesting result was obtained from sequence: $(1 : 2,3,4,5) \rightarrow (2,3 : 4,5) \rightarrow (4 : 5) \rightarrow (2 : 3)$ while uniting the second and third columns into the complex shown in Fig. 8.3b. It turned out that this sequence is better by 30% in terms of expenditures than the best simple sequence.

8.3. Thermodynamically Improved and Thermally Integrated Separation Flowsheets

8.3.1. Thermodynamic Losses and Their Decrease

The comparison of adiabatic and reversible distillation determines several sources of thermodynamic losses leading to increase of energy expenditures compared with separation work at the reversible process (Eq. [8.2]).

1. Thermodynamic losses caused by unequilibrium of liquid and vapor flows mixed at each tray (Δ_1). Two ways to decrease Δ_1 are decrease of excess reflux factor to $\sigma = 1,05 \div 1,1$ and usage of intermediate along the height of the column input or output of heat (see Sections 6.2 and 6.8). The second way is efficient if a wide-boiling mixture is subjected to separation. An example of its application is the usage of pumparounds in petroleum refining (see Section 6.8). A modified variant of this method is stepped condensation of vapor feeding and input of the formed liquid fractions into different cross-sections along the height of the column. The examples can be low-temperature units of obtaining ethylene and propylene from pyrolysis gases (Fig. 8.4).

   A decrease in Δ_1 is always connected with an increase of necessary tray numbers and total capital costs. Therefore, it is justified only in the case of massive energy expenditures.

2. Thermodynamic losses caused by mixing of flows of different composition in the feed cross-section of the column (Δ_2). These losses always arise at separation of multicomponent mixture at any split without distributed components. The losses are absent only at the preferable split when the compositions of the liquid and vapor parts of feeding coincide (in the mode of minimum reflux) or are close (at the reflux bigger than minimum) to the composition of the liquid flow from the top section of the column and to the composition of vapor flow from the bottom section of the column, respectively.

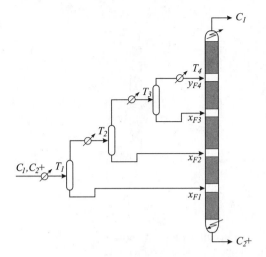

Figure 8.4. The demethanizer with multiple feed streams. $T_1 \div T_4$, temperature since condencers, $x_{F1} \div x_{F3}$, liquid compositions since condencers; y_{F4}, vapor composition since condencer.

Such separation takes place in flowsheets with prefracionator and in Petlyuk flowsheets (Fig. 6.12c $\div$ f,6.13 $\alpha \div$ c).

3. Thermodynamic losses caused by input into the column of unequilibrium flows of reflux from condenser and of vapor from reboiler (Δ_3). To exclude Δ_3, it is necessary to replace condenser and reboiler by the input of liquid and vapor from the other columns (i.e., to turn from the flowsheet in Fig. 6.12d to the flowsheets in Fig. 6.12c,e,f). At such passage, parts of section trajectories $x_D \to S_r^1$ and $x_B \to S_s^1$, at which nonequilibrium of liquid and vapor flows being mixed at the trays is especially big, are excluded. It is very clearly seen in Fig. 8.5 (Petlyuk & Platonov, 1965), which shows working and equilibrium lines for each of three components at the preferable split and mode of minimum reflux ($\alpha_{13} = 5$; $\alpha_{23} = 2$; $x_{F1} = 0.1$; $x_{F2} = 0.6$; $x_{F3} = 0.3$; $x_{B1} = 0.0001$; $x_{D3} = 0.0004$; $L_{min}/F = 0,25$). As is evident from the figure, the nonequilibrium at the end parts of the column $x_D \to S_r^1$ and $x_B \to S_s^1$, if working with a condenser and a reboiler (the shaded regions correspond to them), exceeds many times the nonequilibrium in the middle part of the column at parts $S_r^1 \to x_{f-1}$ and $S_s^1 \to x_f$.

4. Thermodynamic losses caused by hydraulic resistance of the trays (Δ_4). These losses consist of considerable parts of summary thermodynamic losses if the pressure drop along the column is commensurable with the absolute pressure at its top end (vacuum columns, columns for separation of narrow-boiling mixtures – isomers, isotopes). The increase of pressure at the bottom end of the column plays a twofold negative role: it decreases relative volatilities of the components and increases the top-bottom difference of temperatures in the column. Decrease of relative volatilities of the components leads to the necessity of increasing the reflux number and expenditure of heat energy (Q in Eq. [8.2]), and the increase of top-bottom temperature difference means an increase of term ($1/T_{con} - 1/T_{reb}$) in this equation. Both facts lead to an increase of expenditures of energy for

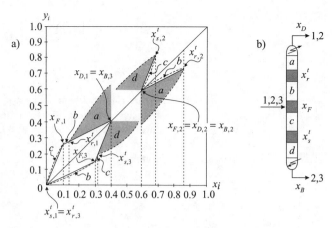

Figure 8.5. McCabe-Thiele plots for all components of a ternary mixture under minimum reflux for the preferable split. Operating (solid) and equilibrium (dotted) lines for parts of column − a,b,c,d. (b) Parts of column a,b,c,d and pinches x_F, x_r^t, x_s^t. $x_{F,1}, x_{F,2}, x_{F,3}, x_{D,1}, x_{D,2}, x_{D,3}, x_{B,1}, x_{B,2}, x_{B,3}$, concentrations of components 1, 2, 3 in feed, overhead product, bottom product, respectively; $x_{r,1}^t, x_{r,2}^t, x_{r,3}^t, x_{s,1}^t, x_{s,2}^t, x_{s,3}^t$, concentrations of components 1,2,3 in pinches, respectively; regions between operating and equilibrium lines for a and d parts of column are shaded.

separation. To decrease Δ_4, it is necessary to apply contact devices with low hydraulic resistance, for example, regular packing.

5. Thermodynamic losses caused by temperature difference between the heat source and flow from reboiler and between flow from condenser and the heat sink (Δ_5). These losses play especially significant role in two cases: in that of separation of narrow-boiling mixtures, and at low-temperature separation. In the first case, the differences of temperatures between the heat source and flow from reboiler and between flow from the condenser and the heat sink can considerably exceed that between the bottom and the top of the column. This leads to an increase of term ($1/T_{sink} - 1/T_{sours}$), which is to replace the term ($1/T_{con} - 1/T_{reb}$) in Eq. (8.2) and to lower thermodynamic efficiency, for example, of units for isotopes separation – 0,01% (London, 1961). In the second case, expensive cold sources (heat sink) are used and, therefore, decrease of losses Δ_5 leads to big economy.

To decrease Δ_5, it is necessary to bring together temperatures of the heat source and the sink with T_{reb} and T_{con}. There are several ways to do this.

One of the ways consists of maximum use of hot or cold flows going out from the unit as heat sources and cold sinks along with heat carriers and refrigerants. The general approach to the solution of this task – "pinch technology" (Linnhoff & Hindermarsh, 1983) – uses construction of curves "temperature – enthalpy" for reboilers, condensers, hot and cold flows going out from the unit, and heat carriers, and refrigerants.

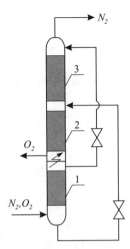

Figure 8.6. Double column for air separation. 1, rectifying section of high pressure; 2, stripping section of low pressure; 3, rectifying section of low pressure.

8.3.2. Thermally Integrated Separation Flowsheets

Another way consists of organization of heat exchange between condenser of one column and reboiler of another column. This way decreases the amount of heat received from heat carriers and/or given to refrigerants. The application of this method usually requires the use of different pressures in the columns. For the purpose that the temperature of condenser of one column would be higher than the temperature of reboiler of the second column, the pressure in the first column should be higher than in the second one. This method is used also at separation of binary mixtures, for example, in low-temperature units of air separation (Fig. 8.6) (Baldus et al., 1983). In this case the first column, working at higher pressure, is used as prefractionator, products of which are concentrated nitrogen and oxygen, and the second column serves to obtain nitrogen and oxygen of the required purity. Reboiler is absent in the first column because its feeding is brought in gas phase; in the second column, condenser is absent because distillate of the first column is used there as a reflux. Thermodynamic efficiency of units of air separation is very high – 18% (Haselden, 1958).

At multicomponent distillation change of pressures in the columns to ensure heat exchange between their condensers and reboilers leads to the increase of total difference of temperatures between the hottest reboiler and the coldest condenser. This increase is limited by temperatures of available heat carriers and refrigerants and in a number of cases also by thermolability of the mixture.

The application of thermally integrated separation flowsheets can give considerable economy of energy expenditures, but runs across the mentioned limitation.

Besides that, usage of this flowsheet usually leads to the increase of capital costs for apparatuses and for the control system.

8.3.3. The Heat Pump

One more method to decrease Δ_S consists of the application of the so-called "heat pump" of mechanical or absorption type (Fonyo & Benko, 1998). Heat, given in

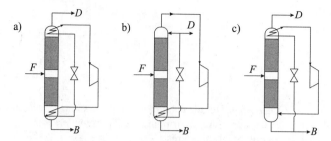

Figure 8.7. (a) Heat pump with supplementary circulation, (b) the overhead product as working medium for heat pump, and (c) the bottom product as working medium for heat pump.

the condenser, and the mechanical work of the compressor for the bringing in of heat into the reboiler are used in the flowsheets with a heat pump of mechanical type. Here are many types of "heat pump", some of which are shown in Fig. 8.7.

Heat can be passed from the condenser to the reboiler with the help of an auxiliary contour, the working body of which can be steam or hot water (Fig. 8.7a). Vapor flow, going out from the column from above (Fig. 8.7b), or liquid, going out from the column from below (Fig. 8.7c), are used as working body in the auxiluary contour in other types of heat pump. Heat pumps of absorption type, having the auxiliary system absorber-desorber, are more complicated.

The application of "heat pump" leads to additional capital and energy expenditures. Therefore, total economy of expenditures on separation can be achieved in such a way only if the difference $(T_{reb} - T_{con})$ is not large (i.e., at separation of a narrow-boiling mixture). Examples of application of "heat pump" in industry are columns for separation of ethane and ethylene, propane and propylene, ethylbenzene and styrene, and isobutane and butane.

In a number of cases, if bubble temperatures of the components of the mixture under separation are very close to each other and the structure of their molecules is different, it is profitable to use *extractive distillation*, even at separation of zeotropic mixtures, to decrease energy and capital expenditures (separation of mixtures of hydrocarbons of different homologous rows). The economy of expenditures on separation is being achieved at the expense of the fact that separation of one narrow-boiling mixture is replaced by separation of two wide-boiling mixtures in two column at extractive distillation.

While synthesizing separation flowsheets, it is necessary to consider the possibility of thermodynamic improvement and thermal integration. Therefore, for each sequence, identified in the process of synthesis, it is necessary to realize possible thermodynamic improvements and thermal integration of the columns. The estimation of expenditures on separation is made taking into consideration these modifications, if these expenditures are smaller than for the sequence without modifications under consideration. This estimation of expenditures is used while comparing the sequence under consideration to all other possible sequences.

At such an approach, the total volume of calculations grows considerably. It can become excessively big at big number of the sequences (big n, big number of distillation complexes). Therefore, to decrease the volume of calculations, various heuristic rules can be used (e.g., to use "heat pump" only if $[T_{reb} - T_{con}] < 20°C$).

8.4. Multicomponent Azeotropic Mixtures: Presynthesis

In contrast to zeotropic mixtures, the main problem for azeotropic ones at synthesis is that of determining possible splits in each potential separation column (the problem of presynthesis). Initially, the analysis of possible splits of homogeneous azeotropic mixtures in simple columns came to nothing more than three-component mixtures and the mode of infinite reflux. This analysis was based on diagrams of residue curves. It was proved that the curvature of boundaries between distillation regions at infinite reflux can be used for separation (Balashov, Grishunin, & Serafimov, 1970; Balashov, Grishumin, & Serafimov, 1984; Laroche et al., 1992). This analysis is of great practical importance for the choice of entrainers with the purpose of separation of binary azeotropic mixtures.

The analysis of possible splits in the mode of infinite reflux with the purpose of synthesis of separation flowsheets was extended to multicomponent mixtures (Petlyuk, Kievskii, & Serafimov, 1977a; Petlyuk, Avetyan, & Inyaeva, 1977; Petlyuk, 1979; Petlyuk, Kievskii, & Serafimov, 1979; Baburina & Platonov, 1990; Safrit & Westerberg, 1997; Rooks et al., 1998; Sargent, 1998), (Doherty & Malone, 2001).

The analysis of possible splits was realized for three-component mixtures not only at infinite, but also at finite, reflux (Petlyuk et al., 1981; Petlyuk, Vinogradova, & Serafimov, 1984; Wahnschaft et al., 1992; Kiva, Marchenko, & Garber, 1993; Poellmann & Blass, 1994; Krolikowski et al., 1996; Davydyan et al., 1997). In some works (Wahnschaft, Le Redulier, & Westerberg, 1993; Bauer & Stichlmair, 1995), various strategies of synthesis are examined, but to determine possible splits it is proposed to use simulation software, which, in the majority of cases, requires too large volumes of calculations and is not systematic for the reason of great number of parameters.

Development of trajectory tear-off theory at sharp separation and finite reflux from the boundary elements of the concentration simplex (Petlyuk & Danilov, 2000a) created a general presynthesis method for multicomponent azeotropic mixtures.

The theory of trajectory tear-off was extended to the section located between the cross-section of entrainer input and that of the main-feeding input in columns of sharp extractive distillation (Petlyuk & Danilov, 1999), which included it into the general method of presynthesis.

Presynthesis on the basis of the theory of trajectory tear-off from the boundary elements of concentration simplex takes into consideration the sharp splits important for practical usage in columns with one and two feedings that are not embraced by the methods of presynthesis on the basis of residue curve structure. In a number of cases, this manages without expensive ways of separation, without entrainers, with a minimum number of columns.

8.4.1. Possible Product Segments at the Edges of Concentration Simplex

In Chapter 5, we examine structural conditions of trajectory tear-off for the top and bottom sections. We now examine these conditions in more detail for multi-component mixtures. We examine edge 1-2 of five-component mixture 1,2,3,4,5 as an example. For trajectory tear-off from edge 1-2 into the concentration pentahedron (for obtaining the mixture 1, 2 as product of the five-component distillation) it is necessary that these trajectories could tear-off into each of adjacent with this edge faces 1-2-3, 1-2-4 and 1-2-5 (that the mixture 1, 2 could be the product of the distillation of 1, 2, 3 mixture, 1, 2, 4 mixture, and 1, 2, 5 mixture). Figure 8.8 shows such a case for the top section at separation of a hypothetical mixture. This graph shows curves $x_{t2} \rightarrow x_{D2}^{rev}$ for the mentioned faces at reversible distillation. The way to obtain these curves is described in Chapter 4.

In this example, the tear-off into face 1-2-4 is possible from all points of edge 1-2, into face 1-2-3 it is possible from the points of segment $[0 - \max x_{t2}^{1-2-3}]$, and into face 1-2-5 it is possible from the points of segment $[\min x_{t2}^{1-2-5} - 1]$. Therefore, tear-off from edge 1-2 into pentahedron is possible only from the points of segment

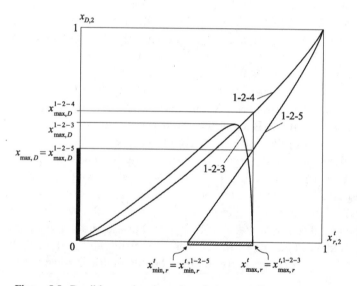

Figure 8.8. Possible overhead product (mixture 1,2) concentrations $x_{D,2}$ of the component 2 for constituent mixtures 1,2,3; 1,2,4; 1,2,5 as function of tear-off point compositions $x_{r,2}^t$ on the edge 1-2 of the concentration simplex for the five-component mixture. Segment possible top product $\text{Reg}_{D \atop 1,2}^{(2)}{}^{3,4,5}$ (solid) and tear-off segment $\text{Reg}_{r \atop 1,2}^{t}{}^{3,4,5}$ (hatched) on the edge 1-2 for the five-component mixture separation; $\text{Reg}_{r \atop 1,2}^{(t)}{}^{4} = [0-1]$, $\text{Reg}_{r \atop 1,2}^{t}{}^{5} = [x_{\min}^{1-2-5} - 1]$, $\text{Reg}_{r \atop 1,2}^{t}{}^{3} = [0 - x_{\max}^{1-2-3}]$, $\text{Reg}_{D \atop 1,2}^{(2)}{}^{3} = [0 - x_{\max,D}^{1-2-3}]$, $\text{Reg}_{D \atop 1,2}^{(2)}{}^{4} = [0-1]$, $\text{Reg}_{D \atop 1,2}^{(2)}{}^{5} = [0-1]$.

$\underset{1,2}{\overset{3,4,5}{\text{Reg}^t_{g_r}}} \equiv [\min x^{1-2-5}_{t2} - \max x^{1-2-3}_{t2}]$ (common part of segments [0-1], [0 – max

x^{1-2-3}_{t2}], and [$\min x^{1-2-5}_{t2} - 1$]) (i.e., $\underset{1,2}{\overset{3,4,5}{\text{Reg}^t_r}} = \underset{1,2}{\overset{3}{\text{Reg}^t_r}} \bullet \underset{1,2}{\overset{4}{\text{Reg}^t_r}} \bullet \underset{1,2}{\overset{5}{\text{Reg}^t_r}}$). For each

tear-off point at segment [$\min x^{1-2-5}_{t2} - \max x^{1-2-3}_{t2}$], there are three prod-

uct points x^{1-2-3}_{D2}, x^{1-2-4}_{D2}, and x^{1-2-5}_{D2} at separation of mixtures 1,2,3; 1,2,4; and

1,2,5, respectively. Possible product segment $\underset{1,2}{\overset{3,4,5}{\text{Reg}_D}}$ at adiabatic distillation of

a five-component mixture is a common part of possible product segments at

adiabatic distillation of mixtures 1,2,3; 1,2,4; and 1,2,5 for tear-off points at seg-

ment $\underset{1,2}{\overset{3,4,5}{\text{Reg}^t_r}} \equiv [\min x^{1-2-5}_{t2} - \max x^{1-2-3}_{t2}]$ (i.e., $\underset{1,2}{\overset{3}{\text{Reg}_D}}$ is function of K_1, K_2, and K_3

at segment $\underset{1,2\,3,4,5}{\overset{3,4,5}{\text{Reg}^t_r}}$ [see for example Fig. 4.11], $\underset{1,2}{\overset{4}{\text{Reg}_D}}$ is function of K_1, K_2, and

K_4 at segment $\overset{5}{\text{Reg}^t_r}$, $\underset{1,2}{\overset{3,4,5}{\text{Reg}_D}}$ is function of K_1, K_2, and K_5 at segment $\underset{1,2}{\text{Reg}^t_r}$ and

$\underset{1,2}{\overset{3,4,5}{\text{Reg}_D}} = \underset{1,2}{\overset{3}{\text{Reg}_D}} \bullet \underset{1,2}{\overset{4}{\text{Reg}_D}} \bullet \underset{1,2}{\overset{5}{\text{Reg}_D}}$). This segment $\underset{1,2}{\overset{3,4,5}{\text{Reg}_D}}$ is located between point $x_{D2} =$

0 and point $x_{D2} = \min[\underset{\text{Reg}^t_r}{\max x^{1-2-3}_{D2}}, \underset{\text{Reg}^t_r}{\max x^{1-2-4}_{D2}}, \underset{\text{Reg}^t_r}{\max x^{1-2-5}_{D2}}]$. In the example un-

der consideration, possible top product segment is $\underset{1,2}{\overset{3,4,5}{\text{Reg}_D}} = [0, \underset{\text{Reg}^t_r}{\max x^{1-2-5}_{D2}}]$.

In the general case, if components of edge are components i_1 and i_2 and the

other components are $j_1, j_2, \ldots j_k$, then possible product segments are $\overset{j_1 \div j_k}{\underset{i_1, i_2}{\text{Reg}_D}} =$

$\underset{i_1, i_2}{\overset{j_1}{\text{Reg}_D}} \bullet \underset{i_1, i_2}{\overset{j_2}{\text{Reg}_D}} \ldots \bullet \underset{i_1, i_2}{\overset{j_k}{\text{Reg}_D}} = [0, \max x_D]$ or $\underset{i_1, i_2}{\overset{j_1 \div j_k}{\text{Reg}_B}} = \underset{i_1, i_2}{\overset{j_1}{\text{Reg}_B}} \bullet \underset{i_1, i_2}{\overset{j_2}{\text{Reg}_B}} \ldots \bullet \underset{i_1, i_2}{\overset{j_k}{\text{Reg}_B}} = [0,$

$\max x_B]$, where the ends of the segments are:

$$\max x_D = \min_j [\underset{\text{Reg}^t_r}{\max x^{i_1-i_2-j}_D}] \tag{8.10}$$

$$\max x_B = \min_j [\underset{\text{Reg}^t_r}{\max x^{i_1-i_2-j}_B}] \tag{8.11}$$

8.4.2. Possible Product Regions at the Boundary Elements of Concentration Simplex

Possible product regions $\underset{i}{\overset{j}{\text{Reg}^{(k)}_D}}$ and $\underset{i}{\overset{j}{\text{Reg}^{(k)}_B}}$ at the faces and hyperfaces $(C^{(k)})$ of the

concentration simplex have its edges as their boundary elements $\underset{i}{\overset{j}{\text{Reg}_{bound,D}}}$ and

$\underset{i}{\overset{j}{\text{Reg}_{bound,B}}}$. Figures 8.9 and 8.10 show possible location of these regions in two-

dimensional faces $\underset{i}{\overset{j}{\text{Reg}^{(3)}_D}}$ and three-dimensional hyperfaces $\underset{i}{\overset{j}{\text{Reg}^{(4)}_D}}$ of concentration

simplex, respectively. As one can see in these figures, possible product regions

$\underset{i}{\overset{j}{\text{Reg}^{(k)}_D}}$ and $\underset{i}{\overset{j}{\text{Reg}^{(k)}_B}}$ are polygons, polyhedrons, or hyperpolyhedrons, part of the

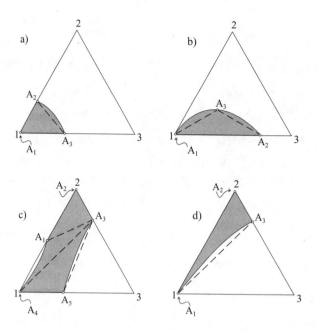

Figure 8.9. Possible product regions Reg_D^j or Reg_B^j (shaded)
for ternary mixtures. $A_1 \div A_5$, vertices of product regions.

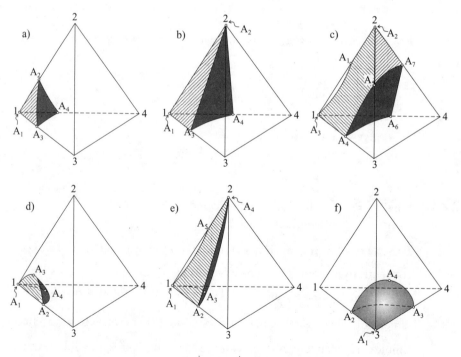

Figure 8.10. Possible product regions Reg_D^j or Reg_B^j (shaded and hatched) for four-component mixtures. $A_1 \div A_7$, vertices of product regions.

boundary elements of which $\mathrm{Reg}_{bound,D}^{j}{}_{i}$ and $\mathrm{Reg}_{bound,B}^{j}{}_{i}$ are curvilinear and vertexes are ends of segments at edges of the concentration simplex.

To determine the location of possible product regions $\mathrm{Reg}_{D}^{(k)}{}_{i}^{j}$ and $\mathrm{Reg}_{B}^{(k)}{}_{i}^{j}$, it is necessary to determine the coordinates of the ends of these segments. The algorithm described in the previous section should be used for this purpose, taking into consideration that we are interested now not in the possible product segments $\mathrm{Reg}_{D}^{(2)}{}_{i}^{j}$ and $\mathrm{Reg}_{B}^{(2)}{}_{i}^{j}$, but in a segments whether are boundary elements of some possible product region $\mathrm{Reg}_{bound,D}^{j}{}_{i}$ or $\mathrm{Reg}_{bound,B}^{j}{}_{i}$.

We examine what conditions should be satisfied by the possible product point x_D or x_B belonging to the k-component hyperface $C^{(k)}$ of the concentration simplex of an n-component mixture C_n and lying at edge $i_1 - i_2$ of this hyperface ($x_D \in [i_1 - i_2] \in C^{(k)}$ or $x_B \in [i_1 - i_2] \in C^{(k)}$). Components $j_1, j_2 \ldots j_{n-k}$ do not enter into the number of components of the k-component hyperface under consideration ($j_1 \notin C^{(k)}$, $j_2 \notin C^{(k)}$, $\ldots j_{n-k} \notin C^{(k)}$). Point x_D or x_B can be product point of the distillation of the n-component mixture if it is a possible product point x_D or x_B of the distillation for all the three-component mixtures $C^{(3)}$ composed of $i_1, i_2, j_1; i_1, i_2, j_2; \ldots; i_1, i_2, j_{n-k}$. Therefore, possible product segment $\mathrm{Reg}_{bound,D}^{j}{}_{i}$ or $\mathrm{Reg}_{bound,B}^{j}{}_{i}$ at edge $i_1 - i_2$ being part of the k-component hyperface $C^{(k)}$ of the concentration simplex C_n is a common part of all possible product segment $\mathrm{Reg}_{D}^{(2)}{}_{i}^{j}$ or $\mathrm{Reg}_{B}^{(2)}{}_{i}^{j}$ at edge $i_1 - i_2$ at separation of all ternary mixtures $C^{(3)}$ composed of i_1, i_2, j; that is,

$$\mathrm{Reg}_{bound,D}{}_{i_1,i_2}^{j_1 \div j_{n-k}} = \mathrm{Reg}_{D}{}_{i_1,i_2}^{j_1} \bullet \mathrm{Reg}_{D}{}_{i_1,i_2}^{j_2}, \ldots, \bullet \mathrm{Reg}_{D}{}_{i_1,i_2}^{j_{n-k}} \text{ or}$$

$$\mathrm{Reg}_{bound,B}{}_{i_1,i_2}^{j_1 \div j_{n-k}} = \mathrm{Reg}_{B}{}_{i_1,i_2}^{j_1} \bullet \mathrm{Reg}_{B}{}_{i_1,i_2}^{j_2}, \ldots, \bullet \mathrm{Reg}_{B}{}_{i_1,i_2}^{j_{n-k}}.$$

The difference between the algorithms of determination of the ends of possible product segments $\mathrm{Reg}_{D}^{(2)}{}_{i}^{j}$ or $\mathrm{Reg}_{B}^{(2)}{}_{i}^{j}$ and of segments that are boundary elements of possible product regions $\mathrm{Reg}_{bound,D}^{j}{}_{i}$ or $\mathrm{Reg}_{bound,B}^{j}{}_{i}$ in the k-component hyperface $C^{(k)}$ consists of the fact that, in the first case, all the components except i_1 and i_2 enter into the number of components j and, in the second case, all components, except the components of the k-component hyperface $C^{(k)}$ under consideration, enter into this number.

Thus, the algorithm of determination of coordinates of vertexes of possible product regions $\mathrm{Reg}_{D}^{(k)}{}_{i}^{j}$ and $\mathrm{Reg}_{B}^{(k)}{}_{i}^{j}$ (for example of points $A_1 \div A_7$ in Fig. 8.10) includes the following steps:

1. Determination of phase equilibrium coefficients of all components in points of each edge, taken with the set step.

2. Determination of the possible product points $\text{Reg}_{D}^{(1)\,j}_{\;\;i}$ and $\text{Reg}_{B}^{(1)\,j}_{\;\;i}$ in the vertexes of the concentration simplex (they are nodes N^- and N^+ of the concentration simplex).

3. Determination of the coordinates of the ends of the possible product segments $\text{Reg}_{D}^{(2)\,j}_{\;\;i}$ and $\text{Reg}_{B}^{(2)\,j}_{\;\;i}$ at the edges of the concentration simplex (see Section 8.4.1).

4. Determination of the coordinates of the ends of the boundary elements possible product regions $\text{Reg}_{bound,D}^{\;\;\;\;\;j}_{\;\;\;\;\;i}$ and $\text{Reg}_{bound,B}^{\;\;\;\;\;j}_{\;\;\;\;\;i}$ at the edges of the concentration simplex for the possible product regions $\text{Reg}_{D}^{(k)\,j}_{\;\;i}$ and $\text{Reg}_{B}^{(k)\,j}_{\;\;i}$ at the two-dimensional faces $C^{(3)}$, three-dimensional hyperfaces $C^{(4)}$, etc., up to $(n-1)$-component hyperfaces $C^{(n-1)}$ (i.e., vertexes of these regions).

8.4.3. Possible Sharp Splits in Columns with One Feed

The splits that meet the following two conditions are feasible: (1) points of products and of feeding lie in one straight line and (2) points of products belong to possible product composition regions $x_D \in \text{Reg}_{D}^{\,j}_{\,i}$ and $x_B \in \text{Reg}_{B}^{\,j}_{\,i}$. Therefore, to determine possible splits, it is necessary to check these conditions.

For splits without distributed components, there is a correspondence between each possible product region $\text{Reg}_{D}^{(k)\,j}_{\;\;i}$ or $\text{Reg}_{B}^{(k)\,j}_{\;\;i}$ containing k components and a possible product region $\text{Reg}_{B}^{(n-k)\,j}_{\;\;\;i}$ or $\text{Reg}_{D}^{(n-k)\,j}_{\;\;\;i}$ containing $(n - k)$ components that is complementary to it (i.e., these elements taken together contain all components). Therefore, to determine feasible splits, one looks over those boundary elements of concentration simplex $C^{(k)}$, which contain possible product regions $\text{Reg}_{D}^{(k)\,j}_{\;\;i}$ and $\text{Reg}_{B}^{(k)\,j}_{\;\;i}$, and checks which of them complement each other. Each pair determines one possible split. If the product points for the set composition of feeding get into possible product regions (i.e., $x_D \in \text{Reg}_{D}^{(k)\,j}_{\;\;i}$ and $x_B \in \text{Reg}_{B}^{(n-k)\,j}_{\;\;\;i}$), then this split is feasible without any recycle flows. If they do not get there, then this split is feasible if recycle flows of separation products of the unit are available, because these flows move the feed point into the concentration simplex. The value of necessary recycle flow rate is proportional to the length of the segment between the product point without recycle and the boundary of possible product region. The segments have to be taken at the secants passing through vertexes of the concentration simplex. The smallest of these segments corresponds to the component the required recycle flow rate of which is the smallest for the split under consideration. To check if product point gets into possible product composition region, one has to divide this region into simplexes, number of vertexes of which and number of product components is equal. Figure 8.9 shows divide of possible composition regions into simplexes (see dotted lines).

To check whether product point with coordinates $(x_1, x_2 \ldots x_k)$ belongs to the simplex of possible product compositions $x_D \in \text{Reg}_{simp,D}^{(k)}$ or $x_B \in \text{Reg}_{simp,B}^{(k)}$, coordinates of vertexes of which are $(x_1^1, x_2^1 \ldots x_k^1)$, $(x_1^2, x_2^2 \ldots x_k^2)$, $\ldots (x_1^k, x_2^k \ldots x_k^k)$, it is sufficient to solve the following system of linear equations:

$$
\begin{aligned}
x_1 &= a_1 x_1^1 + a_2 x_2^1 + \cdots + a_k x_k^1 \\
x_2 &= a_1 x_1^2 + a_2 x_2^2 + \cdots + a_k x_k^2 \\
&\cdots\cdots\cdots\cdots\cdots\cdots\cdots\cdots\cdots \\
x_k &= a_1 x_1^k + a_2 x_2^k + \cdots + a_k x_k^k
\end{aligned}
\tag{8.12}
$$

In this system of equations, the unknown parameters $a_1, a_2, \ldots, a_m$ are proportional to the distance from point $(x_1, x_2, \ldots x_k)$ to the corresponding vertexes of possible product composition simplex. If all parameters $a_1, a_2, \ldots, a_k$ obtained from Eq. (8.12) turn out to be positive, the potential product point being checked belongs to the possible product composition simplex under consideration; otherwise, it does not belong to it. A similar method was used before to determine feasible splits at infinite reflux (Petlyuk, Kievskii, & Serafimov, 1979) (see Chapter 3).

Besides splits without distributed component, splits with one distributed component can be of great practical importance. Therefore, it is necessary to check which splits of this type are feasible.

The check-up is realized in the same way as for the splits without distributed components, taking into consideration the fact that coordinates of the product points depend on the distribution coefficient. Therefore, the check-up is performed for a values 0 and 1 of this coefficient.

8.4.4. Possible Sharp Splits in Columns with Two Feeds

To determine possible splits in columns with two feedings, it is necessary to find trajectory tear-off segments of the intermediate section $\text{Reg}_e^{t(2)j}{}_i$ at the edges of the concentration simplex, while using various autoentrainers. As shown in Chapter 6, the following order of components at decreased phase equilibrium coefficients should be valid for these segments: first comes the group of components of the top product, and last comes the group of components of entrainer, and between them there is the group of the rest of components that are absent at the edge under consideration.

The similar condition should be valid for the trajectory tear-off regions $\text{Reg}_e^{t(k)j}{}_i$ at the faces and hyperfaces of the concentration simplex. Like possible product composition regions, trajectory tear-off regions of the intermediate section are polygons, polyhedrons, and hyperpolyhedrons, the vertexes of which are ends of segments $\text{Reg}_{bound,e}^{t j}{}_i$ at edges.

Therefore, to identify the trajectory tear-off segments and regions, it is necessary at the beginning to calculate the values of phase equilibrium coefficients of all components at all the edges of the concentration simplex in the points taken with a certain step.

Sharp split in the column with two feedings is possible, if the conditions for product compositions are valid: (1) top product point and entrainer point should belong to the same boundary elements of concentration simplex as the trajectory tear-off region of the intermediate section (if $\text{Reg}_e^t \in C_i^j$, then $x_D \in C_i^j$ and $x_E \in C_i^j$); (2) bottom product point should belong to the possible composition region $x_B \in \text{Reg}_B$ in the concentration simplex, containing all components of feed and of entrainer C_{F+E}; (3) top product point should belong to the possible composition region $x_D \in \text{Reg}_D$ in the concentration simplex, containing only the components of top product and of entrainer C_{D+E}. We note that the coordinates of the bottom product point depend on the flow rate and composition of the entrainer.

8.4.5. The Most Interesting Splits of Columns with Decanters

The splits in the columns with decanters are the most interesting for synthesis, if in the concentration simplex there are regions of existence of two liquid phases Reg_{L-L} and a heteroazeotrope that is the unstable node N_{Haz}^- of the concentration simplex C_n or its boundary element $C^{(k)}$ having a trajectory tear-off region of the top or intermediate section Reg_r^t or Reg_e^t.

In the listed cases, one chooses the split at which a vapor close in composition to heteroazeotrope $y_D \approx N_{Haz}^-$ goes from the column to the decanter, one of the liquid phases is taken from the decanter as top product of the complex "column + decanter" $x_D = x_{L2}$, bottom product point of the column belongs to the possible product region of the concentration simplex $x_B \in \text{Reg}_B$, and product points lie in different distillation regions Reg_1^∞ and Reg_2^∞ (see Chapter 6). The second phase from the decanter or a mixture of two phases is used as reflux (at heteroazeotropic distillation) or as entrainer (at heteroextractive distillation, while, besides that, an additional amount of the entrainer is brought into the column).

The above-discussed algorithms of presynthesis for simple columns, columns with two feedings, and columns with decanters do not require visualization and can be executed automatically, which allows them to be used for mixtures with any number of components.

However, visualization can be useful at the ends of the sequence, when ternary and binary azeotropic mixtures are obtained. Visual analysis of distillation diagrams is desirable at this stage of presynthesis and synthesis for application of such separation methods as usage of curvatures of distillation regions boundaries, of different of the pressure in the columns, and of entrainers (see Section 8.5).

8.4.6. Examples of Presynthesis

8.4.6.1. Example 1: Simple Columns

To illustrate the algorithm of presynthesis, we examine two examples of homogenous azeotropic mixtures. The first example is a four-component mixture acetone(1)-benzene(2)-chloroform(3)-toluene(4) with one binary azeotrope 13. Figure 4.13 shows the segments of identical order of the components $\text{Reg}_{ord}^{ijk(2)}$ at the edges of concentration tetrahedron and the regions of identical order of components $\text{Reg}_{ord}^{ijk(4)}$ inside concentration tetrahedron obtained by means of scanning of phase equilibrium coefficients at the edges.

The structural condition of trajectory tear-off determines the vertexes $\text{Reg}_{r,s}^{t(1)}$, the segments at the edges $\text{Reg}_{r,s}^{t(2)}$, and the regions at the faces $\text{Reg}_{r,s}^{t(3)}$ of the concentration tetrahedron from points of which the trajectory tear-off into it is possible (Fig. 5.27a). For the top section, these are $\text{Reg}_r^{t(1)}$ – vertexes $1(\text{Reg}_r^{t}{}_{2,3,4})$ and $3(\text{Reg}_r^{t}{}_{1,2,4})$ (unstable nodes N^-), $\text{Reg}_r^{t(2)}$ – segments at edge 1-2 $(\text{Reg}_r^{t}{}_{3,4})$ and at edge 1-3$(\text{Reg}_r^{t}{}_{2,4})$, and region $\text{Reg}_r^{t(3)}$ – the whole face 1-2-3$(\text{Reg}_r^{t}{}_{1,2,3})$ (for all these boundary elements, the absent components are heavier than present ones). For the bottom section, these are $\text{Reg}_s^{t(1)}$ – vertex $4(\text{Reg}_s^{t}{}_{1,2,3})$ (stable node N^+), $\text{Reg}_s^{t(2)}$ – the whole edge 2-4$(\text{Reg}_s^{t}{}_{1})$, and $\text{Reg}_s^{t(3)}$ – the part of face 2-3-4$(\text{Reg}_s^{t}{}_{2,3,4})$ (for all these boundary elements, the absent components are lighter than present ones).

Figure 5.27b shows possible product composition segments $\text{Reg}_{D,B}^{(2)}$ at the edges of concentration tetrahedron. The whole edge 1-3 is filled up with possible top product points at distillation of three-component constituents $1,2,3(\text{Reg}_D^2)$ and $1,3,4(\text{Reg}_D^{1,3})$, and the whole edge 2-4 is filled up with possible bottom product points for mixtures $1,2,4(\text{Reg}_B^{1}{}_{2,4})$ and $2,3,4(\text{Reg}_B^{3}{}_{2,4})$. Therefore, these edges are entirely filled up with possible points of top and bottom products, respectively, for the whole four-component mixture $([1-3] \equiv \text{Reg}_D^{2,4}{}_{1,3}, [2-4] \equiv \text{Reg}_B^{1,3}{}_{2,4})$. At edge 1-2, there is a small common segment of possible top product compositions for mixtures 1,2,3 and 1,2,4. This segment is a possible top product composition segment $\text{Reg}_D^{3,4}{}_{1,2}$ at distillation of four-component mixture.

The next step of presynthesis is determination of possible product regions $\text{Reg}_{D,B}^{(3)}$ in those two-dimensional faces of concentration tetrahedron that contain trajectory tear-off regions (i.e., in face 1-2-3 for the top product and in face 2-3-4 for the bottom product). We have to find contours of these regions $\text{Reg}_{bound,D,B}^{(2)}$ at the edges of faces 1-2-3, 2-3-4.

In accordance with the described algorithm, we have to examine ternary mixtures composed of each edge of these faces and of the component that is absent at the face (in the given example, only one component is absent: for face 1-2-3, it is component 4; for face 2-3-4, it is component 1). Therefore, for face 1-2-3, it is sufficient to determine possible product segments $\text{Reg}_D^{4}{}_{1,2} \in \text{Reg}_{bound,D}^{4}{}_{1,2,3}$ at edge 1-2 in ternary mixture 1,2,4, $\text{Reg}_D^{4}{}_{1,3} \in \text{Reg}_{bound,D}^{4}{}_{1,2,3}$ at edge 1-3 – in ternary mixture 1,3,4 and $\text{Reg}_D^{4}{}_{2,3} \in \text{Reg}_{bound,D}^{4}{}_{1,2,3}$ at edge 2-3 – in ternary mixture 2,3,4. Edges 1-2, 1-3, and 2-3 are entirely filled up with possible top product points at distillation of the mentioned ternary mixtures ($[1\text{-}2] \equiv \text{Reg}_{D}^{4}{}_{1,2}$, $[1\text{-}3] \equiv \text{Reg}_{D}^{4}{}_{1,3}$, $[2\text{-}3] \equiv \text{Reg}_{D}^{4}{}_{2,3}$). Thus, the whole contour of face 1-2-3 is filled up with possible top product points (i.e., belong $\text{Reg}_{bound,D}^{4}{}_{1,2,3}$). Therefore, we can state that the whole face 1-2-3 is also filled up with possible top product points ($\{1\text{-}2\text{-}3\} \equiv \text{Reg}_{D}^{(3)}$).

For face 2-3-4, it is sufficient to determine possible bottom product composition segment $\text{Reg}_{B}^{1}{}_{2,3} \in \text{Reg}_{bound,B}^{1}{}_{2,3,4}$ at edge 2-3 at distillation of mixture 1,2,3 (this segment joins vertex 2), possible bottom product segment $\text{Reg}_{B}^{1}{}_{2,4} \in \text{Reg}_{bound,B}^{1}{}_{2,3,4}$ at edge 2-4 at distillation of mixture 1,2,4 (the whole edge is entirely filled up with possible bottom product compositions) and possible bottom product segment $\text{Reg}_{B}^{1}{}_{3,4} \in \text{Reg}_{bound,B}^{1}{}_{2,3,4}$ at edge 3-4 at distillation of mixture 1,3,4 (this segment joins vertex 4). If we join with a straight line, the ends of possible bottom product segments at edge 2-3 and at edge 3-4, we obtain at face 2-3-4 the contour $\text{Reg}_{bound,B}^{1}{}_{2,3,4}$ of possible bottom product region (Fig. 5.27b).

Therefore, we obtained the complete set of possible product points in the vertexes $\text{Reg}_{D,B}^{(1)}(1, 3$ and $4)$, at the edges $\text{Reg}_{D,B}^{(2)}$ (edge 1-3, the segment of edge 1-2, edge 2-4), and at the faces $\text{Reg}_{D,B}^{(3)}$ of tetrahedron (face 1-2-3 and the region at face 2-3-4).

If the feed composition is set, it is possible to determine feasible sharp split in the column with one feeding. We accept that the composition of the mixture being separated is equimolar.

We examine complementary boundary elements of concentration tetrahedron that contain possible product regions. Such boundary elements are vertex 1 and face 2-3-4 (direct split), edge 1-3 and edge 2-4 (intermediate split), and face 1-2-3 and vertex 4 (indirect split).

At the direct split 1 : 2,3,4 point x_B (0,333; 0,333; 0,333) is located outside possible bottom product region (see Fig. 5.27b). Therefore, such a split is feasible only if the feed composition is changed with the recycle of component 2, component 4, or their mixture. If the equimolar mixture of components 2 and 4 is used as a recycle, then the rate of recycle flow will be close to minimum. Intermediate 1,3 : 2,4 and indirect 1,2,3 : 4 split will be possible at any feed composition.

The following splits with a distributed component are also feasible: 1,2 : 2,3,4 and 1,2,3 : 2,4. The question of selection of preferable splits is discussed in the section 8.5.1.

8.4.6.2. Example 1: Extractive Distillation

We now examine the possible sharp splits in a column with two feedings. At the beginning, we find the segments $\text{Reg}_{e}^{t(2)}$ of the trajectory tear-off from the edges of the concentration tetrahedron and regions $\text{Reg}_{e}^{t(3)}$ of tear-off from its faces in the intermediate section.

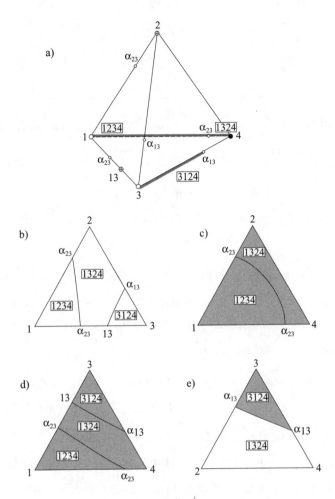

Figure 8.11. Tear-off segments $\mathrm{Reg}_e^{t(2)}$ (thick gray lines) on edges(a) and tear-off regions $\mathrm{Reg}_e^{t(3)}$ (shaded) on faces (b ÷ e) of acetone(1)- benzene(2)- chloroform(3)-toluene(4) concentration thetrahedron for intermediate section: (a) $\mathrm{Reg}_e^{t(2)}$ (entrainer 4) and $\mathrm{Reg}_e^{t(2)}$ (entrainer 4), (c) $\mathrm{Reg}_e^{t(3)}$ (entrainer 4 or 2,4), (d) $\mathrm{Reg}_e^{t(3)}$ (entrainer 4 or 3,4), (e) $\mathrm{Reg}_e^{t(3)}$ (entrainer 2,4).

As one can see in Fig. 8.11, the trajectory tear-off is possible from all points of edge 1-4($[1\text{-}4] \equiv \mathrm{Reg}_e^{t} \in \mathrm{Reg}_e^{t(2)}$), segment $[3, \alpha_{13}]$ of edge 3-4($[3\text{-}\alpha_{13}] \equiv \mathrm{Reg}_e^{t} \in \mathrm{Reg}_e^{t(2)}$), face 1-2-4($\{1\text{-}2\text{-}4\} \equiv \mathrm{Reg}_e^{t} \in \mathrm{Reg}_e^{t(3)}$), face 1-3-4($\{1\text{-}3\text{-}4\} \equiv \mathrm{Reg}_e^{t} \in \mathrm{Reg}_e^{t(3)}$), and region Reg_{ord} in face 2-3-4 ($\mathrm{Reg}_{ord} \equiv \mathrm{Reg}_e^{t} \in \mathrm{Reg}_e^{t(3)}$) (in all these cases, the components absent at the edge or in the face are intermediate).

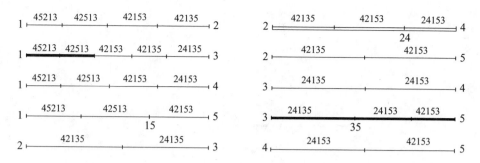

Figure 8.12. Component-order $\text{Reg}_{ord}^{(2)}$ $\overset{i,j,k}{}$ and tear-off (double line for the overhead product $\overset{j}{\text{Reg}_r^{t(2)}}$ and thick line for the bottom product $\overset{j}{\text{Reg}_s^{t(2)}}$) segments on edges of the water(1)-methanol(2)-acetic acid(3)-acetone(4)-pyridine(5) concentration pentahedron.

The following splits are feasible: (1) 1 : 2,3,4 (entrainer 4, $\overset{2,3}{\text{Reg}_e^t}$ – tear-off edge 1-4) because component 1 is possible top product $\overset{4}{\text{Reg}_D}$ at separation of mixture 1,4 and in face 2-3-4 there is possible bottom product region $\overset{1}{\text{Reg}_B}$; (2) 1 : 2,3,4 (entrainer 2,4, $\overset{3}{\text{Reg}_e^t}$ – tear-off region $\overset{1,3,2,4}{\text{Reg}_{ord}}$ in face 1-2-4) because component 1 is possible top product $\overset{1,2,4}{\text{Reg}_D}$ at separation of mixture 1,2,4 and in face 2-3-4 there is possible bottom product region $\overset{1}{\text{Reg}_B}$; and (3) 1,3 : 2,4 (entrainer 4, $\overset{2}{\text{Reg}_e^t}$ – tear-off regions $\overset{3,1,2,4}{\text{Reg}_{ord}}$ and $\overset{1,3,2,4}{\text{Reg}_{ord}}$ in face 1-3-4) because mixture 1,3 is possible top product $\overset{4}{\text{Reg}_D}$ at separation of mixture 1,3,4 and mixture 2,4 is possible bottom product $\overset{1,3}{\text{Reg}_B}$ at separation of the initial mixture.

We note that split 3 : 1,2,4 (entrainer 4, $\overset{2,3}{\text{Reg}_e^t}$ – tear-off segment $[3, \alpha_{13}]$ in edge 3-4) is not feasible in spite of the fact that, as was shown, trajectory tear-off from edge 3-4 is possible in the intermediate section and component 3 can be the top product of separation of mixture 3,4. In this case, the third necessary condition – possible bottom product composition region is absent in face 1-2-4 – is not valid.

Splits 1,2 : 3,4 (entrainer 4, $\overset{3}{\text{Reg}_e^t}$ – tear-off region $\overset{1,2,3,4}{\text{Reg}_{ord}}$ in face 1-2-4) and 3 : 1,2,4 (entrainer 2,4, $\overset{1}{\text{Reg}_e^t}$ – tear-off region $\overset{3,1,2,4}{\text{Reg}_{ord}}$ in face 2-3-4) are not valid for the reasons given above.

8.4.6.3. Example 2: Simple Columns

We now discuss a more complicated task: separation of five-component mixture water(1)-methanol(2)-acetic acid(3)-acetone(4)-pyridin(5) with three binary azeotropes 15, 24, 35 (Petlyuk et al., 1985). Figure 8.12 shows segments $\overset{ijk}{\text{Reg}_{ord}}$ of identical order of components; Figs. 8.13 and 8.14 show

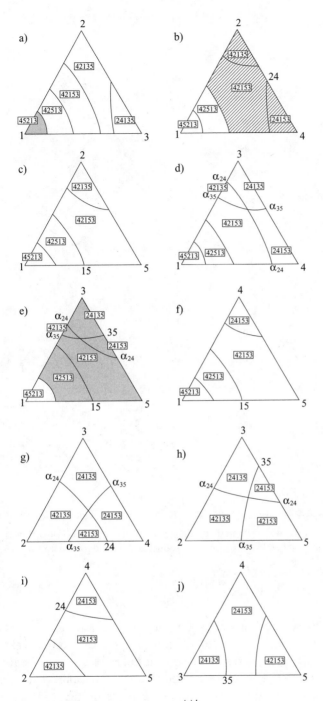

Figure 8.13. Component-order $\text{Reg}_{ord}^{(3)}{}^{i,j,k}$ and tear-off regions $\text{Reg}_r^{t(3)}{}_i^j$ and $\text{Reg}_s^{t(3)}{}_i^j$ on three-component faces of the water(1)-methanol(2)-acetic acid(3)-acetone(4)-pyridine(5) concentration pentahedron (hatched for the overhead product and shaded for the bottom product): (a) $\text{Reg}_s^{t(3)}{}_{1,2,3}^{4,5}$, (b) $\text{Reg}_r^{t(3)}{}_{1,2,4}^{3,5}$, (e) $\text{Reg}_s^{t(3)}{}_{1,3,5}^{2,4}$.

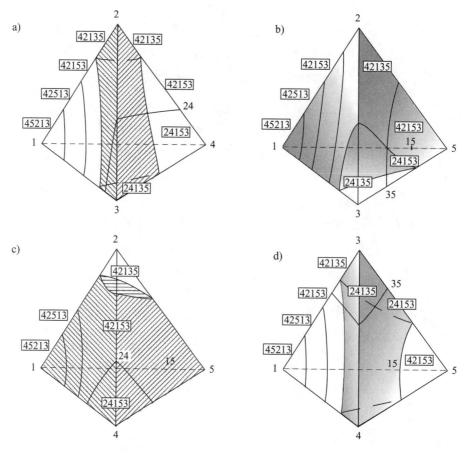

Figure 8.14. Component-order $\mathrm{Reg}_{ord}^{(4)}$ and tear-off regions $\mathrm{Reg}_r^{t(4)}$ and $\mathrm{Reg}_s^{t(4)}$ on four-component hyperfaces of the water(1)-methanol(2)-acetic acid(3)-acetone(4)-pyridine(5) concentration pentahedron (hatched for the overhead product and shaded for the bottom product): (a) $\mathrm{Reg}_r^{t(4)}$, (b) $\mathrm{Reg}_s^{t(4)}$, (c) $\mathrm{Reg}_s^{t(4)}$, (d) $\mathrm{Reg}_s^{t(4)}$.

corresponding to these segments trajectory tear-off regions $\mathrm{Reg}_e^{t(3)}$ in two-dimensional faces and $\mathrm{Reg}_e^{t(4)}$ three-dimensional hyperfaces, respectively. Figure 8.15 shows the direction of residue curves at edges of the pentahedron. It is seen in Fig. 8.15 that all vertexes of the pentahedron are saddles (i.e., cannot be product points in a simple column). As one can see in Fig. 8.12, segments $\mathrm{Reg}_{r,s}^{t(2)}$ of trajectory tear-off from the edges of the pentahedron inside it are available only at edge 2-4 — Reg_r^t (in any point of this edge, the absent components [1,3,5] are the heaviest ones), at edge 1-3 — Reg_s^t (at segment $[1, \alpha_{15}]$ the components absent at the edge [2,4,5] are the lightest ones), and at edge 3-5 — Reg_s^t (in any point of this edge, the absent components [1,2,4] are the lightest ones).

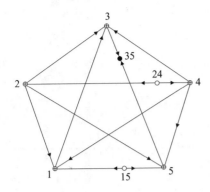

Figure 8.15. Connections between stationary points on the binary edges of the water(1)-methanol(2)-acetic acid(3)-acetone(4)-pyridine(5) concentration pentahedron.

Therefore, top product points $\text{Reg}_D^{(2)}$ can only be at edge 2-4, and bottom product points $\text{Reg}_B^{(2)}$ at edges 1-3, 3-5. Similarly, as Fig. 8.13 shows it, regions $\text{Reg}_{r,s}^{t(3)}$ of trajectory tear-off from two-dimensional faces inside the pentahedron are available only at face 1-2-4 − Reg_r^{t} (in component order regions $\text{Reg}_{ord}^{4,2,1,3,5}$, $\text{Reg}_{ord}^{4,2,1,5,3}$, $\text{Reg}_{ord}^{2,4,1,5,3}$ components 3,5 that are absent at this face are the heaviest), at face 1-2-3 − Reg_s^{t} (in region $\text{Reg}_{ord}^{4,5,2,1,3}$ components 4,5 absent at this face are the lightest) and at face 1-3-5 − Reg_s^{t} (in regions − $\text{Reg}_{ord}^{2,4,1,3,5}$, $\text{Reg}_{ord}^{2,4,1,5,3}$, $\text{Reg}_{ord}^{4,2,1,3,5}$, $\text{Reg}_{ord}^{4,2,1,5,3}$, $\text{Reg}_{ord}^{4,2,5,1,3}$ components 2,4 absent at this face are the lightest). Therefore, bottom product points Reg_B can be located only at faces 1-2-3, 1-3-5, and top product points Reg_D can be located at face 1-2-4.

Finally, as Fig. 8.14 shows, regions of trajectory tear-off from three-dimensional hyperfaces $\text{Reg}_{r,s}^{t(4)}$ inside the pentahedron are available only at hyperface 1-2-3-4 − Reg_r^{t} (in regions $\text{Reg}_{ord}^{4,2,1,3,5}$, $\text{Reg}_{ord}^{2,4,1,3,5}$ component 5 that is absent at this hyperface is the heaviest one), at hyperface 1-2-4-5 − Reg_r^{t} (in regions $\text{Reg}_{ord}^{4,2,1,5,3}$, $\text{Reg}_{ord}^{4,2,5,1,3}$, $\text{Reg}_{ord}^{4,5,2,1,3}$, $\text{Reg}_{ord}^{2,4,1,5,3}$ component 3 absent at this hyperface is the heaviest one), at hyperface 1-2-3-5 − Reg_s^{t} (in regions $\text{Reg}_{ord}^{4,2,1,3,5}$, $\text{Reg}_{ord}^{4,2,1,5,3}$, $\text{Reg}_{ord}^{4,2,5,1,3}$, $\text{Reg}_{ord}^{4,5,2,1,3}$ component 4 absent at this hyperface is the lightest one), and at hyperface 1-3-4-5 − Reg_s^{t} (in regions $\text{Reg}_{ord}^{2,4,1,3,5}$, $\text{Reg}_{ord}^{2,4,1,5,3}$ component 2 absent at this hyperface is the lightest one). Therefore, the top product points Reg_D can be located only at hyperfaces 1-2-3-4, 1-2-4-5, and bottom product points Reg_B can be located at hyperfaces 1-2-3-5, 1-3-4-5.

We now determine segments and possible product regions $\text{Reg}_{D,B}$ at the most interesting boundary elements: at edge 1-3 (it is zeotropic in contrast to edges 2-4, 3-5), at faces 1-2-3 (it is zeotropic) and 1-2-4 (one binary azeotrope), and at hyperfaces 1-2-3-4, 1-2-3-5, 1-2-4-5, 1-3-4-5.

We use the general algorithm of determination of possible product segments and regions $\text{Reg}_{D,B}$. The main element of this algorithm is determination of

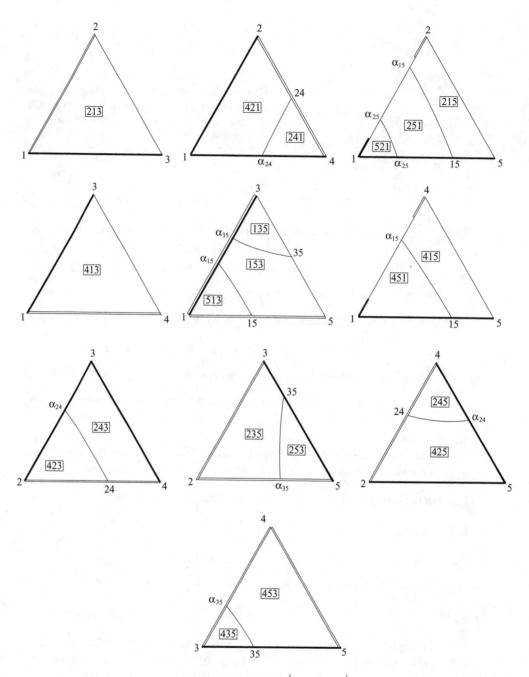

Figure 8.16. Segments of possible product $\mathrm{Reg}_D^{(2)}{}_i^{\,j}$ and $\mathrm{Reg}_B^{(2)}{}_i^{\,j}$ (double line for the overhead product and thick line for the bottom product) on the binary edges of the water(1)-methanol(2)-acetic acid(3)-acetone(4)-pyridine(5) concentration pentahedron for the distillation of ternary constituents.

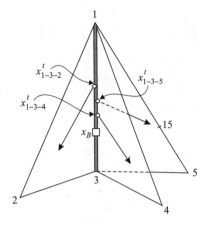

Figure 8.17. The edge 1-3 of the concentration pentahedron is segment of possible bottom product $\operatorname*{Reg}_{B}^{(2)}{}_{1,3}^{2,4,5} = \operatorname*{Reg}_{B}^{(2)}{}_{1,3}^{2} = \operatorname*{Reg}_{B}^{(2)}{}_{1,3}^{4} = \operatorname*{Reg}_{B}^{(2)}{}_{1,3}^{5}$ for the distillation of five-component mixture and of all ternary constituents of five-component mixture. Lines with arrows, section trajectories on three-component faces of the concentration pentahedron.

possible product segments $\operatorname{Reg}_{bound,D,B}^{(2)}$ at the sides of two-dimensional faces of the concentration simplex (according to Fig. 8.8). For the example under consideration, these segments are shown in Fig. 8.16.

For example, faces 1-2-3, 1-3-4, and 1-3-5 are adjacent two-dimensional faces for edge 1-3 (Fig. 8.17). For each face, all points of edge 1-3 are possible bottom product points $\operatorname{Reg}_{1,3}^{2}$, $\operatorname{Reg}_{1,3}^{4}$, and $\operatorname{Reg}_{1,3}^{5}$. Therefore, they are also possible bottom product points at distillation of five-component mixture under consideration ($[1\text{-}3] \equiv \operatorname{Reg}_{B}{}_{1,3}^{2,4,5}$).

We now for example examine face 1-2-3 and its edges 1-2, 1-3, and 2-3 (Fig. 8.18). Components 4,5 are absent in this face. Therefore, it is necessary to examine the segments of possible bottom product $\operatorname{Reg}_{B}{}_{1,2}^{4}$, $\operatorname{Reg}_{B}{}_{1,2}^{5}$ at edge 1-2 at distillation of three-component mixtures 1,2,4 and 1,2,5. The segment joining vertex 1 is common segment of possible bottom product for these two mixtures – $\operatorname{Reg}_{B}{}_{1,2}^{4,5} \in \operatorname{Reg}_{bound,B}{}_{1,2,3}^{4,5}$.

Similarly, we have to examine the segments of possible bottom product $\operatorname{Reg}_{B}{}_{1,3}^{4}$, $\operatorname{Reg}_{B}{}_{1,3}^{5}$ at edge 1-3 at distillation of mixtures 1,3,4 and 1-3-5 (the whole edge $[1\text{-}3] \equiv \operatorname{Reg}_{B}{}_{1,3}^{4,5} \in \operatorname{Reg}_{bound,B}{}_{1,2,3}^{4,5}$) and segments $\operatorname{Reg}_{B}{}_{2,3}^{4}$, $\operatorname{Reg}_{B}{}_{2,3}^{5}$ at edge 2-3 at distillation of mixtures 2,3,4 and 2,3,5 (the segments are absent). Hence, it follows that the region of possible bottom product $\operatorname{Reg}_{B}^{(3)}$ in face 1-2-3 looks like Fig. 8.18d shows it.

The contour of the region of possible top product $\operatorname{Reg}_{bound,B}{}_{1,2,4}^{3,5}$ at face 1-2-4 (Fig. 8.19) is determined in a similar way.

For example, in conclusion, we discuss the determination of the region of possible top product $\operatorname{Reg}_{D}{}_{1,2,4,5}^{3}$ in one of the three-dimensional hyperfaces – in hyperface 1-2-4-5, where component 3 is absent. For this purpose, it is necessary to determine

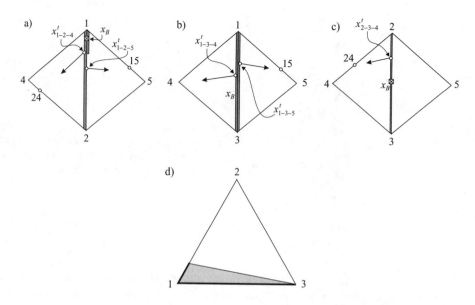

Figure 8.18. Determination of possible bottom product region $\mathrm{Reg}_B^{(3)}\,{}_{1,2,3}^{4,5}$ on three-component face 1-2-3 of the water(1)-methanol(2)-acetic acid(3)-acetone(4)-pyridine(5) concentration pentahedron. Segments of possible bottom product on the binary edges of face 1-2-3: (a) $\mathrm{Reg}_B^{(2)}\,{}_{1,2}^{4,5}$ (double thick segment), (b) $\mathrm{Reg}_B^{(2)}\,{}_{1,3}^{4,5}=[1\text{-}3]$(whole double thick edge), (c) $\mathrm{Reg}_B^{(2)}\,{}_{2,3}^{4,5}=0$ (absent), and (d) region of possible bottom product on the face 1-2-3 $\mathrm{Reg}_B^{(3)}\,{}_{1,2,3}^{4,5}$ (shaded).

the segments of possible top product $\mathrm{Reg}_{bound,D}^{(2)}$ at all the edges of this hyperface: $\mathrm{Reg}_D\,{}_{1,2}^{3}$ at edge 1-2 at distillation of mixture 1,2,3, $\mathrm{Reg}_D\,{}_{1,4}^{3}$ at edge 1-4 at distillation of mixture 1,3,4, $\mathrm{Reg}_D\,{}_{1,5}^{3}$ at edge 1-5 at distillation of mixture 1,3,5, $\mathrm{Reg}_D\,{}_{2,4}^{3}$ at edge 2-4 at distillation of mixture 2,3,4, $\mathrm{Reg}_D\,{}_{2,5}^{3}$ at edge 2-5 at distillation of mixture 2,3,5, and $\mathrm{Reg}_D\,{}_{4,5}^{3}$ at edge 4-5 at distillation of mixture 3,4,5. One can see in Fig. 8.16 that any point of all these edges is a possible top product point. Therefore, we can consider that the whole hyperface 1-2-4-5 is filled up with possible top product points ({1-2-4-5} $\equiv \mathrm{Reg}_D\,{}_{1,2,4,5}^{3} \in \mathrm{Reg}_D^{(4)}$).

It follows from the analysis made that in a column with one feeding only two splits without distributed components are feasible: (1) 2,4 : 1,3,5 and (2) 1,2,4 : 3,5.

Split 1,2,4,5 : 1,3 is the most interesting among splits with one distributed component that is feasible at any feed composition because hyperface 1-2-4-5 and edge 1-3 are entirely filled up with possible product points ({1-2-4-5}$\in \mathrm{Reg}_D^{(4)}$ and [1-3] $\in \mathrm{Reg}_B^{(2)}$).

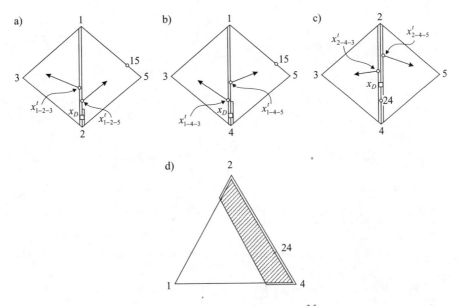

Figure 8.19. Determination of possible overhead product $\text{Reg}_D^{(3)}\overset{3,5}{\underset{1,2,4}{}}$ on three-component face 1-2-4 of the water(1)-methanol(2)-acetic acid(3)-acetone(4)-pyridine(5) concentration pentahedron. Segments of possible overhead product (triple line) on the binary edges of face 1-2-4: (a) $\text{Reg}_D^{(2)}\overset{3,5}{\underset{1,2}{}}$, (b) $\text{Reg}_D^{(2)}\overset{3,5}{\underset{3,5}{}}$, (c) $\text{Reg}_D^{(2)}\overset{3,5}{\underset{1,4}{}} = [2\text{-}4]$ (whole edge), and (d) region of possible overhead product $\text{Reg}_D^{(3)}\overset{}{\underset{1,2,4}{}}$ on the face 1-2-4(hatched).

8.4.6.4. Example 2: Extractive distillation

To analyze variants of autoextractive distillation with one-component entrainer and one-component top product, it is sufficient to examine edges of concentration pentahedron: one of the components of the edge should be the entrainer, and the other one should be the top product. The rest of the components absent at the edge should have intermediate volatilities. The segments $\text{Reg}_e^{t(2)}$ of trajectory tear-off of intermediate section at separation of three- and four-component constituents of five-component mixture are marked out in Fig. 8.20 at edges that do not contain binary azeotropes. As one can see in this figure, one can separate all three-component constituents and some of the four-component constituents of five-component mixture under consideration by means of autoextractive distillation with one-component entrainer and top product, but it is impossible to separate five-component mixtures itself this way.

However, if we use a two-component entrainer or obtain a two-component top product, sharp autoextractive distillation of a five-component mixture becomes possible. It is convenient to use Fig. 8.13 (Reg_{ord}^{ijk} in two-dimensional faces) to analyze the possibility of such splits. Figure 8.21 shows trajectory tear-off regions

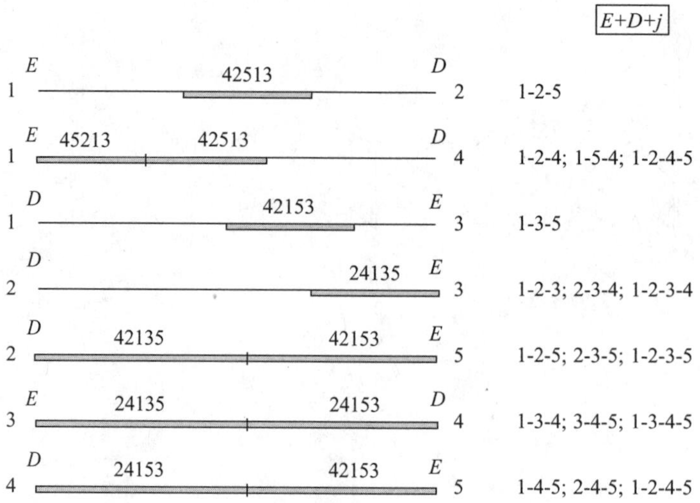

$$\boxed{E+D+j}$$

Figure 8.20. Tear-off segments $\mathrm{Reg}_{e_i}^{t(2)}$ and $\mathrm{Reg}_{e_i}^{t(2)}$ (shaded) for extractive section on edges of the concentration pentahedron for the distillation of three- and four-component constituents of the water(1)-methanol(2)-acetic acid(3)-acetone(4)-pyridine(5) mixture for different entrainer (E) and overhead product (D). To the left, present components i; to the right, all components $i + j$.

$\mathrm{Reg}_e^{t(3)}$ of intermediate sections in two-dimensional faces of the concentration pentahedron. The analysis of possibility of separation for three sections proves that there are several efficient splits meeting the three necessary conditions of sharp separation: (1) $4:1,2,3,5$ (entrainer 1,3), trajectory tear-off regions of intermediate section $\mathrm{Reg}_{e}^{t}{}_{1,3,4} = \mathrm{Reg}_{ord}^{2,5} + \mathrm{Reg}_{ord}^{4,5,2,1,3}$ (Fig. 8.21a); (2) $2:1,3,4,5$ (entrainer 3,5), trajectory tear-off regions of intermediate section $\mathrm{Reg}_{e}^{t}{}_{2,3,5} = \mathrm{Reg}_{ord}^{1,4} + \mathrm{Reg}_{ord}^{2,4,1,5,3}$ (Fig. 8.21c); and (3) $4:1,2,3,5$ (entrainer 3,5), trajectory tear-off region of intermediate section $\mathrm{Reg}_{e}^{t}{}_{3,4,5} \equiv \mathrm{Reg}_{ord}^{4,2,1,5,3}$ (Fig. 8.21e).

8.5.   Multicomponent Azeotropic Mixtures: Automatic Sequencing and Selection

Automatic sequencing is carried out, beginning with the first column and examining it all possible splits. For each possible split in the first column, one obtains composition of two products that for further separation go to the following columns. All possible splits are also examined for each of following columns. The tree of alternative sequences is obtained as a result. The products of each sequence are the set target components or groups of components and, probably, also azeotropes that cannot be separated into target components or groups of

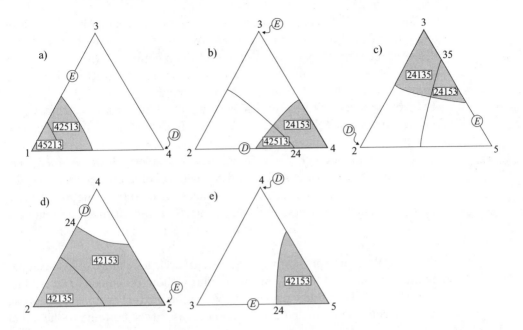

Figure 8.21. Tear-off regions $\mathrm{Reg}_{e}^{t(3)}$ (shaded) for extractive section on faces of the water(1)-methanol(2)-acetic acid(3)-acetone (4)-pyridine(5) concentration pentahedron for different entrainers(E) and overhead products(D): (a) $\mathrm{Reg}_{e}^{t}{}_{1,3,4}^{2,5}$ ($E-$ mixture 1,3; $D-$ component 4); (b) $\mathrm{Reg}_{e}^{t}{}_{2,3,4}^{1,5}$ ($E-$ component 3; $D-$ mixture 2,4); (c) $\mathrm{Reg}_{e}^{t}{}_{2,3,5}^{1,4}$ ($E-$ mixture 3,5; $D-$ component 2); (d) $\mathrm{Reg}_{e}^{t}{}_{2,4,5}^{1,3}$($E-$ component 5; $D-$ mixture 2,4); (e) $\mathrm{Reg}_{e}^{t}{}_{3,4,5}^{1,2}$ ($E-$ mixture 3,5; $D-$ component 4).

components without usage of special methods (entrainers, two levels of pressure) that are discussed in the following section.

8.5.1. Selection of Splits

At the sequencing stage, it is necessary to realize selection of splits in each column to decrease the tree of sequences before calculation of the columns for estimation of expenditures on separation.

As a result of such selection, we obtain several preferable sequences that at the completing stage of synthesis have to be compared with each other by the expenditures in order to choose the optimum one.

Several heuristic rules directed to decrease energy and capital expenditures are used for selection: The first rule is for the mixtures having the region of two liquid phases: it is necessary to use the most interesting splits at heteroazeotropic and heteroextractive distillation described in the section 8.4.5. Such splits separate, in the cheapest way, the mixture into components. The second rule is to exclude splits for which one of the products is binary azeotropic mixture, if other splits exist. This rule excludes, as far as possible, the necessity of expensive methods of separation with application of entrainers and two levels of pressure. The third rule: sequences with recycle flows are less preferable than sequences without them. For example, if it is possible to manage without columns of autoextractive distillation, splits in columns with two feedings have to be excluded.

The rest of heuristic rules are less obvious. The fourth rule: sequences with minimum number of columns are preferable. Hence, it follows that splits with distributed components have to be excluded if there are ones without them. Capital costs at minimum number of columns are the smallest. However, energy expenditures in this case are not always the smallest. For zeotropic mixtures, the situation is the same as for azeotropic: flowsheets with prefractionator are more profitable sometimes than without it (see Section 8.2). Therefore, the designer has to decide him- or herself if he or she should use this rule at the stage of automatic sequences.

The fifth rule: sharp splits with a preliminary recycle for change of feed composition are more preferable than semisharp splits without preliminary recycle. Figure 8.22 shows two possible separation flowsheets of mixture acetone(1)-benzene (2)-chloroform(3) into pure components in the case of the bottom product point

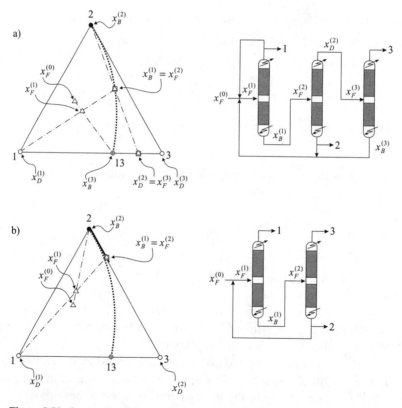

Figure 8.22. Sequences for distillation of the acetone(1)-benzene(2)-chloroform(3) mixture: (a) with best semisharp distillation in the first column, (b) with primary recycle in order that bottom of first column belong to possible bottom product segment $\mathrm{Reg}_{B\ 2,3}^{(2)\ 1}$ (thick line).

not getting into the possible bottom product composition segment at side 2-3 (flowsheet with preliminary recycle (b) and flowsheet with semisharp separation in the first column (a)).

The flowsheet (a) has considerably heavier energy expenditures because of the big necessary reflux number in the third column and big secondary recycle flow rates of azeotrop 13 and components 1 and 2. Besides that, this flowsheet is bigger by one column.

The sixth rule: if a product in the synthesized sequence is the bottom product of one column and the top product of another one, then these two columns have to be made into one four-section column that is the same as the main column in flowsheet with prefractionator (Fig. 6.12d). This situation arises if the split with a distributed component was chosen in one of the previous columns. Unification of two columns leads to decrease of energy and capital expenditures on separation.

8.5.2. Examples of Sequencing and Selection

8.5.2.1. Example I

We discuss two examples, for which presynthesis was made in the previous section. We accept equimolar feed composition for mixture acetone(1)-benzene(2)-chloroform(3)-toluene(4). One of the unstable nodes N^- is component 1. At the opposite face 2-3-4 of concentration tetrahedron, there is possible bottom product region $\text{Reg}_B^{\overset{1}{2,3,4}}$, but the potential product point does not get into this region $x_B^{\overset{1}{2,3,4}} \notin \text{Reg}_B (x_{B,2} = x_{B,3} = x_{B,4} = 0,333)$. Therefore, *direct split* in the first column 1 : 2,3,4 is feasible only with preliminary recycle. Equimolar mixture of components 2,4 can be accepted as a recycle. In this case, the necessary recycle rate is close to minimum possible. Such a split is one of the most preferable. In the first column, we obtain pure component 1 and zeotropic mixture 2,3,4 that can be easily separated in the second and third columns (sequences in Fig. 8.23a,b).

The following split in the first column (*indirect*) 1,2,3 : 4 is feasible without any recycle because component 4 is stable node N^+ and face 1-2-3 is entirely filled up with possible top product points $\text{Reg}_D^{\overset{4}{1,2,3}}$. However, this face contains binary azeotrope 13. Therefore, direct split 1: 2,3 with preliminary recycle of component 2 can be applied in the second column. That separates mixture 1,2,3,4 into pure components in three columns with one preliminary recycle (sequence in Fig. 8.23c).

The last of potential splits in the first column (*intermediate*) 1,3 : 2,4 is feasible because edge 1-3 is filled up with top product points $\text{Reg}_D^{\overset{2,4}{1,3}}$ and edge 2-4 is filled up with bottom product points $\text{Reg}_B^{\overset{1,3}{2,4}}$. However, this split is not expedient because top product is binary mixture with azeotrope.

Splits 1 : 2,3,4 (entrainer is component 4, sequences in Fig. 8.23d,e), 1 : 2,3,4 (entrainer is mixture 2,4, sequences in Fig. 8.23f,g), and 1 : 2,3,4 (entrainer is mixture

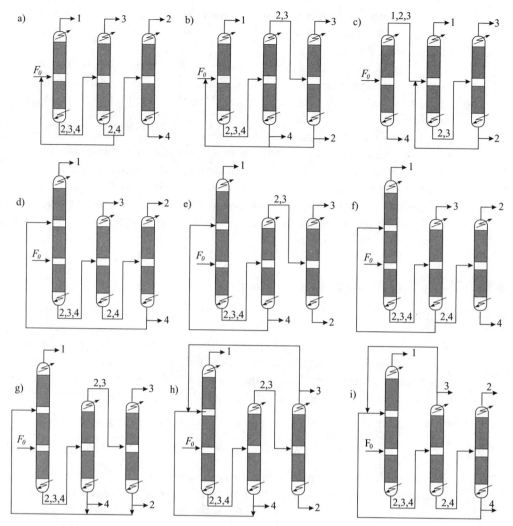

Figure 8.23. Sequences for distillation of the acetone(1)-benzene(2)-chloroform(3)-toluene(4) mixture, obtained by means of automatic sequencing.

3,4, sequences in Fig. 8.23h,i) meet conditions of *autoextractive separation*. These splits are also preferable.

In the given example mixture 2,3,4, that is bottom product of the first column, is zeotropic; therefore, its separation in the second and third columns presents no difficulty. Figure 8.23 shows nine preferable separation flowsheets for this example, obtained by means of synthesis. All these flowsheets separate four-component mixtures using a minimum number (three) of columns without additional entrainers, but with application of autoentrainer. Component 2 or 4 or mixtures 2,4 or 3,4 are used as an autoentrainer. Therefore, *each of nine separation flowsheets contains one column with one or two feedings where recycle flow of autoentrainer is brought in.*

8.5.2.2. Example 2

For the second example, separation of mixture water(1)-methanol(2)-acetic acid(3)-aceton(4)-pyridin(5), both possible splits in the *first column* with one feeding without distributed component 2,4 : 1,3,5 and 1,2,4 : 3,5, are not expedient because one of the products is binary mixture with azeotrope.

Split with distributed component for the *first column* with one feeding (1) 1,2,4,5 : 1,3 and three splits for the first column with two feedings and two-component entrainer; (2) 4 : 1,2,3,5 (entrainer 1,3); (3) 2 : 1,3,4,5 (entrainer 3,5); and (4) 4 : 1,2,3,5 (entrainer 3,5) are preferable splits in the first column.

We turn to the following columns. For *split 1,2,4,5 : 1,3 in the first column*, top product is mixture 1,2,4,5, which can be separated in the *second column with one feeding at split 1,2,4 : 5* because, in face 1-2-4 of the concentration simplex 1-2-4-5, there is a possible top product region $\text{Reg}_D \underset{1,2,4}{\overset{1,2,4}{}}$ joining side 2-4 (Fig. 8.13b), and vertex 5 is a stable node $N^+ \equiv \text{Reg}_B \underset{5}{}$. Therefore, for the part of possible feed composition $x_{F(2)} = x_{D(1)}$ obtained from the previous column, the recycle of component 2 will be necessary. For further separation of the top product of second column (mixture 1,2,4), it is necessary to use in the *third column autoextractive distillation* with entrainer – component 1, top product – component 4, and bottom product – zeotropic mixture 1,2 because, in edge 1-4 of the concentration triangle 1-2-4, there is tear-off region $\text{Reg}_e^t \underset{1,4}{\overset{2}{}} = \text{Reg}_{ord} \overset{4,2,1}{}$ (Fig. 8.13b). Component 1 is the top product at separation of the bottom product of the first column (mixture 1,3), and it simultaneously plays the role of the bottom product at separation of the bottom product of the third column (mixture 1,2). That unites two columns of binary distillation into one *complex column with two feed flows and one side product –* component 1. Therefore, we got a sequence in Fig. 8.24a.

We turn to the following splits in the first column. Mixtures 1,2,3,5 and 1,3,4,5 (bottom products at extractive distillation in first column at splits 2 ÷ 4) can be separated only in the column with one feeding at splits with distributed component.

For separation of *mixture 1,2,3,5 in the second column* (sequences in Fig. 8.24b, d), split 1,2,5 : 1,3 is possible (in face 1-2-5, there is a possible top product region $\text{Reg}_D \underset{1,2,5}{\overset{3}{}}$ [Fig. 8.13c] and, in edge 1-3, there is a possible bottom product segment $\text{Reg}_B \underset{1,3}{\overset{2,5}{}}$ [Fig. 8.12]), and the top product of this column, containing binary azeotrope 15, can be separated by means of distillation in the simple *third column* with preliminary recycle of component 2 at split 1,2 : 5 (there are $\text{Reg}_D \underset{1,2}{\overset{5}{}}$ and $\text{Reg}_B \underset{5}{\overset{1,2}{}}$ [Fig. 8.13c]).

For separation of *mixture 1,3,4,5 in the second column* (sequence in Fig. 8.24c), split 1,4,5 : 1,3 is possible (the whole face 1-4-5 is possible top product region $\text{Reg}_D \underset{1,4,5}{\overset{3}{}}$ [Fig. 8.13f] and, in edge 1-3, there is possible bottom product segment $\text{Reg}_B \underset{1,3}{\overset{4,5}{}}$ [Fig. 8.12]), and the top product of this column, containing binary azeotrope 15,

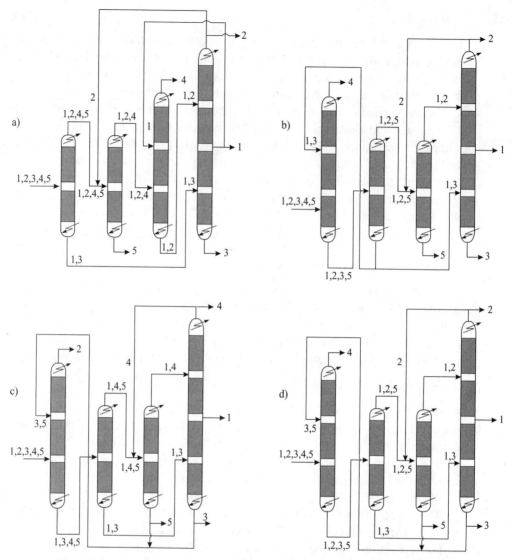

Figure 8.24. Sequences for distillation of the water(1)-methanol(2)-acetic acid(3)-acetone(4)-pyridine(5) mixture, obtained by means of automatic sequencing.

can be separated by means of distillation in the simple *third column* with preliminary recycle of component 4 at split 1,4 : 5 (there are Reg_D $\frac{5}{1,4}$ and Reg_B $\frac{1,4}{5}$ [Fig. 8.13f].)

As a result, we have four best sequences of five-component mixture in four columns (Fig. 8.24). *Each sequence contains one autoextractive column with two feed flows, one autoextractive column with one feed flow (with preliminary recycle), one simple column with one distributed component, and one complex column with two feeds and one side product. Such a set of columns is a consequence of the structure of phase equilibrium coefficients field in concentration pentahedron, (only* of segments $\text{Reg}_{ord}^{(2)}{}^{i,j,k}$ on edges of concentration simplex [Fig. 8.12] not of component names).

In the discussed examples, synthesis of separation flowsheets is carried out automatically on the basis of formal rules. In more complicated cases, in the process of synthesis there can be no other variants other than variants of obtaining azeotropes as some products. These cases are discussed in the following section.

After identification of several preferable sequences, choosing among the optimum sequences, taking into consideration possible thermodinamic improvements and thermal integration of columns, arises. This task is similar to the synthesis of separation flowsheets of zeotropic mixtures (see Section 8.3), and it should be solved by the same methods (i.e., by means of comparative estimation of expenditures on separation). The methods of design calculation, described in Chapters $5 \div 7$ for the modes of minimum reflux and reflux bigger than minimum, have to be used for this purpose. In contrast to zeotropic mixtures, the set of alternative preferable sequences for azeotropic mixtures that sharply decreases the volume of necessary calculation is much smaller.

8.6. Binary and Three-Component Azeotropic Mixtures

If automatic sequencing does not lead to obtaining the prescribed set of target products and the obtained set of products contains azeotropes, then it is necessary to use the special methods to separate the obtained binary and three-component mixtures. These methods were intensively investigated in recent decades. Using curvature of separatrix lines in concentration triangle, and application of entrainers and of two levels of pressure for shift of azeotropes points, belong to these methods. At this stage of synthesis of sequence, the complete automation is not expedient because visualization of separatrix lines, α-lines, binodal lines, and lines liquid–liquid in the concentration triangle can be more profitable.

Some examples of separation of three-component mixtures using curvature of separarix lines are described in Chapter 3. Examples of the application of entrainers forming heteroazeotropes and of columns with decanters for heteroazeotropic and heteroextractive distillation are given in Chapter 6.

We next discuss the examples of application of semisharp extractive distillation (Petlyuk & Danilov, 2000b) for separation of ternary mixture with two binary azeotropes, examples of application of two levels of pressure, and choice of entrainers that do not form heteroazeotrpes.

8.6.1. Application of Semisharp Extractive Distillation

Figure 8.25 shows separation of ternary mixture with two binary azeotropes. The sharp autoextractive distillation at split 2 : 1,3 is possible in the first column (entrainer – component 1) because at side 1-2 there is trajectory tear-off segment $\text{Reg}_e^{t}{}_{1,2}^{3}$ of intermediate section [1,α 23], the whole side 1-3 is possible bottom

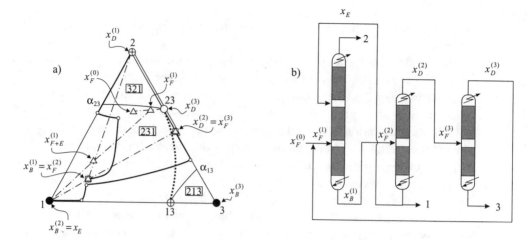

Figure 8.25. Phase equilibria map (a) and sequence for distillation of a ternary mixture with two binary azeotropes in three columns (with semisharp autoextractive distillation in the first column) (b).

product segment $\overset{2}{\mathrm{Reg}_B}$, and binary mixture 1,2 is zeotropic and component 2 is light in it ($2 \equiv \overset{1,3}{\mathrm{Reg}_D}$).

However, sharp extractive distillation is not expedient because bottom product x_B is a binary mixture with azeotrope 13. Split of extractive semisharp distillation $2 : 1,2,3$ ($x_B^{(1)}$) is preferable. Indirect split $2,3 : 1$, at which the top product point of second column $x_D^{(2)}$ gets to the possible top product segment $\overset{1}{\mathrm{Reg}_D}$, has to be applied in the second column. Split $23 : 3$ is applied in the third column, and top product $x_D^{(3)} = 23(\mathrm{Az})$ is used as recycle flow into the first column. The search for such a sequence can hardly be automated because analysis of possible sharp splits in the first column does not give any preferable split. It is necessary to analyze possible splits simultaneously in several columns.

8.6.2. Application of Pressure Change

In some cases, change of pressure very much influences the location of azeotrope points and sometimes even leads to transformation of azeotropic mixture into zeotropic. That uses change of pressure for separation of azeotropic mixtures without entrainers. The example is separation of mixture acetic acid(1)-water(2)-formic acid(3) at two pressures: at atmospheric pressure and at 200 mm Hg pressure (Kuschner et al., 1969).

Figure 8.26a shows the location of azeotropes and separatrix lines at these two pressures. Such location separatrix lines use flowsheet, shown in Fig. 8.26b (in the first and the third columns, there is atmospheric pressure; in the second one, there is pressure 200 mm Hg). The same figure shows lines of material balance and

a)

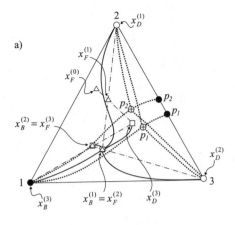

Figure 8.26. Phase equilibria map (a) and sequence for distillation of the acetic acid(1)-water (2)-formic acid(3) mixture with two pressures (p_1 and p_2) in three columns with best semisharp splits.

b)

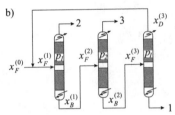

calculation trajectories for each column (Avetyan, Petlyuk, & Serafimov, 1973). Another example is separation of binary azeotropic mixture tetrahydrofurane–water in two columns with recycle at two pressures: at atmospheric pressure and at 8 atmospheres (Abu-Eishuh & Luben, 1985). The large difference in pressures applies thermal integration of the columns (unification of deflegmator of the column of high pressure and of reboiler of the column of low pressure) to decrease expenditures of heat for separation.

Unfortunately, in the majority of cases, change of pressure weakly influence the location of azeotropic points, which makes this method of separation impossible or uneconomical. Possibility or impossibility of usage of two levels of pressure for separation of binary azeotropic mixtures can be easily determined by means of simulation of their phase equilibrium.

8.6.3. Choice of Entrainers

While choosing entrainers for separation of binary azeotropic mixtures, the structure of phase equilibrium diagrams (residue curve maps) of ternary mixtures formed at the addition of entrainer is of great importance.

The most desirable types of residue curve maps were discussed in a number of works (Doherty & Caldorola, 1985; Laroche et al., 1992; Stichlmair & Herguijuela, 1992).

If entrainer – component 3 is intermediately boiling and does not form azeotropics with components of the azeotropic mixture 1,2 under separation, then ternary mixture has a residue curve map shown in Fig. 8.27. In this case,

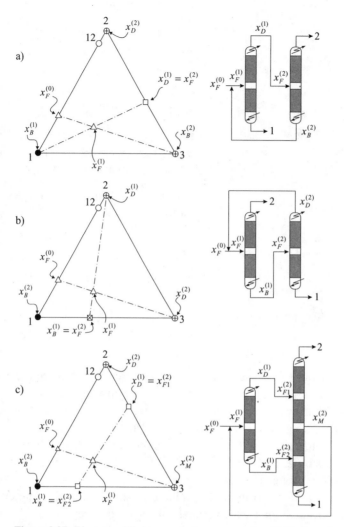

Figure 8.27. Phase equilibria map and sequences for distillation of a binary azeotropic mixture (1,2) with an intermediate boiling entrainer (3): (a) indirect split 2,3 : 1 in the first column, (b) direct split 2 : 1,3 in the first column, and (c) preferred split 2,3 : 1,3 in the first column.

three sequences of two columns with recycle (with indirect split [Fig. 8.27a], with direct split [Fig. 8.27b] and with preferable split [Fig. 8.27c] in first column) (Fig. 8.27a,b,c) can be applied. The main problem consists of choosing a good entrainer.

If entrainer – component 2 is a heavy component that does not form azeotropes with components of the azeotropic mixture 1,3 under separation and the mixture under separation has azeotrope 13 with minimum of bubble temperature, then ternary mixture has a residue curve nap shown in Fig. 8.28 and extractive distillation is used for separation.

If the entrainer is a light component that does not form azeotropes with components of the mixture under separation and the mixture under separation has

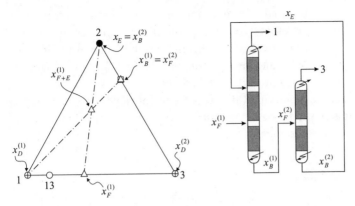

Figure 8.28. Phase equilibria map and sequence for extractive distillation of a binary azeotropic mixture (1,3) with a high boiling entrainer (2); split 1 : 2,3 in the first column.

an azeotrope with a minimum of bubble temperature or the entrainer is heavy component – component 2 and the azeotropic mixture 1,3 under separation has an azeotrope 13 with a maximum of bubble temperature (e.g., acetone(1)-benzene(2)-chloroform(3)) then the ternary mixture has a residue curve map like that shown in Fig. 8.22b or its antipode, and the sequence of two columns with a recycle has to be used for separation.

The main problem consists of the choice of a suitable entrainer that would form a separatrix of big curvature and create a respectively large possible composition product segment at the side of the concentration triangle. If the separatrix is rectilinear or has a small curvature, then it is impossible or uneconomical to apply the sequence in Fig. 8.22b.

If the entrainer – component 2 forms azeotropes with one or two components of the azeotropic mixture 1,3 under separation (Fig. 8.29), then sequence of two columns with recycle of binary azeotrope 12 of entrainer with one of the components can be used for separation.

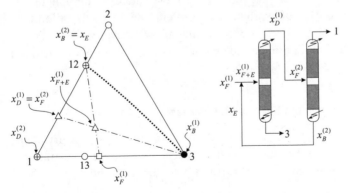

Figure 8.29. Phase equilibria map and sequence for extractive distillation of a binary azeotropic mixture (1,3) with a recycle of azeotropic entrainer (azeotrope 12); split 1,2 : 3 in the first column.

To choose the entrainer among a number of alternative entrainers, it is necessary to carry out comparative estimation of expenditures on separation. For preliminary estimation at extractive distillation, the value of minimum flow rate of entrainer can be used and, at the sequence in Fig. 8.22b, the length of possible product composition segment at the side of the concentration triangle can be used. For more precise estimation, it is necessary to calculate the summary vapor flow in the columns in the mode of minimum reflux at several values of excess factor at the flow rate of entrainer.

8.7.   Petroleum Mixtures

8.7.1. Peculiarities of Petroleum as Raw Material for Separation

A number of significant peculiarities distinguishes petroleum from other kinds of raw materials:

1. The largest world volume of production. It attaches especially great importance to the solution of the task of optimal designing of petroleum refining units.
2. Continuity of properties of components. A large number of components leads to the fact that dependence of properties of components on their normal bubble temperature is continuous. Therefore, in practice, while designing one deals not with the true components but with pseudocomponents (i.e., with groups of components boiling away in a set interval of temperatures), and the quality of the products is characterized not by their purity but by their refinery inspection properties.
3. Wide intervals of bubble temperatures of components from negative temperatures for light hydrocarbon gases up to over 800°C for heavy fractions of mazut and tar.
4. Thermolability–thermal decomposition of components at high temperatures ($360 \div 400°C$). Thermolability limits separability of petroleum and the possibility of recovery of the most valuable components. Therefore, the main task of synthesis of petroleum mixture separation flowsheets is the creation of such flowsheets that allow recovery of a maximum number of valuable components at a permissible temperature (i.e., without thermal decomposition of components).

8.7.2. Methods of Petroleum Separability Increase

To increase recovery of valuable (light) components, it is necessary to increase the amount of the heat brought in without exceeding the set maximum permissible temperature. There are several methods to achieve this goal:

1. It is necessary to bring in heat, not in the reboiler of the column, but in its feeding, bringing in the bottom a light stripping agent (most often, steam serves as one). At such a method of creation, vapor flow maximum

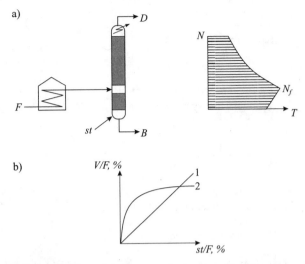

Figure 8.30. A temperature profile of column with steam stripping (a) and fraction of feed vaporized V/F as function of mass flow rate of steam st/F (b). 1, steam-heated reboiler; 2, live steam.

temperature is achieved, not in the bottom, but in feed cross-section (Fig. 8.30).

2. It is necessary to keep in the feed cross-section the lowest possible pressure, which provides the possibility to obtain a maximum fraction of vapor phase in the feeding and a maximum vapor flow rate in the column.

3. It is necessary to increase the number of theoretical trays in sections adjacent to feed cross-section, where the most valuable products are being separated from the less valuable ones.

4. In feeding of the column, where the most valuable products are being separated from the less valuable ones, the fraction of light components that have stripping influence on the rest of components should be as large as possible.

Figure 8.31 shows the calculation dependence of the output of light oil products (benzine + kerosene + diesel oil) on the pressure in the column and on the number of trays in the first section above the feed cross-section (West Siberian petroleum).

The above-listed methods of separability increase require some increase of capital expenditures (an increase in the column diameter – item 1, 2, and 4, an increase in the column height – item 3).

8.7.3. The Best Distillation Complex for Petroleum Refining

Columns with side stripping sections were used for petroleum separation already in the first decades of twentieth century. This choice is quite grounded by the main purposes of designing: increase of separability, and decrease of energy and capital expenditures on separation.

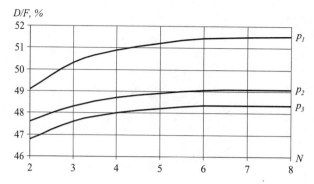

Figure 8.31. The yield of light oil products (<360°C) D/F as function of a pressure p in crude oil distillation column and number of trays N in the first section above feed cross-section (p_1, 0.8 bar; p_2, 1.6 bar; p_3, 2.3 bar, West Siberian petroleum).

Because petroleum is a mixture with a wide interval of bubble temperatures of components and the required purity of products is not very great, reflux and vapor numbers in the sections are not large. Therefore, the heat brought in is used up mostly not for creation of vapor reflux, but for evaporation of those products that are withdrawn above the feed cross-section. Therefore, the split (direct, indirect, intermediate) has but a weak influence over energy expenditures, but it is very important to exclude multiple evaporation and condensation like in multicolumn sequences of simple columns. Columns with side strippings exclude multiple evaporation and condensation.

Replacement of multicolumn sequences with multisection columns also leads to a decrease in capital expenditures because, due to the wide interval of bubble temperatures, there can be but a small number of trays in each section, and it is profitable to unite these sections into one shell.

Among multisection distillation complexes, only columns with side strippings bring practically the whole heat into the feeding and bring live steam into the bottom. Application of pumparounds decreases energy expenditures and recuperates withdrawn heat for heating of petroleum before separation.

At the same time, the application of pumparounds decreases vapor flow rate in the top sections of the column (i.e., to decrease the diameter and capital expenditures).

Therefore, column with side strippings, with live steam into the bottom, and with pumparounds is the best distillation complex for petroleum refining. An optimum way of designing such column is discussed in Section 7.5.2.

8.7.4. Main Succession of Petroleum Refining

Large numbers of products are obtained from petroleum: liquefied gases (propane + butane), benzine and its fractions, kerosene and jet fuel, diesel oil – light one (winter) and heavy one (summer), mazut, gas-oil, lubricant fractions, and tar.

Only part of these products is obtained in each unit. Nevertheless, it is not possible to obtain all the products in one column with side strippings because of the large interval of bubble temperatures of components: vacuum is necessary to obtain heavy products at a permissible temperature, and light products cannot be obtained under vacuum because of the low temperatures that would be required for their condensation, which could not be ensured with the help of cheap cold agents (air or water).

Therefore, the main succession of petroleum refining includes two main units: that of atmospheric distillation and that of vacuum distillation. The lighter products are obtained in the first unit right up to mazut (bottom product of atmospheric column with side strippings) and, in the second unit, the heavier products are obtained right up to tar (bottom product of vacuum column with side strippings).

The first unit usually consists of one column with side strippings (one-column flowsheet) (Fig. 8.32a) or of superatmospheric column of partial recovery of benzine and of main column with side strippings (two-column flowsheet) (Fig. 8.32b).

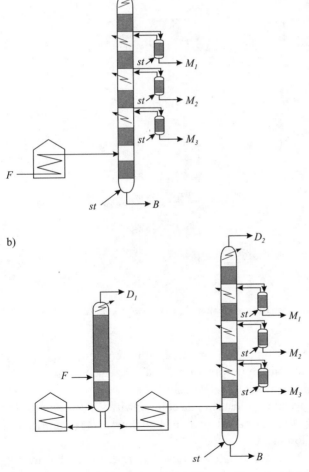

Figure 8.32. Crude oil distillation: (a) in one column, and (b) in two columns.

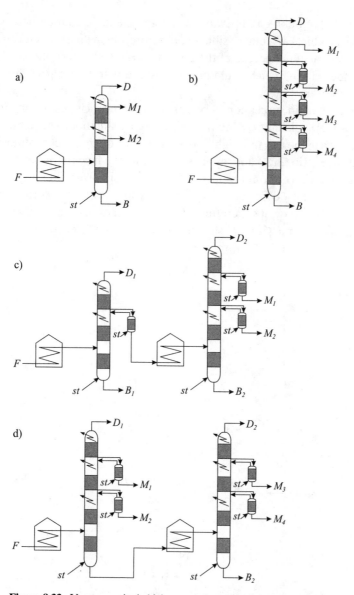

Figure 8.33. Vacuum unit: (a,b) in one column, (c,d) in two columns.

A two-column flowsheet is applied for crude with a large content of gases that are separated in the first column (at pressure 3–5 atm), which keeps sufficiently low pressure in the main column (1,3 ÷ 2,0 atm).

The second unit usually consists of one column with side strippings (Fig. 8.33b) and serves to obtain vacuum gas-oil and tar (fuel flowsheet) or lubricant fractions and tar (lubricant flowsheet).

In some (Fig. 8.33a) cases, steam is not used at separation for strippings ("dry" separation) and deeper vacuum is applied (10 ÷ 20 mm Hg).

To increase the number of theoretical trays, vacuum unit is sometimes made as a two-column one (Fig. 8.33c, d), but according to energy expenditures on separation

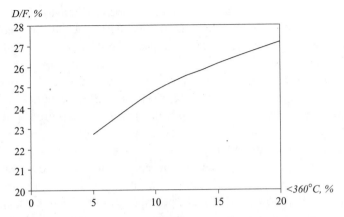

D/F, %

<360°C, %

Figure 8.34. The yield of oil distillates D/F in vacuum column as function of fraction below 360°C in reduced crude (West Siberian petroleum).

it is more profitable to increase the number of trays in a one-column flowsheet by means of applying highly efficient contact devices with small hydraulic resistance (of regular packs).

8.7.5. Modernization of Units for Petroleum Refining

Existing crude units can be reconstructed for a stronger recovery of the most valuable products – light oil products (benzine, kerosene, diesel fuel) and lubricant fractions. The above-listed methods of improving separability have to be used in this reconstruction:

1. Part of diesel fuel has to be recovered not in an atmospheric unit but in a vacuum one. This creates a double effect: (1) in the feeding of a vacuum column the fraction of light components increases, which, due to their stripping influence, increases recovery of lubricant fractions or gas-oil; and (2) the volatilities of components grow under vacuum, which separates diesel fuel from heavier products more sharply and deeply. Figure 8.34 shows the calculation dependence of yield of lubricant fractions in "dry" vacuum column on the fraction in its feeding of components with bubble temperatures below 360°C. However, increase of vapor flow rate in vacuum column requires application of more productive contact devices.

2. If a crude unit contains only a two-column atmospheric unit, then in the main column a moderate vacuum has to be kept ($\approx$0,8 atm) with the help of vacuum-creating equipment joined with c reflux drum.

Increase of the fraction of vapor phase in the feeding at moderate vacuum increases sharpness of separation in the sections of the main column above feeding because of sharp increase of reflux number (in a few times). Besides that, sharpness of separation grows because of increase of relative volatilities of components at lower pressure. This increases output of light oil products (< 360°C) at ensuring their quality (Fig. 8.31) and/or to decrease the flow rate of steam.

In patent literature, a large number of other methods to increase the output of light oil products is introduced, but they require installation of additional columns.

The purpose of modernization of crude unit can be decrease of expenditures of heat and cold.

At two-column atmospheric unit, the simple method to decrease expenditures of heat and cold is application of two feedings with different temperatures in column of partial recovery of benzine (top feeding is at temperature about 100°C and bottom feeding is at temperature achieved by means of heating of crude by hot products and by pumparounds).

Considerable change of main succession of separation with the purpose to decrease energy expenditures at the increase of capital costs was offered in patent (Devos, Gourlia, & Paradowski, 1987). The peculiarity of this flowsheet consists of consequent partial recovery of light fractions (gases, light benzine, heavy benzine, kerosene) in several columns of preliminary separation and obtaining of light end products in several more additional columns. Reboilers of all these columns warm by hot products and by pumparounds, which ensures good recuperation of heat at a large number of heat exchangers. Heavier products are obtained in atmospheric and two vacuum columns with furnace heating and with live steam. Ideas of prefractionation (i.e. split with distributed component) and pinch technologies are used in this flowsheet. The same idea is used in simplified way in the classical two-column flowsheet (Fig. 8.32b). In a more advanced way, this idea was used in the unit recently created in Germany, which contains two prefractionators at pressure three and two atmospheres producing a light naphta with a $80 \div 90°C$ boiling range and a medium naphta with a $90 \div 120°C$ boiling range. This separation flowsheet was named "progressive distillation."

8.8.   Conclusion

While designing and reconstructing separation equipment, one can drastically increase their economy via application of the newest method of scientifically based presynthesis, automatic sequencing, thermodynamic improving, and the use of the best distillation complexes.

The software for automatic sequencing for zeotropic mixtures includes automatic identification of possible splits in simple columns and in distillation complexes, and automatic identification of sequences of these simple columns and complexes, ensuring obtaining from the initial mixture of the prescribed set of products. The degree of complexity of distillation complexes (number of products, number of sections, number of shells of column) can be different and has to be set by the user before the task is solved. After identification of possible sequences, the design calculation by new algorithms of all columns and complexes of each sequence at minimum reflux flow rates or at set excess reflux factor with estimation of energy or summary (energy and capital) expenditures for each sequence is carried out. The degree of simplification of design calculation and of estimation of expenditures can be different and can be set by the user before solution of the task.

For homogeneous azeotropic mixtures, the system of automatic sequencing also includes the stage of determination of possible product compositions (presynthesis) and stage of heuristic selection of splits.

Presynthesis includes the following steps:

1. Scanning of phase equilibrium coefficients of all components in points at all the edges of concentration simplex.

2. Determination of coordinates of ends of segments $\text{Reg}_r^{t(2)}$, $\text{Reg}_s^{t(2)}$, $\text{Reg}_m^{t(2)}$ of trajectory tear-off of top, bottom, and intermediate section at edges of concentration simplex C_n at separation of initial mixture and of all its constituents $C^{(k)}$ with number of components k from three to $(n-1)$.

3. Determination of possible composition regions $\text{Reg}_D^{(2)}$, $\text{Reg}_B^{(2)}$ of top and bottom products at reversible distillation of all three-component constituents $C^{(3)}$ of initial mixture.

4. Determination of coordinates of ends of segments of possible top and bottom products $\text{Reg}_D^{(2)}$, $\text{Reg}_B^{(2)}$ at edges of concentration simplex C_n at separation of initial mixture and of all its constituents $C^{(k)}$ with a number of components k from three to $(n-1)$.

5. Determination of coordinates of vertexes of polygons, polyhedrons, or hyperpolyhedrons of possible top and bottom products $\text{Reg}_D^{(m)}$, $\text{Reg}_B^{(m)}$ at faces and hyperfaces of concentration simplex C_n at separation of initial mixture and of all its constituents $C^{(k)}$ with a number of component k from four to $(n-1)$.

6. Determination of possible sharp splits in columns with one feeding without distributed components and with one distributed component with recycles or without them and in the columns with two feedings with one- or two-component top product and one-component or more component autoentrainer at separation of initial mixture and of all its constituents with number of components from two to $(n-1)$.

For heteroazeotropic mixtures, presynthesis also includes the most interesting variants of heteroazeotropic and heteroextractive distillation.

For petroleum mixtures, their peculiarities predetermine the choice of the main distillation complex (column with side strippings) and of main sequence (atmospheric and vacuum columns), but different modifications are possible for different sets of products and compositions of crude are possible. These modifications may significantly improve the economy of the equipment.

8.9. Questions

1. Let the following segments of component order $\text{Reg}_{ord}^{(2)}$: *12435, 12453, 21453*, and *24153* be located in the direction from vertex 1 to vertex 2 at edge 1-2 of the concentration pentahedron.

   a. Can component 1 be top product?

   b. Can mixture 1,2 be top product?

    c. Which segments of component order are included in the trajectory tear-off segment $Reg_r^{t(2)}$ of the top section?

    d. Can component 2 be the top or the bottom product? If not, then why?

    e. Can mixture 1,2 be the bottom product? If not, then why?

    f. Can trajectory tear-off points $Reg_m^{t(2)}$ of the intermediate section of a column of extractive distillation be located at edge 1-2? If not, then why?

2. Let the following segments of order of components *12435*, *12345*, *12354* (edge 1-2), and *12435*, *21435*, *21345* (edge 1-3) be located in the direction from vertex 1 at edges 1-2 and 1-3.

    a. Can component 1 be the top product?

    b. Can mixture 1,2 be the top product?

    c. Can mixture 1,2,3 be the top product?

    d. Which vertexes of triangle 1-2-3 does the trajectory tear-off region Reg_r^t of the top section join if mixture 1,2,3 is the top product?

3. Let the following segments $Reg_{ord}^{(2)}$ of order of components: *13452*, *14352*, and *14325* be located in direction from vertex 1 to vertex 2 at edge 1-2 of the concentration pentahedron.

    a. Can component 1 be the top product?

    b. Can component 2 be the bottom product?

    c. Can the trajectory tear-off region $Reg_m^{(2)}$ of the intermediate section of the column of extractive distillation be located at edge 1-2? If yes, then why? Which component can be the top product, and which component can be the entrainer if the top product and the entrainer are one-component?

References

Abu-Eishuh, S. I., & Luyben, W. L. (1985). Design and Control of Two-Column Azeotropic System. *Ind. Eng. Chem. Process Des. Dev.*, 24, 132–140.

Agrawal, R. (1996). Synthesis of Distillation Column Configurations for a Multi-component Separation. *Ind. Eng. Chem. Res.*, 35, 1059–1071.

Agrawal, R., & Fidkowski, Z. T. (1998). Are Thermally Coupled Distillation Column Always Thermodynamically More Efficient for Ternary Distillations? *Ind. Eng. Chem. Res.*, 37, 3444–54.

Agrawal, R., & Fidkowski, Z. T. (1999). New Thermally Coupled Schemes for Ternary Distillation. *AIChE J.*, 45, 485–96.

Avet'an, V. S., Petlyuk, F. B., & Serafimov, L. A. (1973). Optimal Design Three-Columns of Separation Unit of Lowest Carbon Acids with Water. In *The Optimal Design in Refining and Chemical Industry*. 3, pp. 127–137. Moscow: VNIPIneft (Rus.).

Balashov, M. I., & Serafimov, L. A. (1984). Investigation of the Rules Governing the Formation of Regions of Continuous Rectification. *Theor. Found. Chem. Eng.*, 18, 360–366.

Balashov, M. I., Grishunin, V. A., & Serafimov, L. A. (1970). The Rules of Con-figuration Boundarys of Regions of Continuous Distillation in Ternary Sistems. *Transactions of Moscow Institute of Fine Chemical Technology*, 2, 121–6 (Rus.).

Baburina, L. V., & Platonov, V. M. (1990). Application of Theory of Conjugate Tie Lines to Structural Analysis of Five-Component Polyazeotropic Mixtures. *Theor. Found. Chem. Eng.*, 24, 382–387.

Baldus, H., Baumgaertner, K., Knapp, H., & Streich, M. (1983). Verfluessigung and Trennung Von Gasen. In *Chemische Technologie*. Bd. 3. Winnacker Kuechler, pp. 567–560 (Germ.).

Bauer, M. H., & Stichlmair, J. (1995). Synthesis and Optimization of Distillation Sequences for the Separation of Azeotropic Mixtures. *Comput. Chem. Eng.*, 19, 515–20.

Davydyan, A. G., Malone, M. F., & Doherty, M. F. (1997). Boundary Modes in a Single-Feed Distillation Column for the Separation of Azeotropic Mixtures. *Theor. Found. Chem. Eng.*, 31, 327–338.

Devos, A., Gourlia, J. P., & Paradowski, H. (1987). Process for Distillation of Petroleum by Progressive Separations. Patent USA No. 4, 664, 785.

Doherty, M. F., & Caldarola, G. A. (1985). Design and Synthesis of Homogeneous Azeotropic Distillation. 3. The Sequencing of Column for Azeotropic and Extractive Distillation. *Ind. Eng. Chem. Fundam.*, 24, 474–485.

Doherty, M. F., & Malone, M. F. (2001). *Conceptual Design of Distillation Systems*. New York: Mc Graw–Hill.

Fonyo, Z., & Benko, N. (1998). Comparison of Various Heat Pumps Assisted Distillation Configurations. *Trans. IChemE.*, 76, Part A, 348–360.

Glinos, K., & Malone, M. F. (1984). Minimum Reflux, Product Distribution and Lumping Rules for Multicomponent Distillation. *Ind. Eng. Chem. Process Des. Dev.*, 23, 764.

Glinos, K., & Malone, M. F. (1988). Optimality Regions for Complex Column Alternatives in Distillation Systems. *Chem. Eng. Res. Des.*, 66, 229–240.

Harbert, W. D. (1957). Which Tower Goes Where? *Petrol. Refin.*, 36(3), 169–74.

Haselden, G. (1958). Approach to Minimum Power Consumption in Low Temperature Gas Separation. *Trans. Inst. Chem. Engrs.*, 36, 123–132.

Hendry, J. E., Rudd, D. F., & Seader, J. D. (1973). Synthesis in the Design of Chemical Processes. *AIChE J.*, 19, 1–15.

Kafarov, V. V., Petlyuk, F. B., Groisman, S. A., & Belov, M. V. (1975). Synthesis of Optimal Sequences of Multicomponent Distillation Columns by Method of Dynamic Programming. *Theor. Found. Chem. Eng.*, 9, 262–9.

Kiva, V. N., Marchenko, I. M., & Garber, Yu. N. (1993). Possible Composition Distillation Products of Three-Component Mixture with Binary Saddle. *Theor. Found. Chem. Eng.*, 27, 373–380.

Krolikowski, L., Davydian, A., Malone, M. F., & Doherty, M. F. (1996). Exact Bounds on the Feasible Products for Distillation of Ternary Azeotropic Mixtures. AIChE Annual Meeting, p. 90F, Chicago.

Kuschner, T. M., Tazievskaya, G. L., Serafimov, L. A., & Lvov, S. V. (1969). Separation of the Lowest Carbon Acids from Oxidate of Benzine. *Chemical Industrie*, 20–33 (Rus.).

Laroche, L., Bekiaris, N., Andersen, H. W., & Morari, M. (1992). Homogeneous Azeotropic Distillation: Separability and Flowsheet Synthesis. *Ind. Eng. Chem. Res.*, 31, 2190–2209.

Linnhoff, B., & Hindmarsh, E. (1983). The Pinch Design Method for Heat-Exchangers Networks. *Chem. Eng. Sci.*, 38, 745–763.

London, G. (1961). *Separation of Isotopes*. London: Newnes.

Malone, M. F., Glinos, K., Marquez, F. E., & Douglas, J. M. (1985). Simple, Analytical Criteria for the Sequencing of Distillation Columns. *AIChE J.*, 31, 683–690.

Modi, A. K., & Westerberg, A. W. (1992). Distillation Column Sequencing Using Marginal Price. *Ind. Eng. Chem. Res.*, 31, 839–48.

Petlyuk, F. B. (1979). Structure of Concentration Space and Synthesis of Schemes for Separating Azeotropic Mixtures. *Theor. Found. Chem. Eng.*, 13, 683–689.

Petlyuk, F. B., Avet'an, V. S., & Inyaeva, G. V. (1977). Possible Product Composition for Distillation of Polyazeotropic Mixtures. *Theor. Found. Chem. Eng.*, 11, 177–183.

Petlyuk, F. B., & Danilov, R. Yu. (1999). Sharp Distillation of Azeotropic Mixtures in a Two-Feed Column. *Theor. Found. Chem. Eng.*, 33, 233–242.

Petlyuk, F. B., & Danilov, R. Yu. (2000a). Synthesis of Separation Flowsheets for Multicomponent Azeotropic Mixtures on the Basis of the Distillation Theory. Presynthesis Prediction of Feasible Product Compositions. *Theor. Found. Chem. Eng.*, 34, 236–254.

Petlyuk, F. B., & Danilov, R. Yu. (2000b). Synthesis of Separation Flowsheets for Multicomponent Azeotropic Mixtures on the Basis of the Distillation Theory. Synthesis: Finding Optimal Separation Flowsheets. *Theor. Found. Chem. Eng.*, 34, 444–456.

Petlyuk, F. B., Kievskii, V. Ya, & Serafimov, L. A. (1977a). Method for the Isolation of the Regions of the Rectification of Polyazeotropic Mixtures Using an Electronic Computer. *Theor. Found. Chem. Eng.*, 11, 1–7.

Petlyuk, F. B., Kievskii, V. Ya, & Serafimov, L. A. (1979). Determination of Distillation Product Composition of Polyazeotropic Mixtures. *Theor. Found. Chem. Eng.*, 13, 643–649.

Petlyuk, F. B., & Platonov, V. M. (1964). The Thermodynamical Reversible Multicomponent Distillation. *Chem. Industry*, (10), 723–725 (Rus.).

Petlyuk, F. B., & Platonov, V. M. (1965). Sequences with Reversible Flows Mixture. In V. M. Platonov & B. G. Bergo (Eds.), *Separation of Multicomponent Mixtures* (pp. 262–78). Moscow: Khemiya (Rus.).

Petlyuk, F. B., Platonov, V. M., & Girsanov, I. V. (1964). Calculation of the Optimal Distillation Cascades. *Chem. Industry*, (6), 445–53 (Rus.).

Petlyuk, F. B., Platonov, V. M., & Slavinskii, D. M. (1965). Thermodynamical Optimal Method for Separating of Multicomponent Mixtures. *Int. Chem. Eng.*, 5(2), 309–17.

Petlyuk, F. B., Serafimov, L. A., Avet'an, V. S., & Vinogradova, E. I. (1981). Trajectories of Reversible Distillation When One of the Components Completely Disappears in Every Section. *Theor. Found. Chem. Eng.*, 15, 185–192.

Petlyuk, F. B., Vinogradova, E. I., & Serafimov, L. A. (1984). Possible Composition of Products of Ternary Azeotropic Mixture: Distillation at Minimum Reflux. *Theor. Found. Chem. Eng.*, 18, 87–94.

Petlyuk, F. B., Zaranova, D. A., Isaev, B. A., & Serafimov, L. A. (1985). The Presynthesis and Determination of the Possible Train of Separation of Azeotropic Mixtures. *Theor. Found. Chem. Eng.*, 19, 514–524.

Poellmann, P., & Blass, E. (1994). Best Products of Homogeneous Azeotropic Distillation. *Gas Separation and Purification*, 8, 194–228.

Rooks, R. E., Julka, V., Doherty, M. F., & Malone, M. F. (1998). Structure of Distillation Regions for Multicomponent Azeotropic Mixtures. *AIChE J.*, 44, 1382–1391.

Safrit, B. T., & Westerberg, A. W. (1997). Algorithm for Generating the Distillation Regions for Azeotropic Multicomponent Mixtures. *Ind. Eng. Chem. Res.*, 36, 1827–1840.

Sargent, R. W. H. (1998). A Functional Approach to Process Synthesis and Its Application to Distillation Systems. *Comput. Chem. Eng.*, 22, 31–45.

Stichlmair, J. B., & Herguijuela, J. R. (1992). Separation Regions and Processes of Zeotropic and Azeotropic Ternary Distillation. *AIChE J.*, 38, 1523–1535.

Stichlmair, J. G., & Fair, J. R. (1998). *Distillation: Principles and Practice*. New York: John Wiley.

Tedder, D. W., & Rudd, D. F. (1978). Parametric Studies in Industrial Distillation. *AIChE J.*, 24, 303.

Thompson, R. W., & King, C. J. (1972). Systematic Synthesis of Separation Schemes. *AIChE J.*, 18, 941–8.

Wahnschafft, O. M., Koehler, J. W., Blass, E., & Westerberg, A. W. (1992). The Product Composition Regions of Single-Feed Azeotropic Distillation Columns. *Ind. Eng. Chem. Res.*, 31, 2345–2362.

Wahnschafft, O. M., Le Redulier, & Westerberg, A. W. (1993). A Problem Decomposition Approach for the Synthesis of Complex Separation Processes with Recycles. *Ind. Eng. Chem. Res.*, 32, 1121–1140.

Westerberg, A. W., & Wahnschafft, O. (1996). Synthesis of Distillation-Based Separation Systems. *Adv. Chem. Eng.*, 23, 63–170.

Short Glossary

Attraction region (Reg_{att}) region of concentration simplex in which the section trajectories at given reflux or reboil ratio come to node product point.

Bond between two stationary points distillation line (c-line at infinite reflux) that comes from the point with less boiling temperature to the point with greater boiling temperature. There should not be any other stationary points between these two points. **Chain of bonds** consists of several unidirectional bonds.

Bundle of section trajectories ($N_r^- \Rightarrow N_r^+$ or $N_s^- \Rightarrow N_s^+$ or $N_e^- \Rightarrow N_e^+$) full set of section trajectories that start from the same point (unstable node) and end in the same point (stable node).

Component order region ($\text{Reg}_{ord}^{i,j,k}$) region of concentration simplex in which points component volatilities are in the same order ($K_i > K_j > K_k$).

Condition for reversible distillation product points points x_D, x_B, x_F, and y_F are located on straight line.

Distillation region at infinite reflux (Reg^∞) region of concentration simplex filled with one bundle of distillation lines (residue curves) at infinite reflux ($N^- \Rightarrow N^+$).

Distillation subregion (Reg_{sub}) part of distillation region contains one bond chain and all feed components $N^- \rightarrow S^1 \rightarrow \cdots \rightarrow N^+$.

Joining condition at minimum reflux for split with distributed component feed cross-section composition point for one column's section must belong to the *separatrix min-reflux region* and the other to the *separatrix sharp distillation region* for given product points and reflux ratio and satisfy the material balance in feed cross-section: $x_{f-1} \in \text{Reg}_{sep,r}^{min,R}$ and $x_f \in \text{Reg}_{sep,s}^{sh,R}$ or $x_{f-1} \in \text{Reg}_{sep,r}^{sh,R}$ and $x_f \in \text{Reg}_{sep,s}^{min,R}$. There is only one possible pair of x_{f-1} and x_f.

Joining condition at minimum reflux for split without distributed components feed cross-section composition points must belong to the *separatrix min-reflux regions* for given product points and reflux ratio and satisfy the material balance in feed cross-section: $x_{f-1} \in \text{Reg}_{sep,r}^{min,R}$ and $x_f \in \text{Reg}_{sep,s}^{min,R}$. There is only one possible pair of x_{f-1} and x_f.

Joining condition for split with distributed components at reflux greater than minimum feed cross-section composition point belong to *separatrix sharp distillation regions* for given product points and reflux ratio and satisfy the material balance in feed cross-section: $x_{f-1} \in \text{Reg}_{sep,r}^{sh,R}$ and $x_f \in \text{Reg}_{sep,s}^{sh,R}$.

Joining condition for split without distributed components at reflux greater than minimum feed cross-section composition points belong to segments located on *separatrix sharp distillation regions* for given product points and reflux ratio and satisfy the material balance in feed cross-section: $x_{f-1} \in [x_{f-1}] \in \text{Reg}_{sep,r}^{sh,R}$ and $x_f \in [x_f] \in \text{Reg}_{sep,s}^{sh,R}$.

Operating condition of trajectory tear-off $L/V > K_j^{t1}$ if only one tear-off point t_1 exists; $K_j^{t2} > L/V > K_j^{t1}$ if both points t_1 and t_2 exist on reversible distillation trajectory.

Pitchfork region (Reg_{pitch}) region of concentration simplex where the section trajectory at certain reflux or reboil ratios leaves its own distillation region and comes into another distillation region.

Possible sharp split split whose product points belong to section product regions $x_D \in \text{Reg}_D$ and $x_B \in \text{Reg}_B$.

Product region ($\text{Reg}_D^{(k)}$ or $\text{Reg}_B^{(k)}$) region of k-component boundary element of concentration simplex in which section product points can be situated.

Product simplex (Reg_{simp}) simplex whose vertexes are stationary points of distillation subregion and their number is equal to the total number of components.

Reversible distillation region of section ($\text{Reg}_{rev,r}^h$, $\text{Reg}_{rev,s}^l$, $\text{Reg}_{rev,e}^m$) unification of several regions of components' order for which one and the same component appears to be (1) the most heavy-volatile for rectifying section, (2) the most light-volatile for stripping section, and (3) middle-volatile between top and entrainer components for extractive section.

Rule of connectedness condition satisfied to product points at $R = \infty$ and $N = \infty$. The stable node N_D^+ of the top product boundary element of distillation region at infinite reflux Reg_D^∞ and the unstable node N_B^- of the bottom product boundary element of distillation region at infinite reflux Reg_B^∞ should coincide ($N_D^+ \equiv N_B^-$), or should be connected with each other by the bond ($N_D^+ \rightarrow N_B^-$) or chain of bonds in direction to the bottom product.

Section composition profile ($x_D \rightarrow x_{f-1}$ or $x_B \rightarrow x_f$) part of working section tra-

jectory from the product point to the feed cross-section.

Section sharp split product of a section contains only the part of feed components.

Segment of possible feed cross-section compositions the set of possible of feed cross-section composition points x_{f-1} or x_f at reflux greater then minimum.

Separatrix element of section trajectories bundle ($\text{Reg}_{sep,r}$ or $\text{Reg}_{sep,s}$ or $\text{Reg}_{sep,e}$) separatrix element that belongs to both working and unworking bundles of section trajectories $\text{Reg}_{sep} \equiv \text{Reg}_w^R \bullet \text{Reg}^R$.

Separatrix min-reflux region ($\text{Reg}_{sep,r}^{min,R}$, $\text{Reg}_{sep,s}^{min,R}$) separatrix element of working bundle of section trajectories for given reflux and product point containing tear-off point S^2 and N^+ and free from points S^1 and N^-: $\{S^2 \rightarrow \cdots \rightarrow N^+\} \equiv S^2 \Rightarrow N^+$ (in abbreviated form $S^2 - N^+$) to which belong the feed cross-section compositions at minimum reflux: $x_{f-1} \in \text{Reg}_{sep,r}^{min,R}$ and $x_f \in \text{Reg}_{sep,s}^{min,R}$.

Separatrix sharp split region ($\text{Reg}_{sep,r}^{sh,R}$, $\text{Reg}_{sep,s}^{sh,R}$) separatrix element of working bundle of section trajectories for given reflux and product point containing tear-off points S^1, S^2, and N^+ and free from point N^-: $\{S^1 \rightarrow S^2 \rightarrow \cdots \rightarrow N^+\} \equiv S^1 \Rightarrow N^+$ (in abbreviated form $S^1 - N^+$) to which belong the feed cross-section compositions at sharp split: $x_{f-1} \in \text{Reg}_{sep,r}^{sh,R}$ and $x_f \in \text{Reg}_{sep,s}^{sh,R}$.

Sharp split split when each of the products contains only the part of feed components. There some kinds of sharp splits: direct split 1 : 2, 3, ..., n; indirect split 1, 2, ..., n – 1 : n; intermediate splits 1, ..., k : k + 1, ..., n; preferred split 1, ..., n – 1 : 2, ..., n; splits with distributed component 1, ..., k : k, k + 1, ..., n.

Sharp split region ($\text{Reg}_{sh,r}^{i:j}$, $\text{Reg}_{sh,s}^{i:j}$, $\text{Reg}_{sh,e}^{i:j}$) region of concentration simplex in which working bundles of section trajectories for given sharp split in a section ($i_D : j_D$ or $i_B : j_B$ or $i_{DE} : j_{DE}$) are situated at any reflux and for any product points. In this region, the following condition is satisfied: $K_{Di} > K_{Dj}$ or $K_{Bi} < K_{Bj}$ or $K_{Di} > K_{DEj} > K_{Ei}$.

Structural condition of trajectory tear-off defined order of component volatilities must be in tear-off points $x_r^t; x_s^t; x_e^t : K_{ri}^t > K_{rj}^t; K_{si}^t < K_{sj}^t; K_{eDi}^t > K_{eDEj}^t > K_{eEi}^t$

Tangential pinch region (Reg_{tang}) region of concentration simplex where there are no pinch points at any reflux or reboil ratios.

Tear-off point $x_{\min,r}^t$ or $x_{\min,s}^t$ **at minimum reflux** (S_r^2, S_s^2) point where the section trajectory at minimum reflux tear-off from boundary element of concentration simplex containing all product components and one impurity component which volatility is niar to product component volatilities.

Tear-off point x_r^t or x_s^t **at sharp split** (S_r^1, S_s^1) point where the section trajectory at given reflux tear-off from product boundary element of concentration simplex.

Tear-off region ($\text{Reg}_r^{t(k)}$ or $\text{Reg}_s^{t(k)}$ or $\text{Reg}_e^{t(k)}$) region of k-component boundary element of concentration simplex contains tear-off points.

Trajectory of distillation section ($\rightarrow$) line in concentration space that connects the points in which the system of distillation equations at given product point and reflux is satisfied.

Working bundle of section trajectories at given reflux ($N_{w,r}^- \Rightarrow N_{w,r}^+$ or $N_{w,s}^- \Rightarrow N_{w,s}^+$ or $N_{w,e}^- \Rightarrow N_{w,e}^+$) bundle contains the distillation **working trajectory** ($x_D \rightarrow$ or $x_B \rightarrow$) (i.e., trajectory going through product point of section).

Working region of section trajectories at given reflux ($\text{Reg}_{w,r}^R, \text{Reg}_{w,s}^R \text{Reg}_{w,e}^R$) region of concentration simplex filled with working bundle of section trajectories $x_D \in \text{Reg}_{w,r}^R$ or $x_B \in \text{Reg}_{w,s}^R$ or $x_e^t \in \text{Reg}_{w,e}^R$.

Index